T0177404

SIGNAL TRANSDUCTION IN THE RETINA

METHODS IN SIGNAL TRANSDUCTION SERIES

Joseph Eichberg, Jr., Series Editor

Published Titles

Lipid Second Messengers, Suzanne G. Laychock and Ronald P. Rubin

G Proteins: Techniques of Analysis, David R. Manning

Signaling Through Cell Adhesion Molecules, Jun-Lin Guan

G Protein-Coupled Receptors, Tatsuya Haga and Gabriel Berstein

Calcium Signaling, James W. Putney, Jr.

G Protein-Coupled Receptors: Structure, Function, and Ligand Screening, Tatsuya Haga and Shigeki Takeda

Calcium Signaling, Second Edition James W. Putney, Jr.

Analysis of Growth Factor Signaling in Embryos Malcolm Whitman and Amy K. Sater

SIGNAL TRANSDUCTION IN THE RETINA

Edited by
Steven J. Fliesler & Oleg G. Kisselev

CRC Press
Taylor & Francis Group
Boca Raton   London   New York

CRC Press is an imprint of the
Taylor & Francis Group, an **informa** business

CRC Press
Taylor & Francis Group
6000 Broken Sound Parkway NW, Suite 300
Boca Raton, FL 33487-2742

© 2008 by Taylor & Francis Group, LLC
CRC Press is an imprint of Taylor & Francis Group, an Informa business

No claim to original U.S. Government works
Printed in the United States of America on acid-free paper
10 9 8 7 6 5 4 3 2 1

International Standard Book Number-13: 978-0-8493-7315-2 (Hardcover)

This book contains information obtained from authentic and highly regarded sources. Reprinted material is quoted with permission, and sources are indicated. A wide variety of references are listed. Reasonable efforts have been made to publish reliable data and information, but the author and the publisher cannot assume responsibility for the validity of all materials or for the consequences of their use.

No part of this book may be reprinted, reproduced, transmitted, or utilized in any form by any electronic, mechanical, or other means, now known or hereafter invented, including photocopying, microfilming, and recording, or in any information storage or retrieval system, without written permission from the publishers.

For permission to photocopy or use material electronically from this work, please access www.copyright.com (http://www.copyright.com/) or contact the Copyright Clearance Center, Inc. (CCC) 222 Rosewood Drive, Danvers, MA 01923, 978-750-8400. CCC is a not-for-profit organization that provides licenses and registration for a variety of users. For organizations that have been granted a photocopy license by the CCC, a separate system of payment has been arranged.

Trademark Notice: Product or corporate names may be trademarks or registered trademarks, and are used only for identification and explanation without intent to infringe.

Library of Congress Cataloging-in-Publication Data

Signal transduction in the retina / [edited by] Steven J. Fliesler and Oleg Kisselev.
　　p. ; cm. -- (Methods in signal transduction series)
　"A CRC title."
　Includes bibliographical references and index.
　ISBN 978-0-8493-7315-2 (alk. paper)
　1. Retina--Physiology. 2. Cellular signal transduction. I. Fliesler, Steven J. II. Kisselev, Oleg. III. Title. IV. Series: Methods in signal transduction.
　　[DNLM: 1. Phototransduction--physiology. 2. Retina--physiology. 3. Rods (Retina)--physiology. 4. Vertebrates. WW 270 S5784 2008]

QP479.S59 2008
612.8'43--dc22   2007027750

Visit the Taylor & Francis Web site at
http://www.taylorandfrancis.com

and the CRC Press Web site at
http://www.crcpress.com

*The Editors dedicate this volume
to Dr. Paul A. Hargrave, in honor of his numerous
and significant contributions to the field of vision science,
particularly in regard to the biochemistry and structure
of rhodopsin and the current understanding of the
phototransduction cascade in retinal rod cells.*

Contents

Section 6   *Lipid Mediators and Signaling in the RPE*

Series Preface

The concept of signal transduction at the cellular level is now established as a cornerstone of the biological sciences. Cells sense and react to environmental cues by means of a vast panoply of signaling pathways and cascades. While the steady accretion of knowledge regarding signal transduction mechanisms is continuing to add layers of complexity, this greater depth of understanding has also provided remarkable insights into how healthy cells respond to extracellular and intracellular stimuli and how these responses can malfunction in many disease states.

Central to advances in unraveling signal transduction is the development of new methods and refinement of existing ones. Progress in the field relies upon an integrated approach that utilizes techniques drawn from cell and molecular biology, biochemistry, genetics, immunology, and computational biology. The overall aim of this series is to collate and continually update the wealth of methodology now available for research into many aspects of signal transduction. Each volume is assembled by one or more editors who are leaders in their specialty. Their guiding principle is to recruit knowledgeable authors who will present procedures and protocols in a critical yet reader-friendly format. Our goal is to assure that each volume will be of maximum practical value to a broad audience, including students, seasoned investigators, and researchers who are new to the field.

The retina has long been a favorite system for the study of signal transduction mechanisms because of its accessibility and the relative abundance of several of the components of the visual phototransduction pathway. Its investigation has yielded valuable insights into the molecular transformation that accompany the transmission of light signals. The editors of this volume have brought together a distinguished group of authors who build on existing knowledge to describe the latest methodological innovations for the exploration of phototransduction at the level of individual molecules. While the chief focus is on vertebrate photo-transduction, aspects of non-visual and invertebrate phototransduction are also covered. Additional chapters deal with techniques used to investigate the rapidly developing area of insulin-mediated signaling in the retina, and with approaches to elucidate the role of signal transduction in the development of retina and supporting structures. A particularly attractive feature is the inclusion of protocols that provide detailed guidance in applying a variety of experimental methods to study retinal signal transduction. Without question, this volume will be a valuable reference to all investigators who are active or interested in this field.

Joseph Eichberg, Ph.D.
Advisory Editor for the Series

Preface

In the current "postgenomic era," there is increasing recognition of the need for integrated approaches to study and understand complex biological systems and signaling networks. The retina—an anatomically and functionally unique part of the central nervous system, responsible for the detection and initial processing of visual information—is illustrative of this. An integrated knowledge of the biochemistry, cell biology, physiology, and physics of phototransduction, as well as postphotoreceptor visual transduction processes, has evolved over the past century, with the finer details becoming apparent particularly within the past decade. The retina is an extremely useful biological system amenable to experimental manipulation in vivo as well as in vitro, affording an accessible model with which to understand individual cellular signaling systems down to the level of molecular interactions at atomic resolution, as well as more complex issues of pathway regulation and the integration of signaling networks that impact cellular and tissue responses, ultimately resulting in visual perception.

The present volume, comprised of fifteen chapters in six sections, brings together a number of internationally recognized authorities in disciplines pertinent to the study of signal transduction in the retina. Each chapter presents a brief overview of the background and current state of knowledge in a particular area relevant to the broader topic of retinal signal transduction, along with detailed information regarding specific methodology for obtaining the primary data necessary to understand the molecular and cellular processes being examined. Because more is known about the rhodopsin-based phototransduction pathway in vertebrate retinal rod cells than in almost any other biological system, and this dominates signaling processes in the retina, a substantial portion of this volume is devoted to that topic. In addition, a diversity of other signaling mechanisms and systems are covered, affording the reader a resource for evaluating the similarities and differences between these systems and the specific research strategies employed for studying them.

Section 1 deals with the molecular mechanisms of vertebrate phototransduction, dissecting the major components of the phototransduction cascade in rod cells and proteins involved in its regulation. The chapters in this section emphasize the breadth of knowledge accumulated in the past decade, especially with regard to determination of the molecular structure of phototransduction cascade components at atomic resolution, as well as the use of transgenic strategies. State-of-the-art approaches for the study of molecular interactions in multiprotein complexes, as well as novel cell-based strategies aimed at understanding the mechanisms of signal shut-off and light adaptation, are presented.

Section 2 focuses on the more recently emerging field of nonvisual phototransduction. Methods for assessing the roles of melanopsin in regulation of the circadian clock and in adaptive photoresponses are described.

Section 3 provides a chapter devoted to essential methods for studying phototransduction in the invertebrate retina, using *Drosophila* as the biological system of choice. Thus, the reader will be able to compare and contrast the juxtaposing processes of visual signaling in vertebrates versus invertebrates.

Section 4 focuses on experimental studies of insulin-based signaling in the retina, both in the outer retina (photoreceptors, per se) as well as inner retinal cells. In addition, insulin receptor structure and ligand-binding specificity as well as mechanisms of downstream signaling are described.

Section 5 presents current methodological approaches relevant to retinal development, including cellular signaling in retinal progenitor cells, and cell–cell communications in developing retina. Because neovascularization is considered an increasingly important factor in various human degenerative retina diseases, particularly those that accompany diabetes and aging, this section also addresses experimental approaches for studying vascular homeostasis.

Section 6 deals with recent developments in the field of lipid-derived mediators, particularly neuroprotectins and the participation of the retinal pigment epithelium in neuronal survival in the retina.

Now in the twenty-first century, we are just beginning to understand the enormous diversity and complexity of signaling processes in the retina. The methodologies and experimental approaches described in this volume have already yielded key fundamental information regarding the cellular and molecular mechanisms that underlie normal retinal physiology. In addition, they have the potential to provide new clues toward elucidating the mechanisms involved in retinal disease processes as well as the development of novel therapeutic approaches for preventing, arresting, and modulating those disease processes and promoting cellular survival and retention of function.

Steven J. Fliesler
Oleg G. Kisselev

Editors

Steven J. Fliesler is a professor and director of research, Department of Ophthalmology (Saint Louis University Eye Institute), as well as a professor in the Department of Pharmacological and Physiological Science at Saint Louis University School of Medicine, in St. Louis, Missouri. He earned a B.A. degree in biochemistry, with a minor in chemistry, from the University of California–Berkeley in 1973, and a Ph.D. degree in biochemistry from Rice University (Houston, Texas) in 1980. Following a three-year postdoctoral research fellowship on retinal lipid metabolism at the Cullen Eye Institute, Baylor College of Medicine (Houston, Texas), Fliesler pursued studies funded by the National Institutes of Health (NIH) on glycoprotein metabolism and photoreceptor membrane assembly in the retina as a research assistant professor at the Cullen Eye Institute. In 1985, Fliesler moved to the Bascom Palmer Eye Institute, with a joint appointment as an assistant professor in the Department of Biochemistry and Molecular Biology and in the Program in Neuroscience at the University of Miami School of Medicine, where he continued his studies on glycoprotein metabolism in the retina. In 1988, he was appointed associate professor at the Bethesda Eye Institute (now Saint Louis University Eye Institute), with a secondary appointment as associate professor in the E.A. Doisy Department of Biochemistry and Molecular Biology and the Cell and Molecular Biology Graduate Program at Saint Louis University School of Medicine (St. Louis, Missouri). Subsequently, Fliesler was promoted (in 1994) to professor in the Department of Ophthalmology, with a secondary appointment (in 2000) as professor in the Department of Pharmacological and Physiological Science at Saint Louis University School of Medicine.

Fliesler's research program encompasses studies on the relationship between cellular metabolism and the establishment and preservation of retinal structure and function, especially in regard to retinal rod photoreceptor cells. A major focus of his research has been cholesterol metabolism and dyslipidemias caused by inborn errors in cholesterol metabolism. In addition, he has a long-standing interest in animal models of human hereditary retinal degenerations, particular those involving defective membrane transport and assembly in retinal photoreceptor cells. Fliesler is the author or coauthor of over 90 peer-reviewed publications, book chapters, and review articles dealing largely with retinal cell biology and glycoprotein and lipid metabolism, and has delivered more than 150 presentations at national and international scientific meetings, colleges and universities, and specialty scientific and biomedical symposia. In addition to editing this volume, Fliesler is the editor of *Sterols and Oxysterols: Chemistry, Biology and Pathobiology*, a multiauthor volume in the "Recent Research Developments in Biochemistry" series, published by Research Signpost (Kerala, India) in 2002. Fliesler is an editorial board member, Retina and Choroid Section editor, and "Focus on Molecules" feature editor for *Experimental Eye Research*, and also serves on the editorial board of the *Journal of Lipid Research*. In addition, he also serves regularly as a reviewer for a variety of federal, state, and private grant-funding agencies, including the National Eye Institute, the Foundation

Fighting Blindness, and Fight for Sight. Fliesler is the recipient of numerous research grant awards, both from federal and private funding agencies, including the National Institutes of Health, the Foundation Fighting Blindness, the March of Dimes Foundation, and Research to Prevent Blindness. In 2007, he was selected as the recipient of a Senior Scientific Investigator Award from Research to Prevent Blindness.

Oleg G. Kisselev is currently an associate professor in the Department of Ophthalmology (Saint Louis University Eye Institute), with a secondary faculty appointment in the Edward A. Doisy Department of Biochemistry and Molecular Biology at Saint Louis University School of Medicine. Kisselev obtained his undergraduate and Ph.D. degrees in biochemistry at Lomonosov Moscow State University, Moscow. In 1992, he emigrated to the United States to pursue postdoctoral training in the laboratory of N. Gautam in the Department of Anesthesiology at Washington University School of Medicine in St. Louis. He later became an American Heart Association Fellow and, in 1997, he received his first faculty appointment as an instructor in that department with a secondary appointment in the Center for Molecular Design headed by Professor Garland R. Marshall at Washington University. In 1999, Kisselev established his own independent research program upon accepting a faculty position in the Department of Ophthalmology (Saint Louis University Eye Institute) of Saint Louis University School of Medicine. He was promoted to his current rank of associate professor in 2007.

Kisselev's research interests are in the biology of sensory signaling utilizing universal mechanisms of transmembrane signal transduction mediated by heterotrimeric GTP-binding proteins, and G-protein-coupled cell surface receptors. He has made seminal contributions to the current state of knowledge regarding the role of individual G-protein subunits in determining the specificity of G-protein signaling and the mechanism of receptor-catalyzed G-protein activation, especially in the vertebrate visual system. His studies of interactions between phototransduction proteins using high-resolution nuclear magnetic resonance (NMR) methods have helped to elucidate the dynamics of the phototransduction machinery at atomic resolution, and have provided essential refinements to the mechanism of visual signal transduction. Kisselev is an author of more than 25 scientific publications including reviews and book chapters dealing with the biochemistry and structural biology of G-proteins. He has received funding for his research program from the National Institutes of Health and the American Heart Association. In addition, he received a William and Mary Greve Scholar Award in 2002 from Research to Prevent Blindness.

List of Contributors

Lingling An
Department of Statistics
Purdue University
West Lafayette, Indiana

Robert E. Anderson
Departments of Ophthalmology and
  Cell Biology, and
Dean A. McGee Eye Institute
Health Sciences Center
University of Oklahoma
Oklahoma City, Oklahoma

Nikolai O. Artemyev
Department of Molecular Physiology
  and Biophysics
College of Medicine
University of Iowa
Iowa City, Iowa

Nicolas G. Bazan
Neuroscience Center of Excellence
  and Department of Ophthalmology
Health Sciences Center
School of Medicine
Louisiana State University
New Orleans, Louisiana

Shawn Beug
Molecular Medicine Program and
  Vision Program
Ottawa Health Research Institute
  and Department of Biochemistry,
  Microbiology and Immunology
University of Ottawa
Ottawa, Ontario, Canada

Terri DiMaio
Department of Ophthalmology and
  Visual Sciences
School of Medicine and Public Health
University of Wisconsin
Madison, Wisconsin

Rebecca W. Doerge
Department of Statistics
Purdue University
West Lafayette, Indiana

Patrice E. Fort
Departments of Ophthalmology and
  Cellular and Molecular Physiology
College of Medicine
Pennsylvania State University
Hershey, Pennsylvania

Thomas W. Gardner
Departments of Ophthalmology and
  Cellular and Molecular Physiology
College of Medicine
Pennsylvania State University
Hershey, Pennsylvania

Eugenia V. Gurevich
Department of Pharmacology
School of Medicine
Vanderbilt University
Nashville, Tennessee

Vsevolod V. Gurevich
Department of Pharmacology
School of Medicine
Vanderbilt University
Nashville, Tennessee

Heidi E. Hamm
Department of Pharmacology
School of Medicine
Vanderbilt University
Nashville, Tennessee

Susan M. Hanson
Department of Pharmacology
School of Medicine
Vanderbilt University
Nashville, Tennessee

Eric L. Harness
Department of Biological Sciences
Purdue University
West Lafayette, Indiana

Byron Hartman
Department of Biological Structure
School of Medicine
University of Washington
Seattle, Washington

Susan Hayes
Department of Biological Structure
School of Medicine
University of Washington
Seattle, Washington

Andreas Helten
Biochemistry Group
Department of Biology and
  Environmental Sciences
Carl von Ossietzky University
Oldenburg, Germany

Song Hong
Neuroscience Center of Excellence and
  Department of Ophthalmology
School of Medicine
Health Sciences Center
Louisiana State University
New Orleans, Louisiana

Fannie Jackson
Neuroscience Center of Excellence and
  Department of Ophthalmology
School of Medicine
Health Sciences Center
Louisiana State University
New Orleans, Louisiana

Kacee Jones
Regulatory Biology Laboratory
The Salk Institute of Biological
  Sciences
La Jolla, California

Eunju Kim
Department of Biological Sciences
Purdue University
West Lafayette, Indiana

Oleg G. Kisselev
Departments of Ophthalmology and
  Biochemistry and Molecular Biology
School of Medicine
Saint Louis University
Saint Louis, Missouri

Junko Kitamoto
Department of Biological Sciences
Purdue University
West Lafayette, Indiana

Karl-Wilhelm Koch
Biochemistry Group
Department of Biology and
  Environmental Sciences
Carl von Ossietzky University
Oldenburg, Germany

Shuji Kondo
Department of Pediatrics
School of Medicine and Public Health
University of Wisconsin
Madison, Wisconsin

Hiep Le
Regulatory Biology Laboratory
The Salk Institute of Biological Sciences
La Jolla, California

Janis Lem
Molecular Cardiology Research
 Institute and Department of
 Ophthalmology
Tufts-New England Medical Center
Tufts University
Boston, Massachusetts
Sackler School of Graduate Biomedical
 Sciences
School of Medicine

Hung-Tat Leung
Department of Biological Sciences
Purdue University
West Lafayette, Indiana

Guohua Li
Department of Biological Sciences
Purdue University
West Lafayette, Indiana

Mandy K. Losiewicz
Departments of Ophthalmology and
 Cellular and Molecular Physiology
College of Medicine
Pennsylvania State University
Hershey, Pennsylvania

Yan Lu
Neuroscience Center of Excellence and
 Department of Ophthalmology
School of Medicine
Louisiana State University
New Orleans, Louisiana

Victor L. Marcheselli
Neuroscience Center of Excellence and
 Department of Ophthalmology

School of Medicine
Louisiana State University
New Orleans, Louisiana

Brian McNeill
Ottawa Health Research Institute
 and Department of Biochemistry,
 Microbiology, and Immunology
University of Ottawa
Ottawa, Ontario
Canada

Hakim Muradov
Department of Molecular Physiology
 and Biophysics
College of Medicine
University of Iowa
Iowa City, Iowa

Michael Natochin
Department of Molecular Physiology
 and Biophysics
College of Medicine
University of Iowa
Iowa City, Iowa

Surendra Kuman Nayak
Genomics Institute of Novartis
 Research Foundation
San Diego, California

Branden Nelson
Department of Biological Structure
School of Medicine
University of Washington
Seattle, Washington

William M. Oldham
Department of Pharmacology
School of Medicine
Vanderbilt University
Nashville, Tennessee

William L. Pak
Department of Biological Sciences
School of Medicine
Purdue University
West Lafayette, Indiana

Satchidananda Panda
Regulatory Biology Laboratory
The Salk Institute of Biological Sciences
La Jolla, California

Victoria Piamonte
Genomics Institute of Novartis
  Research Foundation
San Diego, California

Anita M. Preininger
Department of Pharmacology
School of Medicine
Vanderbilt University
Nashville, Tennessee

Raju V. S. Rajala
Departments of Ophthalmology and
  Cell Biology, and Dean A. McGee
  Eye Institute
Health Sciences Center
University of Oklahoma
Oklahoma City, Oklahoma

Thomas Reh
Department of Biological Structure
School of Medicine
University of Washington
Seattle, Washington

Kibibi Rwayitare
Molecular Cardiology Research
  Institute and Department of
  Ophthalmology
Tufts-New England Medical Center
School of Medicine
Tufts University
Boston, Massachusetts

Elizabeth A. Scheef
Department of Ophthalmology and
  Visual Sciences
School of Medicine and Public Health
University of Wisconsin
Madison, Wisconsin

Nader Sheibani
Departments of Ophthalmology and
  Visual Sciences and Pharmacology
School of Medicine and Public Health
University of Wisconsin
Madison, Wisconsin

Ravi S. J. Singh
Departments of Ophthalmology and
  Cellular and Molecular Physiology
College of Medicine
Pennsylvania State University
Hershey, Pennsylvania

Christine M. Sorenson
Department of Pediatrics
School of Medicine and Public Health
University of Wisconsin
Madison, Wisconsin

Nobushige Tanaka
Regulatory Biology Laboratory
The Salk Institute of Biological
  Sciences
La Jolla, California

Yixin Tang
Department of Ophthalmology and
  Visual Sciences
School of Medicine and Public Health
University of Wisconsin
Madison, Wisconsin

Julie Tseng-Crank
Department of Biological Sciences
Purdue University
West Lafayette, Indiana

Sergey A. Vishnivetskiy
Department of Pharmacology
School of Medicine
Vanderbilt University
Nashville, Tennessee

Dana Wall
Ottawa Health Research Institute
  and Department of Biochemistry,
  Microbiology, and Immunology
University of Ottawa
Ottawa, Ontario, Canada

Valerie A. Wallace
Ottawa Health Research Institute
  and Department of Biochemistry,
  Microbiology, and Immunology
University of Ottawa
Ottawa, Ontario, Canada

Qiong Wang
Department of Biochemistry and
  Molecular Biology
College of Medicine
Baylor University
Houston, Texas

Theodore G. Wensel
Department of Biochemistry and
  Molecular Biology
College of Medicine
Baylor University
Houston, Texas

Sowmya V. Yelamanchili
Regulatory Biology Laboratory
The Salk Institute of Biological
  Sciences
La Jolla, California

Ying Zhou
Department of Biological Sciences
Purdue University
West Lafayette, Indiana

Quansheng Zhu
Regulatory Biology Laboratory
The Salk Institute of Biological
  Sciences
La Jolla, California

1

*Vertebrate Visual
Phototransduction*

1 Biophysical Approaches to the Study of G-Protein Structure and Function

Anita M. Preininger, William M. Oldham, and Heidi E. Hamm

CONTENTS

1.1 INTRODUCTION

Crystallographic studies have revealed much of what we know about G-protein structure, thus allowing us to better understand how these proteins function. The structural
studies of the $G\alpha_i$ and $G\alpha_t$-GDP, GTPγS, GDP-AlF$_4^-$, and Gβγ bound forms (1–6)
have revealed many details about mechanisms of G-protein activation. The crystal
structures of heterotrimeric G_i and G_t (4,6) and activated subunits have allowed the
identification of subunit binding sites and have shed light on how GTP binding leads
to subunit dissociation and effector activation. Comparison of the crystal structures
in the active and inactive states reveal regions of conformational flexibility within the
switch regions that undergo specific changes upon activation and deactivation. When
Gα is activated, Switch I and Switch II regions move inward and stabilize each other
and the active state. Residues within the Switch III region form new salt bridges with
Switch II residues in $G\alpha_t$ (1), which may aid in nucleotide binding.

In $G\alpha_t$, activation brings Trp[207] in Switch II into a more hydrophobic environment, thus shielding tryptophan's electrons and causing an increase in Gα intrinsic
fluorescence. This is the basis for a widely used intrinsic Trp fluorescence assay for
Gα activity (section 1.3.1). Upon activation with the GTP transition state analog,
AlF$_4^-$, Gα-GDP-AlF$_4$ adopts a conformation resembling the transition state for GTP
hydrolysis, and this conformational change is accompanied by an increase in fluo-
rescence of Trp[207] in $G\alpha_t$ (Trp[211] in $G\alpha_i$). Crystal structures of G_t and G_i (4,6) reveal
the plasticity of the Switch II region (disordered in $G\alpha_i$GDP), which adopts a distinct
conformation in the presence of Gβγ. Switch II residues participate in binding to
Gβγ. Gα subunits contact βγ dimers at the switch interface; changes in conformation of residues in this region occur during the activation process and result in the
release of βγ, facilitating the interaction of Gα-GTP and βγ with distinct downstream
effectors.

Solved structures of inactive rhodopsin (7–11) have greatly advanced our understanding of G-protein-coupled receptors, coming some 10 years after the structures
of heterotrimeric G-proteins were solved (1,2,5). The effort to obtain a high resolution structure of an activated GPCR is under intense investigation by a number of
groups. A structure of an activated GPCR bound to G-protein is an even more daunt-
ing undertaking. While these efforts are ongoing, other biophysical methods can
be used to elucidate details of receptor–G-protein interactions. Biophysical studies
complement structural information, and there is still much to learn about conformational changes associated with GPCR signaling. Crystallographic studies are static
by nature, whereas G-protein signaling is a dynamic process. Hence, the need exists
for assays to probe conformational changes associated with signal transduction from
an activated GPCR to G-protein in solution. Both fluorescence and electron paramagnetic resonance (EPR) are tools that can be used to examine changes in specific
regions of G-proteins in solution. This can be achieved by attaching probes in a site-
specific manner and directly examining changes in the dynamics and environment
of the probes as G-proteins undergo conformational changes during the G-protein
cycle. These techniques are especially useful in examining changes in regions of
the protein that are typically absent or disordered in crystal structures, such as the

extreme amino and carboxy termini, both of which are known to interact with receptors and undergo activation-dependent changes.

Biophysical studies of G-proteins have provided a wealth of information about dynamic changes that occur in G-protein signaling that cannot be gleaned from current crystal structures. Recent EPR studies reveal that the carboxy termini of G-proteins provide a critical link from receptor activation to GDP release through a rotation and translation of the α-5 helix (12). The α-5 helix in the carboxy region adjoins the $\beta6/\alpha5$ loop that contains a TCAT motif involved in guanine nucleotide binding (3), coupling receptor activation to nucleotide release, which is the slow step in G-protein activation. Fluorescence and EPR studies have demonstrated that residues in the amino terminus of myristoylated $G\alpha_i$ proteins are relatively immobile and reside in a more solvent-excluded environment than the nonmyristoylated forms, consistent with an intramolecular binding site for the amino terminus of $G\alpha$ proteins (13). Together with crystallographic information, these biophysical studies enlarge our understanding of the mechanisms underlying G-protein signaling.

1.1.1 OVERALL STRATEGY

Because both fluorescence and EPR rely on a uniquely labeled protein, this requires creation of a parent protein lacking solvent-exposed cysteines, which can be accomplished with site-directed mutagenesis (SDM). Labeling of specific cysteine residues requires conservative replacement of solvent-exposed cysteines that would react with thiol directed labels. SDM can be performed easily, aided by numerous commercial kits available, to engineer specific labeling sites (in this case, cysteines) into a protein for dynamic fluorescence and EPR studies. This is done against a background lacking solvent-accessible cysteines to achieve labeling specificity. After the expression of the mutant protein containing cysteines at sites of interest, the protein is modified by fluorescent and spin labels, a dual strategy for the study of protein dynamics (scheme 1.1). Although EPR reveals changes in mobility and conformation at the backbone level as well as information regarding tertiary contacts, fluorescence can be used to gauge the overall environment of a specific residue. Together, these complementary approaches reveal information about the mobility and environment of a specific residue in regions undergoing conformational changes during G-protein signaling.

Fluorescence has long been used to examine changes in protein dynamics and conformation. There is an ever-increasing variety of commercially available fluorescent probes, many with superior spectral characteristics. Advances in EPR technology allow distance measurements between labeled residues during the G-protein cycle. Although mutagenesis and expression of GPCRs are beyond the scope of this chapter, these techniques can be used in conjunction with similarly modified receptors to examine interactions between G-proteins and receptors during the G-protein cycle, using both fluorescence resonance energy transfer (FRET) and EPR double electron–electron resonance EPR spectroscopy (DEER).

Once the parent protein lacking solvent-exposed cysteines (termed Hexa I) is generated, it becomes the background against which further mutations are made, allowing attachment of thiol-reactive labels at specific sites of interest. Expression and purification results in protein quantities sufficient for both fluorescent and EPR studies, and provides two independent and complementary measures of conformational changes that occur during the G-protein cycle. Although fluorescence reveals information about the relative hydrophobicity of the probe's overall environment, EPR can resolve populations with distinct mobilities. For example, a highly dynamic region may demonstrate both a mobile and immobile component and the timescale of such motions (14). A mutational strategy that places these labels at a series of strategic points within the protein allows construction of an overall picture of protein conformational changes, as the labeled Gα proteins are interrogated throughout the G-protein cycle. Regions undergoing conformational changes during the cycle can be characterized by a gradient of dynamics, mobility, and environment using fluorescence and EPR approaches.

1.2 PROTEIN EXPRESSION AND PURIFICATION

$G\alpha_{i1}$ protein was selected as the basis of our studies of Gα proteins, because it is more amenable to expression in bacteria than $G\alpha_t$ and interacts nearly as well with rhodopsin (15). These two Gα subunit family members share an 80% sequence similarity (16), and the conformational changes observed in the $G\alpha_{i1}$ background are likely to be a good approximation of those in $G\alpha_t$. Prior to generating cysteine mutants at sites of interest for conformational studies, we created the $G\alpha_i$ Hexa I parent protein (15) by conservative replacement of solvent-exposed cysteines in $G\alpha_i$ protein. Native $G\alpha_{i1}$ contains 10 cysteine residues (C3, C66, C139, C214, C224, C254, C286, C305, C325, and C351); however, only six of these are solvent-exposed, making them sensitive to thiol-reactive labels. We conservatively replaced these six cysteine residues (C3S, C66A, C214S, C305S, C325A, and C351I) using SDM, following the manufacturer's protocol (QuickChange, Stratagene, La Jolla, California). The amino acids used to conservatively replace native cysteines were chosen based on the specific environment of the native cysteine, where serine was placed at more polar sites and alanine at less polar sites. As a result, most of the cysteines were replaced by serine, but metarhodopsin II stabilization assays of $G\alpha_i$ indicated that alanine was a better substitute for native serines at residues 66 and 325, whereas isoleucine was a better substitute for residue 351. The isoleucine substitution residue 351 was critical for maintaining normal interactions with the receptor, as serine and alanine substitutions prevented receptor-catalyzed GTPγS binding. $G\alpha_i$ Hexa I protein expresses well, is folded properly, and has rates of receptor-catalyzed (^{35}S) GTPγS binding similar to wild-type $G\alpha_{i1}$ (15). A hexahistidine tag was inserted between residues M119 and T120 of the $G\alpha_{i1}$ coding region, which aided in purification of expressed protein without perturbing amino, carboxy, or GTPase domains, all of which undergo activation-dependent conformational changes (13,15,17,18). The site of the hexahistidine tag is located in a solvent-exposed region of the helical domain, far removed from these critical regions of the protein. The resulting $G\alpha_i$ Hexa I protein has properties functionally similar to that of $G\alpha_t$ (15), and provides the cysteine-less background against which later mutations were made.

Following creation of the cys-less Hexa-I parent protein, selected individual residues in the parent $G\alpha_i$ Hexa I protein can be mutated to cysteine, for the purpose of attaching fluorescent or nitroxide spin labels at the selected site. After SDM to

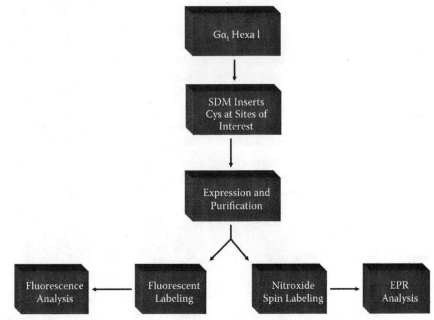

SCHEME 1.1 Overall strategy.

specifically engineer in cysteine residues, proteins are expressed and assayed for function in the same manner as the Gα$_i$ Hexa I parent protein. This is then followed by labeling for biophysical studies to probe conformational changes at each position along a region of interest. An advantage of this approach is that expression and purification of each protein typically yields sufficient quantities for both fluorescent and EPR studies (scheme 1.1). Mutations that disrupt protein folding and function are often poorly expressed, therefore several adjacent residues in a region of interest may need to be mutated in order to build a picture of conformational changes within a region.

1.2.1 EXPRESSION AND PURIFICATION OF Gα$_i$ HEXA I PROTEINS

BL21DE3 Gold *E. coli* (Stratagene) are transformed according to the manufacturer's instruction with the expression vector containing the Gα$_{i1}$ Hexa I cDNA (15), which was created by SDM of a Gα$_{i1}$ expression vector (courtesy of M. Linder, Washington University, St. Louis, Missouri) with a hexahistidine coding region after residue 119 in the helical domain for ease of purification. Site-directed mutagenesis is commonly used to generate mutations, and the prevalence of commercially available kits and reagents allow for introduction of a mutation in one day. It is helpful to keep in mind that primers should be gel-purified, generally 25–45 bases in length, and with a melting temperature of at least 75–80°C. Best success is found with primers that consist of at least 40% GC content. Following manufacturer's instructions, high-quality DNA is obtained that is suitable for sequencing and subsequent transformation into *E. coli*. The Gα$_{i1}$ Hexa I parent protein and cysteine mutants for labeling experiments were expressed and purified according to the protocol below. Although a great many mutant proteins were generated (12,13,15), not all express well or result in a functional protein. Therefore, evaluation of proteins in functional assays is a critical part of the experimental plan. Note that $G\alpha_i$ Hexa I proteins can also be coexpressed with pbb131 plasmid (courtesy of M. Linder) encoding for N-myristoyl transferase in order to express proteins in their N-myristoylated form. The $G\alpha_i$ Hexa I vector encodes ampicillin resistance, whereas the NMT plasmid encodes kanamycin resistance, allowing for selection of expressed proteins in appropriately supplemented media.

After sequence verification and transformation into *E. coli*, the protein is expressed and purified. As our system utilizes a hexahistadine tag, Ni-NTA purification is used to obtain a semipurified protein. An additional round of purification using anion-exchange chromatography is required in order to obtain purity of greater than 85–90%, suitable for labeling and functional assays. The protocol 1 generates quantities sufficient for functional assays and thiol-directed labeling, as summarized in scheme 1.1.

PROTOCOL 1: EXPRESSION AND PURIFICATION OF $G\alpha_i$ HEXA I PROTEINS

1. Transfected *E. coli* cells are grown to OD_{600} of 0.3 units in 2YT broth containing 100 µg/mL ampicillin. For myristoylated protein, 2YT broth is additionally supplemented with kanamycin, 50 µg/mL, and myristic acid, 50 µM. Typically, 50 mL overnight culture is then added to 500 mL of media, and grown to 0.5 OD_{600} over a period of 4–8 h.
2. Expression is induced with 30 µM isopropyl-β-D-thiogalactopyranoside at room temperature for 12–16 h with shaking.
3. Cell pellets from a 1 L culture are resuspended in 50 mM NaH_2PO_4, pH 8.0, 300 mM NaCl containing 5 mM imidazole and 1 µg/mL of the protease inhibitors pepstatin, leupeptin, and aprotinin.
4. Cells are lysed by sonication on ice, generally delivered in 30 s pulses with 1 min cooling time between pulses. Care should be taken to prevent sonicated extracts from excessive heat due to sonication.
5. The lysed extracts are subjected to 50,000 rpm centrifugation for a period of 1 h, and the supernatant from a 1 L culture is combined with 5 mL of a 50% slurry of NiNTA agarose (Qiagen, Valencia, California) and rotated at 4°C for 1 h to allow for complete binding to resin.
6. The slurry is loaded into a gravity filtration column, and flow-through is discarded. The remainder of unbound proteins is removed by washing with 30 mL of resuspension buffer (step 3), followed by an additional 30 mL wash with resuspension buffer containing 10 mM imidazole.
7. $G\alpha_i$ Hexa I proteins are eluted with 7 mL of resuspension buffer containing 40 mM imidazole.
8. The semipurified eluates are dialyzed overnight into a 1 L solution of 50 mM Tris, pH 8.0, 50 mM NaCl, 1 mM $MgCl_2$, 20 µM GDP, 10 mM 2-mercaptoethanol, 100 µM PMSF, and 20% glycerol.
9. After dialysis, proteins are purified using anion exchange chromatography (MonoQ, Amersham, or other strong anion exchanger). Proteins are applied to column in a low-salt buffer consisting of 50 mM Tris, 50 mM NaCl,

and 1 mM $MgCl_2$ at pH 7.8. Purified protein is eluted in a linear gradient of 50–500 mM NaCl, with proteins eluting between 150–250 mM NaCl. Typically, 0.5 mL fractions are collected into tubes containing 300 μL of low-salt buffer supplemented with 50 μM GDP; including GDP in running buffer may obscure absorbance of Gα protein, as they both bind to anion exchange columns and elute at a similar salt concentration.

10. Purified protein fractions demonstrating at least 40% increase in intrinsic fluorescence upon AlF_4^- activation (see functional assays) are pooled, concentrated in a 10 kDa cut-off concentrator (Sartorius, Goettingen, Germany), and washed with low-salt buffer supplemented with 10 μM GDP.

11. Proteins are quantitated by the Coomassie blue method using bovine serum albumin as a standard, and further confirmed by Coomassie blue staining of SDS-PAGE gels, migration consistent with a molecular weight of 37 kDa (figure 1.1).

12. Proteins are concentrated and stored in low-salt buffer supplemented with 10 μM GDP and 10% glycerol prior to storage at −80°C. For $Gα_i$ proteins containing native, solvent-exposed cysteines, 1 mM DTT is included in the storage buffer, and must be completely removed prior to labeling with any thiol-reactive reagents by extensive buffer exchange or use of P6 (Biorad) desalting gel.

   $Gα_i$ Hexa I parent protein and $Gα_i$ Hexa I proteins with cysteine mutations at the sites of interest (hereafter referred to as Hexa I proteins) are ready for labeling if stored in the absence of DTT or BME. Alternatively, proteins for storage can be supplemented with Tris(2-carboxyethyl) phosphine hydrochloride (TCEP), a reducing

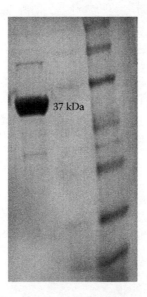

FIGURE 1.1   SDS-PAGE of $Gα_{i1}$ Hexa I protein after anion exchange purification, 5 μg, 10–20% Tris-Glycine gel, visualized by Coomassie blue staining.

agent that is reported to be compatible with thiol-reactive reagents, eliminating the need for their removal prior to the labeling process. However, the presence of TCEP may also reduce labeling efficiency, so one may need to consider this carefully and optimize labeling conditions in the presence of this reducing agent (section 1.4.1). In Hexa I proteins, substitution of native solvent-exposed cysteines considerably reduces the propensity for dimerization between adjacent molecules.

1.2.2 RHODOPSIN AND Gβγ PREPARATION

Rhodopsin and $G\beta_1\gamma_1$ ($G\beta\gamma$) are obtained from native retinas as detailed previously in Reference 18. Unwashed rod outer segments (ROS) are stored at −80°C and in buffer containing 10 mM MOPS, pH 8.0, 90 mM KCl, 30 mM NaCl, 2 mM $MgCl_2$, 0.1 mM EDTA, 1 mM DTT, and 50 µM PMSF. To obtain rhodopsin free of endogenous G-proteins, ROS membranes were stripped of G-proteins with 4 mM urea as described in (19). Rhodopsin concentration was determined by measuring the absorbance of solubilized ROS membrane suspensions at 500 nm before and after photobleaching (protocol 2). $G\beta\gamma_t$ was prepared from ROS, as summarized in protocol 3 below.

PROTOCOL 2: MEASUREMENT OF RHODOPSIN CONCENTRATIONS IN UREA-WASHED ROS SAMPLES

1. Because quantitation of rhodopsin requires comparison of absorbances between dark- and light-activated rhodopsin, the assay must be performed under dim red light, using a filter that passes light greater than 650 nm. A water-jacketed cuvette holder is used to keep the temperature of the cuvette between 5–10°C.
2. Under dim red light (Kodak, Filter 1), 25 µL of rhodopsin is added to 500 µL of 20 mM hexadecyltrimethyl ammonium chloride, mixed well and placed in the cuvette chamber. Absorbances are recorded at 500 nm and 650 nm (reference).
3. The sample is then illuminated with three successive flashes of bright white light.
4. For each dark and light measurement at 500 nm, subtract the absorbance of reference measured at 650 nm. To calculate rhodopsin concentration from Beer's law, the difference in absorbances between light-activated rhodopsin measurement (less 650 absorbance) and dark-adapted rhodopsin (less 650 absorbance) is calculated, and multiplied by the dilution factor used. With the extinction coefficient (ε) of 42,000 $cm^{-1}M^{-1}$ for rhodopsin, and a path length specific for the instrument (L), concentration is calculated from absorbance, where absorbance = ε C L.

PROTOCOL 3: $G\beta_1\gamma_t$ PREPARATION

First, transducin is extracted from ROS, followed by chromatography to isolate $G\beta\gamma$ from $G\alpha$ as described in detail in Mazzoni et al. (18). Briefly, the process is:

1. ROS membranes are washed four times with isotonic buffer (5 mM Tris, pH 8.0, 120 mM KCl, 0.6 mM $MgCl_2$, 1 mM DTT, 0.1 mM EDTA, and 0.1 mM PMSF) and twice with hypotonic buffer lacking KCl.
2. Transducin is eluted from the membranes by washing twice with hypotonic buffer containing 0.1 mM GTP.
3. To isolate Gβγ from transducin, transducin is applied to a HiTrap™ Blue HP blue sepharose column (Amersham Biosciences, Piscataway, New Jersey) at 4°C using a peristaltic pump in 10 mM Tris, pH 7.5, 150 mM NaCl, 20 mM $MgSO_4$, 1 mM EDTA, 10% glycerol, and 14.3 mM 2-mercaptoethanol.
4. $MgSO_4$ encourages subunit dissociation, and as Gβγ does not bind the blue Sepharose column, it is obtained in the flow-through fraction, whereas $Gα_t$ remains bound to the column.
5. The flow-through fractions from chromatography are concentrated and evaluated on SDS-PAGE to identify fractions containing pure Gβγ, which can be pooled and stored in buffer containing 10 mM Tris, pH 7.5, 100 mM NaCl, 5 mM 2-mercaptoethanol, and 10% glycerol at −80°C.
6. Just prior to use, $Gβ_1γ_1$ samples are buffer exchanged into 20 mM MES, pH 6.8, 100 mM NaCl, 2 mM $MgCl_2$, and 10% glycerol.

1.3 FUNCTIONAL ASSAYS

1.3.1 Intrinsic Tryptophan Fluorescence

A wide variety of functional assays exist to examine functional integrity of Gα proteins. The intrinsic tryptophan fluorescence assay demonstrates the ability of Gα proteins to undergo an activation-dependent conformational change in solution. In this assay, aluminum fluoride (AlF_4^-) complexes with bound GDP to form a transition state that mimics the conformation of Gα during the transition state associated with GTP hydrolysis. A tryptophan residue within the Switch II region acts as a reporter of this activation-dependent conformational change. This conformational change is reflected by an increase in tryptophan fluorescence emission for properly folded, active Gα proteins. The increase in emission is typically expressed relative to its emission in the GDP-bound state; minimally, a 40% increase over basal is exhibited by functional Gα subunits. This assay allows efficient screening of fractions for activity after HPLC purification. Oftentimes, only selected fractions from the peak eluting at retention times that correspond to Gα has functional activity, and it is necessary to discard inactive fractions prior to pooling and concentrating the purified Gα protein.

Protocol 4: Intrinsic Tryptophan Fluorescence

1. Using a fluorescence spectrophotometer, set excitation/emission wavelengths to 280/340 nm.
2. Zero emission reading from buffer consisting of 50 mM Tris, 50 mM NaCl, 1 mM $MgCl_2$, and 10 μM GDP.
3. Add 100 nM of Gα subunit, mix well, and measure emission in the GDP-bound state.

4. To form activated GαGDP-AlF$_4$⁻ complexes, add NaF to 10 mM and AlCl$_3$ to 50 μM, and mix.
5. Measure emission after addition of AlF$_4$⁻; for properly folded, active Gα subunits, the emission should be at least 40% greater in the GDP-AlF$_4$⁻-bound state compared to the Gα-GDP-bound form (figure 1.2a).

1.3.2 BASAL NUCLEOTIDE EXCHANGE

G-proteins undergo GDP release under basal conditions at a relatively low level, as compared to rates in the presence of activated receptor. Furthermore, this rate is

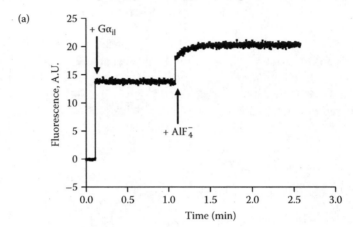

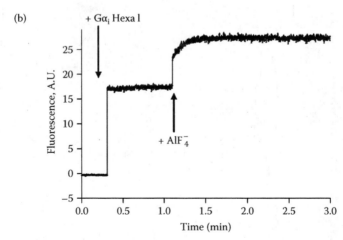

FIGURE 1.2 Intrinsic tryptophan activation of Gα proteins. Gα$_{i1}$ (a) or Gα$_{i1}$ Hexa I (b), 100 nM, is added to buffer consisting of 50 mM Tris, 50 mM NaCl, 1 mM MgCl$_2$, and 10 μM GDP. Tryptophan emission is monitored at excitation/emission 280/340 nm, both before and after addition of NaF to 10 mM and AlCl$_3$ to 50 μM. Proteins exhibit ≥ 40% increase in fluorescence in GDP-AlF$_4$⁻-activated form, relative to inactive, GDP-bound form, as would be expected for properly folded and fully functional Gα proteins.

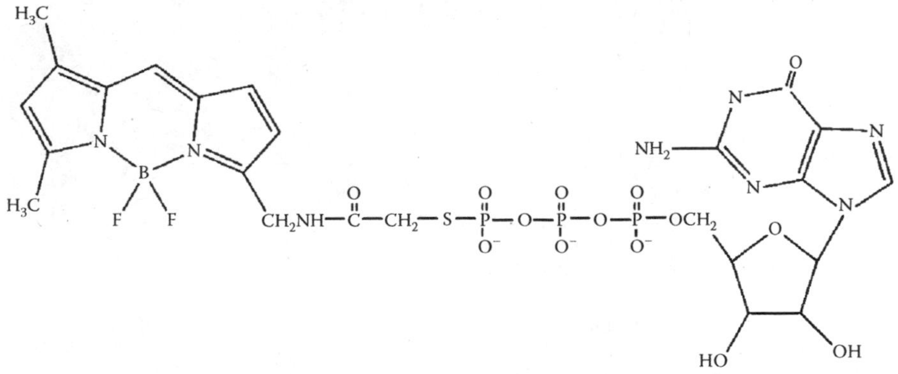

FIGURE 1.3   Structure of BD-GTPγS.

somewhat increased for Gα$_i$ relative to Gα$_t$. The low rate of intrinsic GDP exchange for Gα$_t$ subunits supports fast, efficient signal termination required for these subunits during the visual transduction process.

In order to measure intrinsic nucleotide exchange in the absence of a receptor, a fluorescently modified GTPγS can be used to monitor GDP release (protocol 5), as can intrinsic tryptophan fluorescence (protocol 6). Both of these methods avoid the use of radioactivity, and results from these assays correlate well with published radioactive assays for Gα$_i$ subunits. Bodipy FL-GTPγS (Invitrogen, Madison, Winconsin), has a fluorescent label on the γ-phosphate of GTPγS (figure 1.3). The emission of Bodipy GTPγS is quenched in aqueous solution, and increases upon binding to a number of GTP-binding proteins, including Gα$_i$ proteins (20). Bodipy-GTPγS provides a convenient tool to measure GDP release. As the spectral properties of the probe overlap with rhodopsin, this GTP analog is more suited to intrinsic nucleotide exchange assays than receptor-mediated assays. Intrinsic tryptophan fluorescence can also be used to monitor GTPγS binding (both basal and receptor-stimulated), increasing options for nonradioactive monitoring of nucleotide binding.

PROTOCOL 5: BASAL NUCLEOTIDE EXCHANGE ASSAY BY EXTRINSIC FLUORESCENCE

1. Emission of 5 μM Bodipy-GTPγS (BD-GTPγS, figure 1.3) is monitored for approximately 1 min at excitation/emission 490/512 nm with stirring at 18°C in 50 mM Tris, 50 mM NaCl, and 1 mM MgCl$_2$ at pH 7.8.
2. 500 nM Gα$_i$ (with or without Gβγ in a 2 molar excess) is added to the cuvette with BD-GTPγS, and emission is monitored over a 5 min period (figure 1.4a) using a Varian Cary Eclipse (Mulgrave, Australia).

(a)

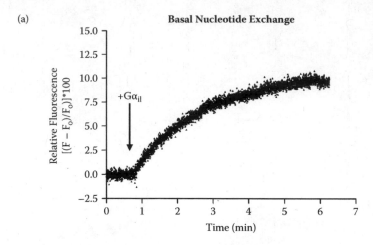

(b)

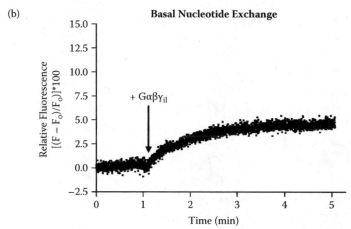

FIGURE 1.4   Basal nucleotide exchange of $G\alpha_{il}$ subunits, measured in the presence of 5 μM BD-GTPγS in buffer consisting of 50 mM Tris, 50 mM NaCl, 1 mM $MgCl_2$, and pH 7.8. Excitation/emission set at 490/512 nm using a Varian Eclipse fluorescence spectrophotometer. (a) 550 nM Gα or (b) reconstituted heterotrimer formed from preincubation of 550 nM Gα and 1 μM Gβγ for 5 min at 4°C prior to addition to a cuvette containing 5 μM BD-GTPγS. Data are normalized to basal BD-GTPγS emission measured in the absence of G-protein, and represented as percent increase over basal. Basal nucleotide exchange can also be measured by monitoring conformational changes in the Switch II region by monitoring Trp emission, as described in the intrinsic fluorescence assay (protocol 4).

PROTOCOL 6: BASAL NUCLEOTIDE EXCHANGE ASSAY USING INTRINSIC TRYPTOPHAN FLUORESCENCE

1. Using a fluorescence spectrophotometer, set excitation/emission wavelengths to 300/345 nm.
2. Zero the emission on buffer alone (10 mM MOPS, pH 7.2, 150 mM NaCl, and 2 mM $MgCl_2$) at 15°C.
3. Add 500 nM Gα subunit to cuvette, mix, and monitor emission for 1 min.

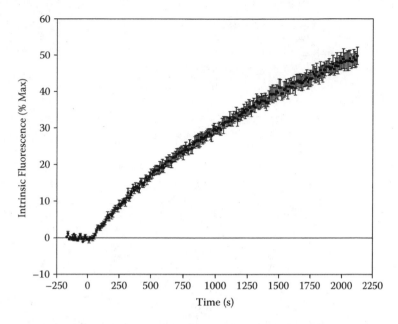

FIGURE 1.5   Increase in the intrinsic fluorescence of $G\alpha_{i1}$ upon the addition of GTPγS (t = 0). Data are the average ±SEM of three independent experiments.

4. Add GTPγS to 10 μM and continuously monitor the emission over a period of 40 min (figure 1.5). Add NaF (10 mM) and AlCl$_3$ (50 μM) to fully activate the subunits at the end of 40 min.

1.3.3   RECEPTOR-MEDIATED GTPγS BINDING

Although radiolabeled GTPγS-binding assays have long been used in determination of GTPγS binding to Gα subunits, nonradiative methods are often preferable when available. Using nonradioactive nonhydrolyzable GTP analogs (protocols 7 and 8), receptor-mediated GTP binding of Hexa I proteins can be conveniently measured in a cuvette or 96-well plate format. The 96-well format allows for screening a large number of recombinant proteins quickly, with minimum sample volumes required.

PROTOCOL 7: RECEPTOR-MEDIATED GTPγS
BINDING USING EXTRINSIC FLUORESCENCE

1. The Eu-GTP binding assay is based on the DELFIA Europium-GTP binding kit (PerkinElmer) using a Victor fluorescence spectrophotometer (PerkinElmer).
2. Under dim red light, incubate 1 nmol rhodopsin with 1 nmol G-protein in a 2 mL microcentrifuge tube for 10 min on ice.
3. Add 1.8 mL of buffer containing 100 mM HEPES (pH 8.0), 1 mM EDTA, 10 mM MgSO$_4$, and 10 mM DTT and mix well to resuspend the membrane.
4. Wash each well of a Pall-Gelman AcroWell 96-well filter plate with 0.45 μm GHP nitrocellulose membrane once with 300 μL cold dilute GTP Wash Buffer from the kit.

5. Photolyze rhodopsin and add 200 μL of 100 nM Eu-GTP (10 nM final concentration) to initiate the reaction.
6. Filter 2 × 100 μL reaction at each time point (1, 3, 5, 7, 10, 15, 20, and 25 min). Wash with 3 × 300 μL cold GTP Wash Buffer.
7. After the final time point, disconnect the vacuum and add 50 μL of GTP Wash Buffer to each well.
8. Generate a Eu-GTP standard curve by adding 10 μL of serial dilutions from 0–5 nM to the filter plate. Allow dilutions to absorb into the filter, then add 50 μL of cold GTP Wash Buffer.
9. Read the plate in the Victor fluorescence spectrophotometer using the manufacturer's preset protocol for Europium time-resolved fluorometry.
10. Convert counts to moles of Eu-GTP-bound using the standard curve (figure 1.6).

Protocol 8: Receptor-Mediated GTPγS Binding Using Intrinsic Fluorescence

1. Using a Varian Cary Eclipse (or any other standard fluorescence spectrophotometer), zero the instrument on buffer alone containing 10 mM MOPS, pH 7.2, 150 mM NaCl, 2 mM $MgCl_2$ at 15°C, using excitation/emission wavelengths of 300/340 nm.
2. In a separate tube, reconstitute heterotrimer by incubation of 500 nM Gα and 500 nM Gβγ on ice for 15 min prior to assay.
3. Add heterotrimer plus 100 nM rhodopsin to the cuvette and monitor the emission over 1 min to obtain baseline.
4. Add GTPγS to 10 μM and monitor the emission over 40 min at 15°C.
5. Normalize data to the baseline (set to 0%) and the maximum fluorescence at 100%, set to the value obtained by activation by AlF_4^-.

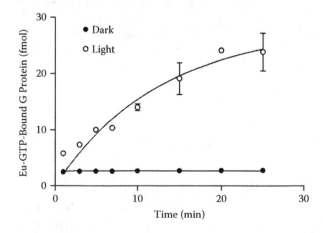

FIGURE 1.6 Basal (closed circles) and rhodopsin-catalyzed (open circles) binding of Eu-GTP to transducin under conditions described in protocol 7. Data are the average ±SEM of three independent experiments.

6. The rate of receptor catalyzed nucleotide exchange can be calculated by fitting the data to the equation $F = Fmax(1-e^{-kt})$, where F is the percent of maximum fluorescence (Fmax) at time t s and k is the catalytic activation rate constant for GTPγS binding in s^{-1}.

1.3.4 RECEPTOR BINDING

G-proteins couple to activated receptors, which catalyze release of GDP and uptake of GTP, and both GαGTP and Gβγ then act on distinct downstream effectors. A traditional rhodopsin-binding assay is used to confirm the ability of Hexa I proteins to bind to rhodopsin, a prerequisite to measuring receptor-catalyzed changes in the Hexa I proteins using EPR and fluorescence methods. In the rhodopsin binding assay (protocol 9), Gα is incubated with Gβγ and rhodopsin under three conditions: dark, light, and light with GTPγS. The samples are centrifuged, and the membrane and supernatant fraction are analyzed by SDS-PAGE. Upon light activation, the G-protein moves from the supernatant to the membrane, and it is released upon the addition of GTPγS (figure 1.7).

PROTOCOL 9: RHODOPSIN-BINDING ASSAY

1. Incubate 10 μM Gα hexa I protein with an equimolar amount of Gβγ at 4°C for 15 min.
2. Incubate urea-washed ROS membranes (100 μM) with 10 μM reconstituted heterotrimer at 20°C with 100 μM GTPγS for 30 min under light conditions.
3. Following this 30 min incubation, pellet membranes by centrifugation at $20,000 \times g$ for 1 h.
4. Pellet and supernatant fractions are then analyzed by SDS-PAGE (figure 1.7), visualized with Coomassie blue and quantified by densitometry (Molecular Imager ChemiDoc™, XRS System, Biorad, Hercules, California).
5. The percent bound can be calculated from the equation below, where GαL and GαG are the amount of Gα in the light and GTPγS supernatants, respectively.

$$\% \text{ Bound} = 1 - (G\alpha_L/G\alpha_G)$$

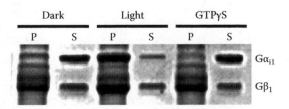

FIGURE 1.7   Rhodopsin-binding assay. Gα is incubated with Gβγ and rhodopsin in dark (left), light (center), and light with addition of GTPγS (right). Pellet (P) versus supernatant (S) are analyzed by SDS-PAGE.

1.4 FLUORESCENT PROBES FOR ANALYSIS OF CONFORMATIONAL CHANGES IN GPCR SIGNALING

The engineering of cysteine residues into sites of interest (in the background of the $G\alpha_i$ Hexa I parent protein with solvent-exposed cysteines removed) provides a valuable tool to study changes in specific regions of $G\alpha$ proteins during the activation cycle by both fluorescence and EPR. Understanding the data obtained from these two techniques together provides a fuller, more comprehensive picture of protein dynamics than either technique alone.

There is a great variety of commercially available fluorescent probes to choose from when labeling $G\alpha$ subunits. Some probes such as bimane and 1-anilino-8-naphthalene sulfonate (ANS) exhibit environmentally dependent spectral shifts, whereas other probes report changes in environment by changes in emission intensity. It is important to choose a probe or probes that reflect conformational changes without perturbing protein function and do not conflict with spectral characteristics of other components used in the assay. For example, rhodopsin absorption maxima change as it moves through the G-protein cycle. As probes can be charged or uncharged, hydrophobic or hydrophilic, it is important to consider their potential effects on the labeled protein and its ability to interact with other proteins. To control for this, conformational changes measured by fluorescence are best measured with several probes of varying size and chemistry. This is generally not difficult to achieve, given the variety of fluorescent probes currently available (Invitrogen, Madison, Wisconsin; Toronto Research Chemicals, Toronto, Canada).

1.4.1 FLUORESCENT LABELING OF $G\alpha$ SUBUNITS

In this section we describe various methods to label $G\alpha$ subunits for fluorescence assays. First, one must choose a suitable fluorescent label or labels. For example, if one is conducting a protein–protein interaction assay with labeled $G\alpha$ subunit as reporter of the interaction, Lucifer yellow and bimane are good choices, as they are relatively small in size and moderately water soluble. Bodipy, although somewhat more hydrophobic, has a higher quantum yield, making it a sensitive reporter of changes in emission intensity. Bimane and other probes demonstrate shifts in their emission maxima as a result of changes in conformation or binding that exclude solvent from the probe's environment. If the labeled proteins will be used in conjunction with rhodopsin, then probes with spectral characteristics in the red region are preferred in order to avoid spectral interferences from rhodopsin. It is also important to ensure the probe is photostable (resists bleaching) and does not perturb protein function. Therefore, all labeling should be accompanied by functional assays to ensure that the functional integrity of the labeled protein is maintained.

The low degree of background labeling in the $G\alpha_i$ Hexa I parent protein promotes labeling at the cysteine residue of interest. However, cysteine mutations in regions of the protein with reduced solvent accessibility may result in poor labeling, despite good functional activity of the unmodified protein; therefore, labeling efficiency at different locations within the protein may vary considerably. Labeling conditions may need to be optimized for specific cysteine mutants. For example, labeling at the extreme termini of the protein is very efficient, requiring less labeling time than other regions of the protein. Labeling efficiency also depends on the nature of the probe used; probes of a more hydrophobic nature, which exhibit poor solubility in water, tend to label less efficiently than water-soluble probes. This can be ameliorated to some extent by labeling in the presence of DMSO, which can be used at varying concentrations up to 25% for short periods without adversely affecting $G\alpha_i$ protein's ability to undergo activation-dependent conformational changes. Because only labeled residues act as reporters in these fluorescence assays, a relatively modest degree of labeling is generally sufficient (<50%), and even desirable, as excessive labeling may perturb protein function if secondary labeling is an issue. For example, iodoacetamides can react with residues other than cysteines, albeit at a much slower rate, and generally only when targeted cysteines have reduced solvent accessibility. Malemides, commonly used to label cysteines (figure 1.8) are more selective for cysteine residues. In general, if the inserted cysteines are in a solvent-accessible environment, these will be successfully labeled using a relatively short labeling protocol of less than 4 h.

Intermolecular disulfide bonding is predicted to lower labeling efficiency, therefore labeling at high protein concentrations is not generally recommended (above 3 mg/mL). If the purified proteins have been stored in reducing agents such as DTT or 2-mercaptoethanol, these need to be removed prior to modification with thiol-reactive probes, by gel filtration, or dialysis. As mentioned previously, TCEP (tris-(2-carboxyethyl)phosphine) is a mild reducing agent reported to selectively reduce disulfides with aqueous exposure (21). Because this reagent does not contain thiols, it should not require removal prior to labeling. Nevertheless, its presence may reduce labeling efficiency; therefore, it is worthwhile to do a pilot labeling experiment to determine the optimum ratio of probe:protein in the presence of TCEP.

Once the labeling reaction has been carried out, the reaction is quenched, and the free probe must be removed from the labeled protein. This step is highly dependent on the nature of the probe used, but is a critical step, as contamination by the free probe will decrease the ability to measure fluorescent changes in the specific, labeled residue undergoing conformational changes. The purified, labeled protein is then

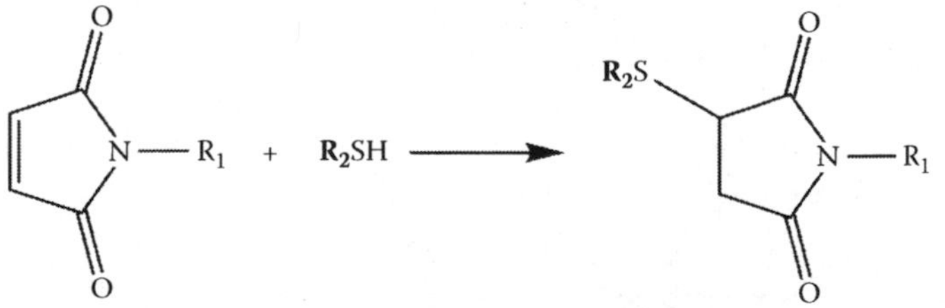

FIGURE 1.8 Fluorescent malemide (probe, R1) labels thiol group of inserted cysteine residue in a $G\alpha_i$ Hexa I protein (R2) to produce a specifically labeled $G\alpha_i$ Hexa I protein. The $G\alpha_i$ Hexa I cysteine mutant used in this reaction is a product of SDM of the parent, cys-less $G\alpha_i$ Hexa I protein, which exhibits very little background labeling and retains functional qualities similar to wild type. (From Medkova, M. et al., Conformational changes in the amino-terminal helix of the G protein α_{i1} following dissociation from $G\beta\gamma$ subunit and activation, *Biochemistry* 41, 9962, 2002. With permission.)

subjected to repeated functional assays to ensure that the protein's folding and activity is maintained. $G\alpha_i$ Hexa I can be labeled at cysteine residues by the following general method (protocol 10), using malemides or iodoacetamide derivatives, which form stable thioether bonds (see figure 1.8).

PROTOCOL 10: GENERAL FLUORESCENT LABELING PROTOCOL

1. Prepare a 10 mM stock solution of probe in DMSO.
2. For a water-soluble probe, add fluorescent probe in a 5:1 molar ratio to $G\alpha$ protein at 1–3 mg/ml protein concentration in 50 mM Tris, 50 mM NaCl, 1 mM $MgCl_2$, and 100 µM GDP.
3. Rotate at 4°C for 1–4 h, protected from light. This time should be determined from labeling trials to determine optimal conditions for labeling. Addition of a small amount of DMSO to the protein solution prior to introduction of the probe may be required to enhance the solubility of probes which are susceptible to precipitation in aqueous solution.
3. At end of the incubation, quench the reaction with an equimolar quantity of β-mercaptoethanol.
4. Excess probe may be removed by overnight dialysis or alternatively, washing in a suitable centrifugal device with a 10 kDa molecular weight cut-off, followed by gel filtration using a P6 gel spin column (Bio-Rad Laboratories, Hercules, California), according to the manufacturer's directions. HPLC gel filtration chromatography using a TosoHaas SW2000 column (Toso-Haas, Montgomeryville, Pennsylvania) can be used to eliminate traces of unreacted probe from labeled protein. Care must be taken to filter the sample prior to application to the column if unreacted probe precipitates. Use of a guard column is also recommended.
5. If the crude labeling mixture contains precipitated (unreacted) probe, the combination of P6 gel spin column (Bio-Rad Laboratories, Hercules, California) and a final passage through a 0.45 µM cellulose acetate syringe filter (Nalgene, Rochester, New York) effectively removes unbound, precipitated probe from labeled protein. This labeled protein may then be further purified by HPLC as described earlier.
6. Determine labeling stoichiometry by the ratio of $G\alpha_i$ protein to probe as measured by the Coomassie blue binding method to determine molar concentration of protein and the concentration of probe in the labeled protein as measured by Beer's law (using the extinction coefficient specific for the label and path length specific for the spectrophotometer used).
7. Labeled proteins can be visualized by UV illumination of SDS-PAGE gels (figure 1.9) and confirmed by Coomassie blue staining.
8. Store labeled protein in 50 mM Tris, 50 mM NaCl, 1 mM $MgCl_2$, 10 µM GDP, and 10% glycerol at −80°C.

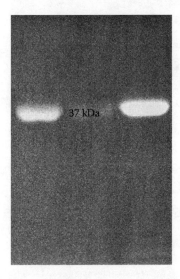

FIGURE 1.9 Bodipy-labeled Gα$_i$ Hexa I proteins after SDS-PAGE on 10–20% Tris-glycine gel, using 5 μg of Hexa I protein with fluorescent label at residue 13 (left) and 351 (right), visualized by UV illumination.

1.5 FLUORESCENCE MEASUREMENT OF PROTEIN DYNAMICS

1.5.1 ANALYSIS OF CONFORMATIONAL CHANGES IN Gα
SUBUNITS USING FLUORESCENT TECHNIQUES

Fluorescence is a convenient tool for measuring dynamic changes in G-proteins upon activation. For example, Lucifer yellow (LY), acrylodan, and a variety of other commercially available probes have been used to look at activation-dependent changes within the amino and carboxy termini of proteins (protocol 11) (13,15,17). These are relatively small probes with excitation and emission wavelengths in the ultraviolet range. Changes in the emission from these probes indicate a change in the environment for the particular probe upon activation and deactivation. Changes in the emission of a series of residues in a selected location in the protein can be used to build a picture of the changes that occur within that region during activation and deactivation. In general, activation-mediated changes or binding events that bring a probe into a more solvent excluded environment are reflected in enhanced emission, blue-shifted wavelength, or both, depending on the probe. Therefore, excitation and emission scans should be performed for individual fluorophores to determine how changes in the probe's environment are reflected in the spectral characteristics of the probe. The experimental design will determine additional controls to use, such as a quenched probe in place of labeled protein in a binding assay. In some cases, (unlabeled) interacting proteins may have some intrinsic fluorescence at the probe's wavelength, and this should be examined routinely, as it could lead to an extraneous signal. Changes in specific regions of the protein (reported by changes in fluorescence) can be measured upon activation, as well as binding to effectors such as Gβγ. These changes can often be quantified; for a Gα-Gβγ subunit association, increasing amounts of unlabeled Gβγ can be incubated with specifically labeled Gα$_i$ Hexa I proteins (15) to demonstrate dose-dependent binding of these subunits. This can be done with steady-state fluorescence emission (15) or by using fluorescence polarization as described in protocol 11.

PROTOCOL 11: MEASUREMENT OF SITE-SPECIFIC ACTIVATION-DEPENDENT CHANGES BY FLUORESCENCE

1. Gα$_i$ Hexa I proteins with cysteines at sites of interest are labeled according to the general protocol 10.
2. Excitation and emission is set to probe specific wavelengths. (For LY-labeled protein, an emission at 520 is monitored with excitation at 430.) Emission is set to zero on buffer alone, containing 50 mM Tris, 50 mM NaCl, 2 mM MgCl$_2$, and 10 μM GDP.
3. Gα$_i$ Hexa I proteins (200 μM) are added to 800 μL buffer, and the emission is monitored at the probe-specific wavelength.
3. Proteins are activated by AlF$_4^-$ (10 mM NaF and 50 μM AlCl$_3$), and emission is monitored for several minutes until the reading is stabilized (figure 1.10).
4. Changes in emission in AlF$_4^-$-activated subunits are typically expressed as a relative increase over basal (Gα-GDP).

1.5.2 SUBUNIT INTERACTIONS AS MEASURED BY FLUORESCENCE RESONANCE ENERGY TRANSFER (FRET)

Changes in the emission from the labeled Gα protein as a result of binding to another interacting protein, such as Gβγ, can be measured by fluorescence. For a Gα Hexa I

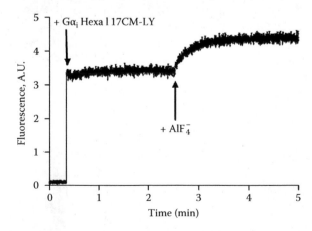

FIGURE 1.10 Gα$_i$ Hexa I 17CM-LY (200 nM) exhibits changes in emission upon AlF$_4^-$ activation (10 mM NaF and 50 μM AlCl$_3$). Excitation/emission measured at 430/520 nm.

cysteine mutant labeled along the amino terminus, this binding is detected as a dose-dependent increase in emission from $G\alpha_i$ Hexa I 13C-LY upon binding $G\beta\gamma$ subunits (15). In this protein, the amino terminus is relatively unstructured, whereas binding to $G\beta\gamma$ orders it, a result seen in both fluorescence and EPR assays (15). For the myristoy-lated counterpart, both EPR and fluorescence support the idea that the amino terminus retains a relatively immobile and solvent excluded environment in the GDP- and $G\beta\gamma$-bound form; therefore, a more sensitive readout of binding is required. To enhance detection of $G\alpha$–$\beta\gamma$ interactions, both $G\alpha$ and $G\beta\gamma$ can be labeled with fluorophores that overlap in the emission and excitation spectra. This allows detection of binding by FRET, which is very sensitive to changes in the distance between two probes in differ-entially labeled proteins. FRET relies on the nonradiative transfer of energy via long-range dipole–dipole coupling from one fluorescent molecule to another, which occurs when the two molecules come within Förster distance (the distance at which energy transfer is 50% efficient). Typically this requires molecules to be within 100 Å or less, and the excitation spectrum of one molecule must overlap the emission spectra of the other. As FRET is proportional to the inverse sixth power of the distance between the two molecules, it is highly sensitive to interactions that bring molecules into the prox-imity required for energy transfer to occur.

Interaction of labeled $G\alpha$ Hexa I proteins to $G\beta\gamma$ can be measured using FRET between LY-$G\alpha_i$ Hexa I proteins and 2-(4'-maleimidylanilino)naphthalene-6-sulfonic acid (MIANS) labeled $G\beta\gamma$ proteins. MIANS has emission maxima at 417 nm, whereas LY excitation and emission maxima are at 420/530 nm. Therefore, binding of these can be monitored by excitation at MIANS specific wavelength, and monitoring the emis-sion increase at 530 nm using a small amount of LY-labeled $G\alpha_i$ Hexa I protein and increasing amounts of MIANS-labeled $G\beta\gamma$ protein (protocol 12). If desired, a binding curve can be constructed using increasing amounts of $G\beta\gamma$ until saturation is reached, as evidenced by a plateau in the emission intensity increases (13).

PROTOCOL 12: INTERACTIONS BETWEEN LABELED G-PROTEIN SUBUNITS MEASURED BY FRET

1. To label $G\beta\gamma$ subunits with MIANS at sites distinct from that involved in binding $G\alpha$ subunits, holotransducin is incubated with a 10 M excess of MIANS at 4°C for 2 h in buffer containing 50 mM Tris-HCL, 50 mM NaCl, 1 mM $MgCl_2$, and 50 µM GDP at pH 8.0.
2. The reaction is quenched with 2-mercaptoethanol.
3. The reaction mixture is purified in labeling buffer supplemented with AlF_4^- (10 mM NaF and 50 µM $AlCl_3$), which allows recovery of $G\beta\gamma$-MIANS free of active $G\alpha$-GDP-AlF_4^- in late fractions from gel filtration chromatography on an SW2000 column (TosoHaas, Montgomeryville, Pennsylvania). This results in $G\beta\gamma$ labeled at sites distinct from the α–$\beta\gamma$ binding interface.
4. A 1.5 molar excess of MIANS-$G\beta\gamma$ is added to LY-labeled $G\alpha$ subunits. Binding of labeled subunits is detected as a fluorescence emission increase when the complex emission is scanned at acceptor emission wavelengths, with excitation at the donor-specific excitation maxima, less any contribu-tions in the absence of labeled protein (figure 1.11).

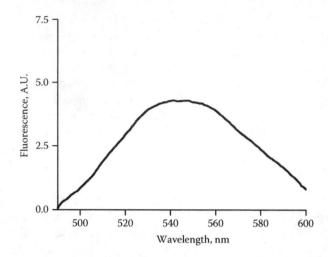

FIGURE 1.11   Binding of LY-labeled Gα Hexa I 21 CM to MIANS-labeled Gβγ, as detected by FRET. MIANS-labeled Gβγ is donor, with emission maxima at 425 nm. LY-labeled $Gα_i$ Hexa 21CM protein serves as acceptor, with excitation maxima at 420 nm. Emission scanned from 495–620 nm, with sample excitation set at 320 nm (bandpass 5 nm); emission represents energy transfer from labeled Gβγ to labeled $Gα_i$ Hexa I protein, after subtraction of emission in the absence of interacting protein.

1.5.3   FLUORESCENCE POLARIZATION MEASUREMENT OF PROTEIN DYNAMICS

Steady-state fluorescence anisotropy can also be used to examine protein dynamics using a spectrophotometer equipped with polarizers. In general, polarization reflects the rotation of a molecule during the period between excitation and emission. Small molecules rotate quickly, hence have a low polarization value, as opposed to larger molecules, with a larger degree of polarized emission due to slower rotation in solution. Fluorescence anisotropy can be used to examine protein–protein or protein–ligand interactions, such as binding of Gα to Gβγ subunits. In this case, one can measure changes in polarization due to reduction of global tumbling of a labeled Gα subunit bound to Gβγ, or vice versa. By monitoring the value of fluorescence polarization over a range of concentrations, one can construct a binding curve. In the case of protein–ligand interactions, the optimal strategy would be to fluorescently label the smaller molecule (ligand) and measure polarization upon binding to the larger protein molecule.

In order to measure fluorescence polarization, emission is measured parallel and perpendicular to the vertically polarized excitation light. Polarization (p) is defined in terms of emission intensity from parallel (S) and perpendicular (P) directions; polarization values are typically reported in mp, which is 1000 times the polarization value.

$$mp = 1000 \times \frac{(S-P)}{(S+P)}$$

where S and P are background subtracted fluorescence counts.

Furthermore, this value can be expressed in terms of anisotropy (A), a unitless number, by measurement of a grating correction factor, G. This instrument-dependent factor normalizes for disparities in the detection efficiencies in the parallel and perpendicular directions by comparison of intensity of emission of a pure fluorophore in solution in the parallel and the perpendicular directions.

Results can also be expressed as anisotropy (A) by the following method, where G is ratio of emission intensity of a pure fluorophore in the parallel and perpendicular directions.

$$G = (S/P)$$

$$A = [(S - GP)/(S + 2PG)]$$

Many screening methods of protein–protein or protein–ligand interactions compare the polarization values in terms of mp, which can provides a relative comparison of interactions for identical systems. Anisotropy is a unitless number that takes into account changes in light intensity due to differential efficiencies in emission between the parallel and perpendicular directions (making polarization values more instrument dependent). For screening purposes, mp values may be entirely sufficient to describe the relative differences in binding of ligands within one receptor system, using a specified instrument and experimental protocol. For comparison of values under differing experimental conditions, anisotropy may be the preferred readout. For a totally depolarized molecule, the limiting anisotropy is 0.4; values in excess of this limit may represent scatter, and a closer examination of experimental conditions is required.

In highly viscous solutions, such as 50% glycerol, global tumbling is minimized, and anisotropy values more closely reflect mobility of isolated residues (protocol 13). To determine mobility of residues in selected regions of $G\alpha$, LY-labeled $G\alpha$ subunits were examined by fluorescence polarization using a Victor V multilabel plate reader in highly viscous solution (figure 1.12). Results were mathematically converted to

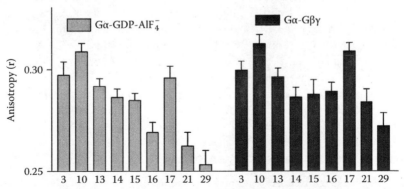

FIGURE 1.12 Anisotropy of myristoylated $G\alpha_i$ Hexa I-LY proteins labeled at the indicated amino-terminal residues (ex/em 430/520). Anisotropy of $G\alpha_i$ subunits (200 nM) in 50 mM Tris, 50 mM NaCl, 1 mM MgCl$_2$, 5 μM GDP, and 50% glycerol is calculated from fluorescence polarization values obtained with the Victor multilabel plate reader (G = 1.143) upon addition of (left) AlF$_4^-$ (10 mM NaF and 50 μM AlCl$_3$) or (right), in the G$\beta\gamma$ (400 nM) bound state. Data are average of three or more independent experiments.

anisotropy by taking into account differences in efficiency of light transmission in the parallel and perpendicular directions.

PROTOCOL 13: MOBILITY OF SELECTED Gα$_i$ HEXA I RESIDUES AS MEASURED BY FLUORESCENCE POLARIZATION

1. Using a Victor V multilabel plate reader (PerkinElmer, Waltham, Massachusetts) set excitation/emission at probe specific wavelengths, using appropriate filters for the fluorescent label excitation and emission specifications.
2. Measure the emission in parallel and perpendicular directions of buffer alone and buffer containing labeled Gα$_i$ Hexa I protein, either activated with AlF$_4$ or in the presence of Gβγ. Buffer consists of 50 mM Tris, 100 mM NaCl, 1 mM MgCl$_2$, 5 μM GDP, pH 7.5, and 50% glycerol. This is conveniently performed in 96- or 384-well plates, which can be specified from the software pull-down menu using the protocol wizard.
3. Readings are taken every 1.5 min over a 20 min period and averaged over this time period.
4. It is necessary to subtract buffer counts in parallel and perpendicular directions from experimental samples. Polarization may be reported directly (in mp) or converted to anisotropy (r) by measurement of the efficiency of light transmission in the parallel and perpendicular directions (G factor) for pure fluorophore in solution. Use buffer subtracted values and G factor to mathematically convert polarization to anisotropy as described above.

It is common in high-throughput screening assays under rigorously controlled conditions to report polarization in mp units, which does not require measurement of a G factor, because the experimental conditions under which these assays are performed are held constant. For monochromator-driven spectrometers equipped with polarizers, the wavelengths of interest can be selected from the instrument control panel, and using a Varian Cary Eclipse, G factors can be measured automatically or results can be reported in mp units. Manual polarizers will perform similarly, but one must be careful to set them appropriately before each measurement.

Fluorescence anisotropy also provides a convenient measure of binding between G-protein subunits, using a fluorescently labeled Gα or Gβγ as a reporter of the interaction (protocol 14). As binding between these subunits reduces global tumbling of the macromolecule, the degree of polarized emission will increase.

PROTOCOL 14: SUBUNIT ASSOCIATION MEASURED BY STEADY-STATE FLUORESCENCE ANISOTROPY

1. Using a Varian Eclipse equipped with automated polarizers, set excitation at 490 nm and emission at 512 nm. Select the G factor measurement from menu options.
2. Measure anisotropy from Bodipy-labeled Gα$_i$ subunit in 50 mM Tris, pH 7.8, 50 mM NaCl, 2 mM MgCl$_2$, 10 μM GDP, 1 mM DTT, and 0.05% Chaps.
3. Incubate labeled Gα$_i$ subunits with increasing amounts of Gβγ for 5 min at 18°C, and record anisotropy for at least three replicates.

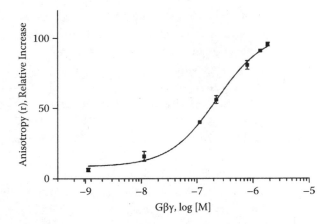

FIGURE 1.13   Binding of BD-Gα_i Hexa I 305C to unlabeled G$\beta\gamma$ results in a dose-dependent increase in anisotropy. Anisotropy of 30 μM Gα_i Hexa I 305C-BD incubated with increasing amounts of unlabeled G$\beta\gamma$ is measured on a Varian Cary Eclipse, excitation/emission 490/512 nm, G factor 1.721, in 50 mM Tris, pH 7.8, 50 mM NaCl, 2 mM MgCl$_2$, 10 μM GDP, 1 mM DTT and 0.05% Chaps. Polarized emission is relative to emission of labeled Gα_i protein in the absence of G$\beta\gamma$, and shown as percent of maximum, resulting in an apparent K$_d$ of 250 nM for the interaction.

4. Increases in anisotropy were plotted as a percent of maximum anisotropy versus the G$\beta\gamma$ concentration (figure 1.13), and results were analyzed using GraphPad Prism (GraphPad Software, San Diego, California).

1.6   ANALYSIS OF CONFORMATIONAL CHANGES USING EPR

EPR spectra reveal information about the mobility of nitroxide spin-labeled residues, including the time scale of such motion, and these measurements can be extended to determine distance changes between two spin-labeled residues. EPR relies on the interaction of an applied magnetic field with the magnetic dipole moment of an unpaired electron. The nitroxide label provides the unpaired electron that acts as the reporter of conformational changes. Electron paramagnetic resonance spectroscopy measures the absorption of microwave radiation by this unpaired electron when it is oriented in a magnetic field. This is very similar to nuclear magnetic resonance spectroscopy, which measures the absorption of radiofrequency radiation by spin-active nuclei. In either method, the electron or nucleus is extremely sensitive to its local environment, enabling the spectrum to provide detailed structural information. In an extension of this technique, continuous-wave EPR spectrum provides information about mobility, solvent accessibility, and distances between paramagnetic centers (reviewed in (22)).

EPR is extremely sensitive; it can measure concentrations well below 0.1 mM, and reports changes in the spin-labeled residue without interference from the rest of the protein. In EPR spectroscopy, the magnetic field is varied, whereas the frequency of electromagnetic radiation is held constant. Typically, the magnetic field range used is 0–1000 Gauss (up to about 1 T), and frequency is about 9 GHz, within the microwave region. With a series of single nitroxide-labeled residues within a protein, the nitroxide side chain is scanned through the protein sequence, and a picture of global structure and dynamics can be constructed (23). The mobility of a nitroxide side chain reflects internal motions of the side chain (bond rotations) and fluctuations of the backbone. EPR spectra from proteins of known structure have resulted in a large database of EPR structures (24), which correlate to secondary structure sites, buried sites, loop sites, and tertiary contact sites on the basis of their EPR structure. The nitroxide label does not perturb structure, as has been shown extensively for T4 lysozyme (25). Thus, EPR, combined with SDM, provides a powerful method to explore backbone dynamics of both soluble and membrane proteins. By interrogation of a series of sites along a specified region of the Gα protein using this technique, structural and dynamic changes can be characterized at specific times within the G-protein cycle (scheme 1.2).

1.6.1   SPIN-LABELING OF Gα$_i$ HEXA I PROTEINS

The thiol-reactive methanethiosulfonate spin label is the most commonly used paramagnetic probe in the biophysical studies of proteins (scheme 1.2). This label reacts with cysteine residues in proteins to generate the nitroxide side chain R1 (protocol 15), and the dynamics of the nitroxide on the nanosecond time scale can be determined by spectral line shape analysis (26).

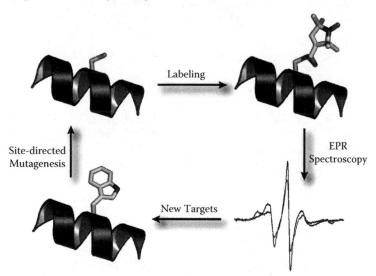

SCHEME 1.2   EPR Strategy. Site-directed cysteine mutagenesis provides a reactive thiol group for the specific incorporation of the paramagnetic nitroxide probe. Local structural changes in the labeled region can be identified by comparing the EPR spectra recorded for different conformations of the protein.

PROTOCOL 15: NITROXIDE SPIN-LABELING OF $G\alpha_i$ HEXA I PROTEINS

1. Methanethiosulfonate spin label reagent, S-(1-oxy-2,2,5,5-tetramethylpyr-rolinyl-3-methyl) methanethiosulfonate (MTSS, Toronto Research Chemicals, Inc., Ontario, Canada) is added to 2 mg/ml $G\alpha_i$ Hexa I proteins solution in a 1:1 molar ratio in 20 mM MES, pH 6.8, 100 mM NaCl, 2 mM $MgCl_2$, 50 μM GDP, and 10% glycerol at room temperature for 5 min.
2. Unreacted probe is removed by extensive washing using a 10 kDa molecular weight cutoff filtration device (Millipore).
3. Gβγ and rhodopsin are also buffer exchanged into the labeling buffer (without additional GDP) prior to use in these experiments. For rhodopsin, the concentration apparatus is foil-wrapped during this process to prevent light activation of rhodopsin.

1.6.2 EPR ANALYSIS OF CONFORMATIONAL CHANGES IN $G\alpha_i$ HEXA I PROTEINS

A typical EPR experiment begins by recording an EPR spectrum for the purified, labeled, GDP-bound Gα subunit. This spectrum can then be correlated to the location of the labeled side chain in the G-protein structure, and compared to an extensive library of spectra of known mobility and solvent accessibility. For example, sharp peaks have been shown to correlate to highly disordered positions, such as solvent-exposed loop regions, whereas peak broadening represents a more restricted mobility. Next, each limit conformation of Gα is examined by sequential recording of EPR spectra after addition of (1) Gβγ, (2) dark (inactive) rhodopsin, (3) light, and (4) GTPγS (protocol 16). Relative changes in conformation are identified by comparison of the EPR spectra recorded for any two states in the G-protein cycle. For example, the receptor activation-dependent conformational changes in Gα leading to GDP release can be observed by comparing the EPR spectra recorded for the reconstituted heterotrimer before and after photoactivation of the receptor (figure 1.14). In this way, dynamics of each structural transition along the activation pathway of Gα can be mapped for numerous sites in the protein.

PROTOCOL 16: EPR ANALYSIS OF CONFORMATIONAL CHANGES

Urea-washed ROS samples are buffer exchanged into 20 mM MES, pH 6.8, 100 mM NaCl, 2 mM $MgCl_2$, and 10% glycerol prior to use in assays with labeled $G\alpha_i$ Hexa I proteins.

1. First, cysteine mutants (30 μM) were loaded into a sealed quartz flat cell.
2. Spectra are recorded at room temperature on a Bruker E580 spectrometer using a high-sensitivity resonator (HS0118) at X-band microwave frequencies (Bruker BioSpin Corp., Billerica, Massachusetts).
3. Each spectrum is collected using a 100 Gauss field scan at a microwave power of 19.92 mW.

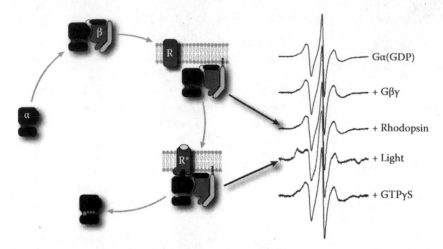

FIGURE 1.14 Electron paramagnetic resonance spectroscopy (performed in collaboration with Wayne L. Hubbell, Ph.D., and Ned Van Eps, Ph.D., in the Department of Chemistry and Biochemistry and Department of Ophthalmology at the Jules Stein Eye Institute, University of California–Los Angeles). Beginning with the purified, labeled GDP-bound Gα subunit, each limit conformation of Gα is explored by EPR. Conformational changes are identified by comparing the EPR spectra recorded for any two conformations. In this example, there is a receptor activation-dependent decrease in the mobility of the spin-labeled side chain.

4. Optimal field modulation amplitudes are selected to give maximal signal intensity without lineshape distortion.
5. 20–50 scans are averaged to obtain spectra.
6. After collecting each Gα spectrum, Gβγ is added to the sample in a 1:1 molar ratio to form heterotrimers. The diluted samples were concentrated to the same concentration as the initial Gα mutants.
8. The EPR spectra of each heterotrimer is recorded both alone in solution and upon addition of urea-washed ROS in the dark (150 μM).
9. For each mutant, a dark spectrum was recorded.
10. Following recording of dark spectrum, the sample is irradiated for 30 s using a tungsten lamp (cutoff filter; λ > 500 nm), and the EPR spectra is collected immediately after bleaching.
11. Finally, GTPγS (200 μM) was added to the samples to form activated Gα.
12. Data analysis: Fitting of spectra to the MOMD model of Freed and coworkers follows previously published methods (Columbus et al., 2001) using principal values for the A and g tensors of Axx = Ayy = 6 G, Azz = 37 G, gxx = 2.0078, gyy = 2.0058, and gzz = 2.0023.
13. Qualitative analysis of spectra in terms of nitroxide mobility is as described in the literature (20,27). The fitting procedure gives values for the order parameter and the diffusion tensor of the nitroxide, Dxx, Dyy, and Dzz.

Biophysical approaches to the study of G-proteins can be used to elucidate key elements of the G-protein cycle. Recent EPR studies have revealed the C terminus of Gα subunits provide a critical mechanistic link between receptor and G-protein activation (12). Advances in the use of DEER spectroscopy in combination with FRET technology will extend our reach, enabling distance measurements between two labeled regions of a protein, or between labeled protein and receptor during receptor-mediated G-protein activation. These technologies, combined with other biophysical studies, can be harnessed to reveal detailed information regarding activation dynamics that occur in solution during the G-protein cycle.

REFERENCES

1. Noel, J.P., H.E. Hamm, and P.B. Sigler, The 2.2 Å crystal structure of transducin-α complexed with GTPγS, *Nature* 366, 654, 1993.
2. Coleman, D.E. et al., Structures of active conformations of G_{ia1} and the mechanism of GTP hydrolysis, *Science* 265, 1405, 1994.
3. Lambright, D.G. et al., Structural determinants for activation of the α-subunit of a heterotrimeric G protein, *Nature* 369, 621, 1994.
4. Lambright, D.G. et al., The 2.0 Å crystal structure of a heterotrimeric G protein, *Nature* 379, 311, 1996.
5. Sondek, J. et al., GTPase mechanism of G proteins from the 1.7-Å crystal structure of transducin α•GDP•AlF$_4^-$, *Nature* 372, 276, 1994.
6. Wall, M.A. et al., The structure of the G protein heterotrimer $G_{ia1}\beta_1\gamma_2$, *Cell* 83, 1047, 1995.
7. Palczewski, K. et al., Crystal structure of rhodopsin: A G protein-coupled receptor, *Science* 289, 739, 2000.
8. Teller, D.C. et al., Advances in determination of a high-resolution three-dimensional structure of rhodopsin, a model of G-protein-coupled receptors (GPCRs), *Biochemistry* 40, 7761, 2001.
9. Okada, T. et al., Functional role of internal water molecules in rhodopsin revealed by x-ray crystallography, *Proc Natl Acad Sci USA* 99, 5982, 2002.
10. Okada, T. et al., The retinal conformation and its environment in rhodopsin in light of a new 2.2 Å crystal structure, *J Mol Biol* 342, 571, 2004.
11. Li, J. et al., Structure of bovine rhodopsin in a trigonal crystal form, *J Mol Biol* 343, 1409, 2004.
12. Oldham, W.M. et al., Mechanism of the receptor-catalyzed activation of heterotrimeric G proteins, *Nat Struct Mol Biol* 13, 772, 2006.
13. Preininger, A.M. et al., The myristoylated amino terminus of $G\alpha_{i1}$ plays a critical role in the structure and function of $G\alpha_{i1}$ subunits in solution, *Biochemistry* 42, 7931, 2003.
14. Van Eps, N. et al., Structural and dynamical changes in an alpha-subunit of a heterotrimeric G protein along the activation pathway, *Proc Natl Acad Sci USA* 103, 16194, 2006.
15. Medkova, M. et al., Conformational changes in the amino-terminal helix of the G protein α_{i1} following dissociation from Gβγ subunit and activation, *Biochemistry* 41, 9962, 2002.
16. Gilman, A.G., G proteins: Transducers of receptor-generated signals, *Annu Rev Biochem* 56, 615, 1987.
17. Yang, C.S. et al., Conformational changes at the carboxyl terminus of G-alpha occur during G protein activation, *J Biol Chem* 274, 2379, 1999.
18. Mazzoni, M.R., J.A. Malinski, and H.E. Hamm, Structural analysis of rod GTP-binding protein, G_t. Limited proteolytic digestion pattern of G_t with four proteases defines monoclonal antibody epitope, *J Biol Chem* 266, 14072, 1991.
19. Yamanaka, G., F. Eckstein, and L. Stryer, Stereochemistry of the guanyl nucleotide binding site of transducin probed by phosphorothioate analogues of GTP and GDP, *Biochemistry* 24, 8094, 1985.

20. Adhikari, A. and S.R. Sprang, Thermodynamic characterization of the binding of acti-
    vator of G protein signaling 3 (AGS3) and peptides derived from AGS3 with G alpha i1,
    J Biol Chem 278, 51825, 2003.
21. Graham, D.E., K.C. Harich, and R.H. White, Reductive dehalogenation of monobromo-
    bimane by tris(2-carboxyethyl)phosphine, *Anal Biochem* 318, 325, 2003.
22. Hubbell, W.L., D.S. Cafiso, and C. Altenbach, Identifying conformational changes with
    site-directed spin labeling, *Nat Struct Biol* 7, 735, 2000.
23. Mchaourab, H.S. et al., Motion of spin-labeled side chains in T4 lysozyme. Correlation
    with protein structure and dynamics, *Biochemistry* 35, 7692, 1996.
24. Isas, J.M. et al., Structure and dynamics of a helical hairpin and loop region in Annexin
    12: A site-directed spin labeling study, *Biochemistry* 41, 1464, 2002.
25. Langen, R. et al., Crystal structures of spin labeled T4 lysozyme mutants: Implications for
    the interpretation of EPR spectra in terms of structure, *Biochemistry* 39, 8396, 2000.
26. Budil, D.E. et al., Nonlinear-least-squares analysis of slow-motion EPR spectra in one
    and two dimensions using a modified Levenberg–Marquardt algorithm, *J Magn Reson
    Ser A* 120, 155, 1996.
27. Budil, D.E., Lee, S., Saxen, A., and Freed, J.H. *J. Magn. Reson., Ser A* 120(2), 155–189,
    1996.

2 Photoactivation of Rhodopsin and Signal Transfer to Transducin

Oleg G. Kisselev

CONTENTS

2.1 INTRODUCTION

The mechanism of the phototransduction cascade in mammalian rod cells is of great interest because of the direct relevance to virtually hundreds of closely related GPCR systems involved in transmission of signals from drugs, hormones, neurotransmitters, growth factors, odors, and light.

The light receptor rhodopsin relies on interactions with the membrane-associated heterotrimeric GTP-binding protein, transducin (Gt) to initiate response to light in retinal rod photoreceptor cells (figure 2.1) (1–5). Both the receptor and the G-protein are molecular switches (6–11). Rhodopsin is inactive while occupied by the chromophore 11-*cis*-retinal. Light-induced isomerization of the 11-*cis*-retinal to all-*trans*-retinal triggers discrete conformational changes in rhodopsin, ultimately leading to the active "on" state, Metarhodopsin II, R* (12,13).

Activation of transducin, consisting of the α-subunit (Gtα) and a tight βγ-subunit complex (Gtβγ), is a multistep process. It first involves binding of the GDP-liganded transducin to Metarhodopsin II. R* catalyzes release of GDP by Gtα, and formation of a stable nucleotide-empty complex R*-Gtαβγ (14). Binding

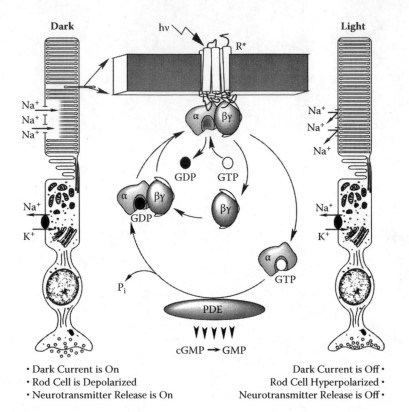

• Dark Current is On	Dark Current is Off •
• Rod Cell is Depolarized	Rod Cell Hyperpolarized •
• Neurotransmitter Release is On	Neurotransmitter Release is Off •

FIGURE 2.1   The prototypical G-protein activation cycle in rod cells.

of the activating cofactor GTP leads to the activation of Gtα, and dissociation of both subunits from R*, and from each other. Gtα-GTP relays the signal to cyclic-GMP phosphodiesterase, an effector molecule that lowers the concentration of the intracellular cGMP. The cGMP-gated cation channels close in response to low cGMP, interrupting the dark current and leading to hyperpolarization of the membrane and the generation of the nerve impulse (figure 2.1).

The G-protein switch turns off because of the intrinsic GTP-ase activity of Gtα. The activation cycle is complete when the subunits of transducin reassociate into the Gtαβγ heterotrimer. At this point transducin is ready to interact with active rhodopsin again. A single active molecule of rhodopsin is capable of activating hundreds of transducin molecules (15). This effect underlies the first step in the intracellular amplification of the light signal. Rhodopsin continues activation of transducin until it is shut off by phosphorylation with rhodopsin kinase (16,17) and binding of visual arrestin (18).

The molecular role of Gtα, which confers transducin's enzymatic activity, is fairly well understood both in vitro (6) and in vivo (19). Much less is known about the functions of the second half of Gt, the Gtβγ subunit complex (20). This chapter discusses several competing models of transducin activation, and focuses on various methodological aspects of rhodopsin photoactivation and signal transfer to transducin, with an emphasis on the role of Gtβγ in these processes.

2.2 ISOLATION OF RHODOPSIN AND TRANSDUCIN

The great success of biochemical and biophysical studies of the phototransduction mechanism is in no small part due to the fact that native components of the phototransduction cascade in rods, rhodopsin, and transducin can be isolated from bovine retinas at approximately 0.8–1 mg of rhodopsin and 0.05–0.1 mg of transducin/retina. Each procedure starts by obtaining fresh bovine eyes from a local slaughterhouse, transporting them in the dark and on ice, then additionally dark-adapting them before the removal of retinas for at least 2 h, or overnight, in a buffer containing 10 mM HEPES, pH 7.5, 120 mM NaCl, 0.2 mM $CaCl_2$, 0.2 mM $MgCl_2$, 0.1 mM EDTA, 10 mM glucose, 1 mM DTT, and 0.2 mM PMSF. Alternatively, dark-adapted frozen bovine retinas can be obtained from sources such as W.L. Lawson Co., Lincoln, Nebraska. Purification of rod outer segment (ROS)-specific proteins relies on the procedure developed by Papermaster and Dreyer in the early 1970s, which elegantly takes advantage of a unique anatomical feature of the rod cell, the fragile cilium connecting the inner segment and the body of the cell with the outer segment. The outer segments can be selectively broken off by applying uniform and gentle shear force such as during vigorous shaking, instead of full retina homogenization using mechanical devices. ROSs are then further enriched on a discontinuous sucrose gradient. This procedure is described in Protocol 1, and can be scaled from just 10–20 retinas to mega preps utilizing 500–1,000 retinas. Protocol 2 deals with the isolation of urea-washed ROS membranes (UMs), a highly enriched preparation of rhodopsin in native ROS membranes depleted of peripherally bound membrane proteins. Protocol 3 describes the transducin purification procedure, which can be used for isolation of the intact heterotrimeric Gt, as well as the individual Gtα and Gtβγ subunits.

For additional information about purification and biochemical characterization of the phototransduction proteins, we would like to direct readers to two excellent books, *Photoreceptor Cells* (21) (ed. Hargrave, P.A.) and *Signal Transduction in Photoreceptor Cells* (22) (eds. Hargrave, P.A., Hofmann, K.P., and Kaupp, U.B.).

PROTOCOL 1: ISOLATION OF ROD OUTER SEGMENTS (ROSs)

ROSs are prepared by the method of Papermaster and Dreyer (23). Except as noted otherwise, all procedures are performed in a dark room under dim red light conditions (overhead light equipped with a >620 nm pass filter).

1. Resuspend retinas in buffer ROS1 (1 mL/retina; 10 mM Tris-HCl, pH 7.4, 2 mM $MgCl_2$, 65 mM NaCl, 1 mM DTT, 0.5 mM PMSF, 34% (w/w) sucrose) directly in JA-14 rotor bottles and shake vigorously for 3 min to shear out most of the ROS.
2. Centrifuge at 4,000 rpm for 5 min (JA-14 rotor, Beckman). Keep the first supernatant and resuspend the pellet in buffer ROS1 and centrifuge again at 4,000 rpm for 5 min. Keep the second supernatant.
3. Combine the first two supernatants and dilute in buffer ROS2 1:2 (10 mM Tris-HCl, pH 7.4, 1 mM $MgCl_2$, 1 mM DTT, 0.5 mM PMSF) and centrifuge at 15,000 rpm for 15 min. Keep the pellet.

4. Resuspend the ROS pellet in buffer ROS2 containing 26.3% (w/w) sucrose, and homogenize with a 26-gauge needle attached to a syringe. Layer the suspension onto a discontinuous sucrose density gradient (29%, 35%, 40% sucrose steps).
5. Centrifuge at 24,000 rpm for 45 min in a Beckman SW-50 rotor. Collect ROS disks at the 29%–35% sucrose interface. Wash ROS disks thrice with buffer ROS-Iso (10 mM Tris-HCl, pH 7.4, 100 mM NaCl, 5 mM MgCl$_2$, 1 mM DTT, 0.1 mM PMSF) followed each time by centrifugation at 18,000 rpm for 30 min (SW-50 rotor, Beckman).
6. Finally, resuspend ROS disk membranes in buffer ROS-Iso, aliquot, and store at −70°C.

PROTOCOL 2: ISOLATION OF UREA-WASHED ROS MEMBRANES (UMs)

We routinely prepare UM using the procedure adapted from Yamazaki et al. (24) and Willardson et al. (25) essentially as we described (26). ROS disk membranes are subjected to a series of washes followed each time by centrifugation at 50,000 rpm for 20 min (Ti-60, Beckman).

1. Wash 1: ROS disk membranes are washed thrice with buffer ROS-Hypo+GTP (10 mM Tris-HCl, pH 7.4, 0.5 mM MgCl$_2$, 1 mM DTT, 0.1 mM PMSF, 1 mg/mL leupeptin, 1 mg/mL pepstatin, and 0.1 mM GTP).
2. Wash 2: ROS disk membranes are incubated in buffer ROS-Hypo containing 5 M Urea for 20 min to remove peripheral proteins, and centrifuged as described earlier.
3. Wash 3: Urea-washed ROS membranes are washed thrice with buffer ROS-Iso, resuspended in buffer ROS-Iso (5 mg/mL), and stored at −70°C.
4. Assess purity of rhodopsin by SDS-PAGE and protein staining with EZBlue Coomassie Brilliant Blue G-250 (Sigma, St. Louis, MO) (figure 2.2a). Do not boil the sample after the addition of a gel sample buffer, to avoid rhodopsin aggregation.
5. Measure rhodopsin concentration as ΔA498 before and after bleaching in the presence of 20 mM hydroxylamine, based on the molar extinction coefficient at 498 nM of 42,700 M^{-1} cm^{-1} (27). Record rhodopsin absorbance spectra using a Cary-50 UV/Visible spectrophotometer (Varian, CA), at 4°C, cuvette path-length 10 mm, essentially as we described earlier (28,29). Samples typically contain 2.5–10 μM of urea-washed ROS membranes. Run scans from 700 to 250 nm before and after photoactivation for 30 s by the Fiber-Light PL-900 equipped with a 480 ± 5 nm pass filter. Representative UV/Visible scan of urea-washed ROS membranes is shown in figure 2.2b.

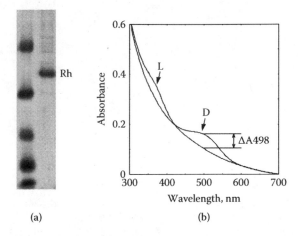

(a)   (b)

FIGURE 2.2   (See accompanying color CD.) (a) SDS-PAGE of 0.5 µg of urea-washed ROS membranes (right lane) and protein markers (left lane). Marker MW from the top: 46 kDa, 30 kDa, 21.5 kDa, 14.3 kDa, and 6.5 kDa. (b) Representative UV-Visible scan of urea-washed ROS membranes before (D) and after (L) light activation in the presence of hydroxylamine.

PROTOCOL 3: PURIFICATION OF TRANSDUCIN

Purify transducin (Gt) by GTP elution from the isotonically washed ROS disks as follows:

1. Resuspend ROS disks under bright light in the buffer ROS-Hypo for 10 min. Centrifuge the suspension at 50,000 rpm for 20 min (Ti-60, Beckman). Keep the pellet and wash it with ROS-Hypo by the same procedure once more.
2. Release Gt from the light-activated ROS membranes by three additional washes with buffer ROS-Hypo containing 100 µM GTP, followed each time by centrifugation at 50,000 rpm for 20 min (Ti-60, Beckman).
3. Combine supernatants of the GTP-containing washes (figure 2.3, Lane 1), and separate Gα and Gtβγ using AKTA FPLC on a 25 mL Blue-Sepharose Cl-6B column (Pharmacia) equilibrated in 10 mM HEPES, pH 7.5, 5 mM $MgSO_4$, 1 mM EDTA, 1 mM DTT, 10 µM PMSF.
4. Apply Gt on the Blue-Sepharose column at 0.5 mL/min at 4°C. Collect unbound fractions and analyze by SDS-PAGE. Combine unbound fractions containing Gtβγ (figure 2.3, Lanes 2–4), and concentrate to 1–2 mL using a mini (0.5 mL) DEAE-Sephacel column. Reapply unbound fractions containing holotransducin (figure 2.3, Lanes 5–10) onto a fresh Blue-Sepharose column and follow exactly the same separation procedure.
5. Wash the Blue-Sepharose column with 20 mL of equilibration buffer.
6. Elute Gtα bound to the Blue-Sepharose by a 75 mL linear gradient of KCl (0–1 M) in a column equilibration buffer. Analyze fractions by SDS-PAGE/

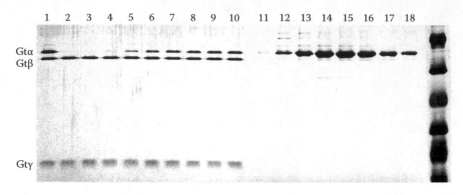

FIGURE 2.3 SDS-PAGE of fractions obtained after the affinity chromatography of purified transducin on the Blue-Sepharose column, silver staining. Lane 1 is an aliquot of transducin applied to the column. Lanes 2–10 are the flow-through fractions. Fractions 2–4 represent purified Gtβγ. Lanes 11–18 are fractions after applying a linear 0–1 M gradient of KCl on the column.

    silver staining (figure 2.3, Lanes 11–18). Combine fractions containing Gtα and concentrate to 1–2 mL by ultrafiltration using an Amicon PM-10 filter.

7. Desalt Gtα and Gtβγ separately on a Superose-12 gel filtration column using AKTA FPLC (flow rate 1 mL/min; final buffer is 10 mM Tris-HCl, pH 7.5, 1 mM MgCl$_2$, 1 mM DTT).
8. Analyze fractions by SDS-PAGE/silver staining. Determine protein concentration by BioRad Protein Assay, using BSA as a standard. Aliquot the separated subunits of transducin and store at −20°C in the presence of 20% glycerol. Estimated protein purity obtained by this procedure is 95–99%.

2.3 THE ROLE OF Gtβγ IN PHOTOTRANSDUCTION

All aspects of Gtα functions, especially its primary interactions with rhodopsin and rhodopsin-catalyzed nucleotide exchange, require the Gtβγ complex. More than 20 years of research into the molecular role of Gtβγ in phototransduction have supported, in general, the original reports by Shinozawa (30) and Kühn (31), who identified a G (for GTPase, Gtα) and an H (for Helper, Gtβγ) components of transducin. In fact, the Gtβγ role was defined to "provide an allosteric conformational advantage for the PDEase reaction, perhaps by more efficient coupling of rhodopsin, G, and PDEase" (30). It is now firmly established that some of the major functions of Gtβγ are: (1) facilitation of Gtα–GDP interactions with R* and formation of the transient nucleotide-empty R*–Gt complex (14); (2) dissociation from Gtα–GTP, allowing Gtα–GTP to activate PDE (32–36); (3) reassociation with Gtα–GDP to form an inactive form of transducin, ready for the next activation cycle (Gtβγ acts as a guanine nucleotide dissociation inhibitor (GDI)) (37); (4) interactions with phosducin, which regulates the concentration of the free form of Gtβγ, allowing fine modulation of the phototransduction cascade (38–40); and finally, but not less important, (5) Gtβγ facilitates membrane attachment of Gt via its hydrophobic farnesylated and carboxymethylated C-terminus (41–45). The weaker membrane attachment of the

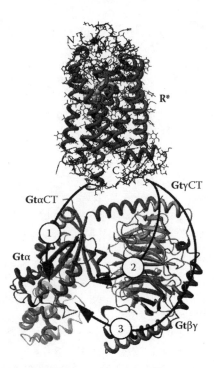

FIGURE 2.4 (See accompanying color CD.) Three-dimensional structures of rhodopsin (orange) and the G-protein, transducin (grey—Gtα, blue—Gtβγ). Rhodopsin-interacting domains of Gt at the C-terminal regions of Gtα and Gtγ are shown in yellow, and modeled based on the Tr-NOESY NMR studies (85,86). Arrows indicate proposed mechanisms of signal propagation from R* to the nucleotide-binding site of Gtα. The Gtγ C-terminal region is involved in an R*-regulated conformational switch.

farnesylated Gtβγ, compared to the many other Gβγ subtypes that are geranyigeranylated, has been shown to be crucial for proper subcellular translocation of Gtβγ and the mechanism of light adaptation (46). Nonretinal Gβγ complexes can also signal through direct interaction with effectors, but similar functions of Gtβγ remain poorly understood.

The primary role of Gtβγ in R* interactions is strongly supported by the mutational, peptide competition and NMR studies from our and other labs (20). Available data suggest that at the farnesylated C-terminal region of Gtγ is the molecular domain of Gtβγ involved in Metarhodopsin II interactions (figure 2.4). Identification of Gtγ sites involved in R* interactions (28,29,47–49) reveal an apparent distance paradox: the footprint from R* on Gt is almost exactly twice the diameter of the cytoplasmic surface of the model of the active state of rhodopsin (50,51). Because direct structural data on the R*–Gt complex are still unavailable, this discrepancy has been addressed by a number of hypotheses, including interactions with a rhodopsin dimer proposed by Palczewski (52), the lever-arm hypothesis proposed by Bourne (53), and the two-step sequential fit hypothesis that we (and others) have proposed (29,54) (figure 2.5). Building on our two-step sequential fit model, Chabre recently proposed the gear-shift model (55). Three putative gear-shift steps are a hallmark of this model:

1. Docking of Gt with R* using the C-terminal domains of Gtγ and the C- and N-terminal domains of Gtα.
2. A hinge motion of the Gtα N-terminus pushes Gtβγ into close contact with the Switch II region to help stabilize the transient nucleotide-empty state after dissociation of GDP.
3. The small hinge movement at the Gtα N-terminus/Gtβγ interface translates through the rigid-body motion of Gtβγ into the large pushing movement of the Gtγ N-terminus to displace the helical domain of Gtα and to trigger GDP release. Thus, Gtγ is viewed as the main gear-shift for coupling R* to Gtα.

The gear-shift model and the sequential fit model assume interactions of Gt with R* monomer, but differ substantially in explaining the mechanism of nucleotide release from Gtα. The sequential fit model shares common features with the lever-arm model. In short, the activating signal from R* is proposed to propagate via well-defined sites on Gtα, such as the N-terminal helix, the C-terminal domain, the α5-helix, and also via the C-terminal domain of Gtγ, shown by arrows 1 and 2 on figure 2.4. The end-result is the pulling motion by Gtβγ on the Switch II region of Gtα and release of GDP. This view was originally proposed by Bourne in his lever-arm model, which is based on an analysis showing that Gtβγ occupies the space of a nucleotide exchange factor for the Gtα-subunit, similar to the position of the nucleotide exchange factor EF-Ts, crystallized in complex with the nucleotide-empty

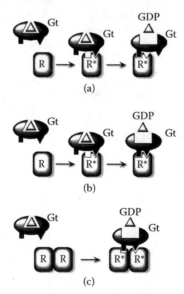

FIGURE 2.5   Two-site sequential scheme of signal transfer from light-activated rhodopsin (R*) to the G protein (Gt). Two domains representing the carboxy termini of Gtα and Gtγ subunits are symbolized with a small black triangle and a square. In model (a), two sites interact sequentially. Model (b) postulates that binding of the first site triggers binding of the second one. Model (c), interactions of Gt with a rhodopsin dimer. All models explain a large footprint on the G-protein from R*.

elongation factor Tu (Ef-Tu) (56). The gear-shift model incorporates this step as the engagement of the second gear, but it does not view this step as being responsible for the nucleotide exchange. The gear-shift model proposes the engagement of the third gear, the third route, shown by arrow 3 on figure 2.4, and argues that a pushing motion by the N-terminal domain of Gtγ partially displaces the helical domain of Gt and triggers GDP release.

The third competing model, interaction of a single molecule of Gt with a rhodopsin dimer, has been in the center of considerable debate from the moment of its discovery by atomic-force microscopy (AFM) (57,58). Additional data in support of this model have been recently reviewed (59). Perhaps the most important feature of this model is the almost perfect match in size between R*R* and Gt, which explains the paradox between the large footprint on all three subunits of Gt from a relatively compact cytoplasmic surface of R*.

Recently, the debate has shifted to demonstrating the physiological significance of rhodopsin dimerization. It has been shown that detergent solubilization conditions that favor rhodopsin oligomers also favor faster rates of Gt activation (60). This observation is consistent with earlier, as well as more recent, reports from direct binding experiments showing that R*–Gt interactions display significant positive cooperativity, with binding curves modeled with Hill coefficient 2 (25,49). Both rhodopsin and transducin oligomerization, induced by the initial binding, can contribute to this effect, which may play an important role in determining the kinetics of the visual response.

Computer modeling studies, which integrate the majority of the available biochemical and biophysical data on R*–Gt interactions, also favor the R*R*–Gt model (61), whereas other modeling attempts argue that monomeric rhodopsin contains all necessary determinants for transducin recognition (62). The latter argument is supported strongly by recent biochemical data obtained under controlled in vitro reconstitution conditions showing that a monomeric form of rhodopsin is fully capable of activating Gt (63). It is quite possible that both arguments are correct, and that rhodopsin does exist in the dimeric or higher oligomeric form, but that a single photoactivated molecule is necessary and sufficient for signal transfer to Gt and for effective nucleotide exchange.

In summary, in all current models of R*–Gt complex, Gtβγ occupies a central position as a potential conduit of the activating conformational changes from Metarhodopsin II to Gtα, but the mystery of the unified mechanism of the heterotrimeric G-protein activation remains to be resolved.

2.4 FUNCTIONAL ANALYSIS OF RHODOPSIN–TRANSDUCIN INTERACTIONS

Gt interacts with rhodopsin in a light- and nucleotide-dependent fashion (31). In the dark, these interactions are weak. Photoactivation leads to the binding of Gt–GDP to R*, dissociation of GDP, and formation of a high-affinity R*–Gt nucleotide-empty complex. Addition of GTP results in dissociation of active Gt–GTP from R* and lipid membranes, and of Gtα–GTP from Gtβγ. Both Gtα and Gtβγ are dependent on each other for these high-affinity interactions with R*. This high-affinity binding of

Gt to light-activated rhodopsin membranes and subsequent release by GTP serves as a quantitative assay of rhodopsin–transducin interactions, which can be readily monitored by quantitative immunoblotting described in Protocol 4.

In the nucleotide-empty state complex, Gt interacts with and stabilizes the active conformation of rhodopsin, R*, Metarhodopsin II, which is characterized by the deprotonated state of the Schiff-base linkage between Lys-296 of rhodopsin and all-*trans* retinaldehyde. Meta II is also identified by the absorbance maximum (A_{max}) of 380 nm, compared to A_{max} 500 nm for dark-adapted rhodopsin, and A_{max} 495 nm for Metarhodopsin I. Under the assay conditions of 4°C and pH 8.0, Meta I is the predominant photoproduct formed. Addition of Gt, or specific synthetic peptides from Gt, leads to the shift of the Meta I—Meta II equilibrium toward Meta II, producing the so-called extra-Meta II (64). Extra Meta II is measured spectrophotometrically as an A380–A417 absorbance difference from full dark/light UV/visible scans, or is monitored by real-time two-wavelength spectroscopy (Protocol 5). The key step of Gt activation is R*-catalyzed nucleotide exchange on the Gtα subunit, which can be monitored by a number of methods, including fluorescence spectroscopy (65). (Protocol 6) and uptake of [^{35}S]-GTPγS (Protocol 7).

For additional in-depth coverage of various biophysical and biochemical methods employed in the studies of the phototransduction and regulatory proteins, we would highly recommend *Molecular Mechanisms in Visual Transduction* (66) (eds. Stavenga, D. G., De Grip, W. J. and Pugh, E. N., Jr.) and *Vertebrate Phototransduction and the Visual Cycle* (67) (eds. Abelson, J. N., Simon, M., and Palczewski, K.).

PROTOCOL 4: GT BINDING TO ROS MEMBRANES

At a fixed concentration of Gtα, binding to R* occurs in a Gtβγ-concentration-dependent manner. Various amounts of the Gtβγ (0–10 μg) are reconstituted with 10 μg of Gtα and 30 μg of urea-washed ROS membranes, UMs, in 100 μl of buffer ROS-ISO (10 mM Tris, pH 7.4, 100 mM NaCl, 5 mM MgCl$_2$, 1 mM DTT, 0.1 mM PMSF) on ice.

1. Mix rhodopsin, Gtα, and Gtβγ under dim red light on ice and initiate the reaction by exposure to light for 30 s by a Fiber-Light PL-900 equipped with a 480 ± 5 nm pass filter.
2. Centrifuge the sample at 109,000 × g, 4°C for 10 min in a TLA-100.3 rotor on a Beckman TL-100 Ultracentrifuge.
3. Wash the pellet twice with buffer ROS-ISO on ice by resuspending the UM pellet and then centrifuging at 109,000 × g, 4°C for 10 min in a TLA-100.3 rotor on a Beckman TL-100 Ultracentrifuge.
4. Resuspend UM with Gt bound in 50 μl of buffer 10 mM Tris-HCl, pH 7.4, 0.5 mM MgCl$_2$, 1 mM DTT, 0.1 mM PMSF, and 250 μM GTPγS; incubate on ice for 30 min and centrifuge as before.
5. Analyze the supernatant for the presence of G-protein subunits by immunoblotting, using antibodies specific to one of the Gt subunits (figure 2.6). We prefer rabbit polyclonal antibodies generated against the N-terminal portion of the Gtβ subunit for their specificity, high-affinity interactions, and linearity of binding.

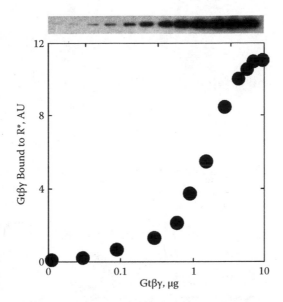

FIGURE 2.6 Transducin binding to urea-washed rhodopsin membranes and its subsequent release by GTPγS washes. Immuno-blot using Gtβ1-specific antibodies (top) and quantitation of the results by densitometry (bottom graph).

6. Quantify the protein bands on ECL films using a densitometer with Image Gauge (FujiFilm). Analyze data, do curve fitting, and plot graphs in Kaleida-Graph 3.6.2.

PROTOCOL 5: UV/VISIBLE SPECTROSCOPY

The amount of extra Meta II is measured on a Cary-50 UV/visible dual beam spectrophotometer (Varian, CA), at 4°C, cuvette path-length 10 mm, essentially as we described before (28,29).

1. Prepare samples that typically contain 2.5–10 μM of urea-washed ROS membranes (UMs) in buffer MII (20 mM Tris-HCl, pH 8.0, 130 mM NaCl, 1 mM MgCl$_2$, 1 mM EDTA) and various concentrations of Gt or G-protein peptides. Sonication of the UM stock solution by three sonic pulses (5 sec, medium power setting, Virsonic 100) improves the reproducibility of scans.

2. Record 700 nm to 250 nm spectra before and after activation of the sample for 30 s with light produced by a Fiber-Light PL-900 equipped with a 480 ± 5 nm pass filter. Calculate the amount of Meta II as the absorbance difference A380 − A417 before and after photoactivation. Consider the amount of Meta II without Gt added to be zero (figure 2.7a).

3. Single- and two-wavelength time-resolved experiments are recorded as described earlier, except that they utilize Cary-50's rapid scanning mode at 80 data points/s and 4 data points/s, respectively. In most situations where light-scattering properties of the sample are stable, rapid single-wavelength mode produces excellent results. Add buffered peptides or Gt to the cuvette

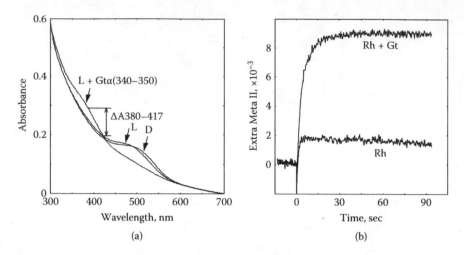

FIGURE 2.7 Extra Meta II measured by UV/visible spectroscopy. (a) Full spectra of the dark-adapted rhodopsin (D), light-activated rhodopsin (L), and light-activated rhodopsin in the presence of 3 mM of synthetic peptide representing Gtα (340–350) C-terminal amino acid sequence of Gtα at pH 8.0. The amount of extra Meta II formed is shown as the A380–A417 absorbance difference. (b) Time-resolved dual wavelength (A380/A417) UV/visible spectroscopy of rhodopsin sample (10 µM) alone and in the presence of Gt (1.5 µM), pH 7.1, Rh bleaching estimated at 15%.

in the dark, if required, before the start of the experiment. Record in kinetic mode at A380 or as the A380–A417 absorbance difference.

4. Start collecting the time-resolved data in the dark. Cary-50's scanning beam produces no noticeable bleaching of the sample for up to 10 min. Photoactivate the sample with a shutter-controlled Fiber-Light PL-900 equipped with a 480 ± 5 nm pass filter, and record the traces for up to 3–5 min. Process the data offline using KaleidaGraph 3.6.2. Normalize full spectra scans to zero at 700 nm. Single- and two-wavelength scans are normalized to zero using baseline absorbance values immediately before light activation (figure 2.7b). For the analysis of rates of Meta II decay, normalize the traces to the peak maximum of Meta II.

PROTOCOL 6: FLUORESCENCE SPECTROSCOPY

Fluorescence spectroscopy measurements are made using a Varian Eclipse Fluorescence spectrophotometer. Traces of transducin activation represent a percentage change of fluorescence emission at 340 nm recorded after exciting the sample at 295 nm essentially as we described before (29).

1. Prepare the samples by reconstituting 50 nM rhodopsin in urea-washed membranes with 330 nM Gt in a final buffer: 20 mM BTP, pH 7.5, 130 mM NaCl, 1 mM $MgCl_2$.
2. Photoactivate the samples and initiate Gt activation by injection of GTPγS into the cuvette to a final concentration of 10 µM. Total typical recording

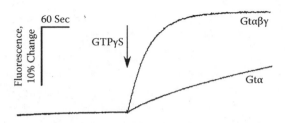

FIGURE 2.8   Activation of transducin is measured by time-resolved changes of intrinsic fluorescence. The point of GTPγS addition is shown with an arrow. R*-catalyzed activation of Gtα (lower trace) and of Gtαβγ heterotrimer (upper trace) are shown.

times are 1–5 min. All recordings are under continuous stirring at 20°C for membrane and vesicle samples, or 10°C for the samples in detergents.

3. Evaluate data as the change in fluorescence relative to the initial fluorescence intensity (figure 2.8).

PROTOCOL 7: [35S]-GTPγS UPTAKE

1. Prepare a 50 µL reaction mixture, which typically contains 1.5 pmol of light-activated rhodopsin, 3 pmol Gtα , 0.1–5 pmol Gtβγ, and 20 mM MOPS (pH 7.5), 10 mM MgCl$_2$, 1 mM EDTA, 1 mM DTT, 100 mM NaCl, 0.5 mg/ mL BSA, and 0.1% octylglucoside, on ice.

2. Start the reaction by adding [35S]GTPγS to a final concentration of 0.1 µM from the [35S]GTPγS stock solution (10 µCi/mL). Incubate the reaction mixture at 22°C for 15 min under continuous illumination.

3. Withdraw 10 µL aliquots at selected intervals of time. Dilute each aliquot tenfold with ice-cold buffer B containing 20 mM Tris HCl (pH 8.0), 100 mM NaCl, and 25 mM MgCl$_2$ and immediately filter through 0.45 µm nitrocellulose filters (GIBCO BRL, PK48).

4. Wash filters five times with 0.85 mL aliquots of ice-cold buffer B. Measure the amount of [35S]GTPγS bound to the filters in the presence of scintillation fluid.

2.5 CONSTRUCTION, EXPRESSION, AND PURIFICATION OF Gtβγ MUTANTS

The exact role of the Gtβγ subunit complex in the R*-catalyzed activation of Gt is not completely understood, which warrants targeted mutagenesis experiments and subsequent studies of how mutations affect protein functions. There is compelling evidence that Gtγ, and especially the farnesylated C-terminal region of Gtγ, is involved in R* interactions and serves as a R*-regulated switch (figure 2.4). So far, the only viable system for expression of recombinant Gtβγ has been the baculovirus/ insect cell system. In protocols 8 and 9, we describe individual steps to construct Gtγ mutants, express them in the presence of the Gtβ subunit, and purify the Gtβγ complexes for functional studies.

PROTOCOL 8: INSECT CELLS—INSECT CELLS AND CONSTRUCTION OF BACULOVIRUSES CONTAINING MUTANT Gtγ

1. Obtain *Spodoptera frugiperda* Sf9 cells from ATCC (ATCC CRL 1711) and maintain in suspension at 27°C at 120 rpm in Sf-900 II SFM medium (Gibco BRL) or in Grace's complete medium. Use Erlenmeyer flasks (Gibco BRL) with cell suspension cultures and shake in INNOVA model 4000 (New Brunswick Scientific, Edison, New Jersey) equipped with a refrigeration unit to maintain temperature at 27°C. Maintain Sf9 cells between 1.5×10^6 and 3×10^6 cells/mL in log phase. Do not overdilute cells.

2. Construct Gtγ1 (bovine) mutants using PCR by introducing the desired nucleic acid substitutions into the 3′ primer, which also has an Xba I restriction site introduced immediately after the TAA stop codon for easy subcloning. The 5′ primer should contain the engineered Pst I site, but should otherwise be identical to the wild-type nucleic acid sequence in the untranslated region of Gtγ immediately preceding the ATG start codon.

3. Construct the baculoviruses using the Bac-to-Bac Baculovirus Expression System (Gibco/BRL). Subclone the mutant cDNA into the pFastBac vector (Invitrogen) for transforming DH5-α competent cells (Gibco).

4. Grow transformants on LB-ampicillin plates for selection. Isolate the DNA from positive colonies using a Fastplasmid miniprep kit (Eppendorf) and analyze by digestion with appropriate restriction endonucleases and DNA sequencing.

5. For transposition into the bacmid, add pFastBac/γ1 mutant DNA to DH10Bac competent cells (Invitrogen) on ice and incubate for 30 min. Heat-shock at 42°C for 45 s, and place on ice for 2 min. Add 300 μL SOC media, and then incubate in a 37°C shaker for 4 h.

6. Prepare serial dilutions of transposition mix in SOC media, and plate 100 μL of undiluted and serially diluted mix onto Luria Agar plates containing 50 μg/mL kanamycin, 7 μg/mL gentamicin, 10 μg/mL tetracycline, 100 μg/mL Bluogal, and 40 μg/mL IPTG and incubate for 48 h at 37°C for color development.

7. Restreak white colonies on Luria Agar complete plates and subculture. Isolate the recombinant bacmid and transfect Sf9 cells as per the Bac-to-Bac Baculovirus Expression System protocol using Cellfectin.

8. Confirm the nucleotide sequence of Gtγ in each bacmid by DNA sequencing. Amplify each virus four times to produce high-titer stock.

PROTOCOL 9: PROTEIN EXPRESSION IN SF9 CELLS AND PURIFICATION

Purification of the recombinant Gtβγ subunits is essentially as before (68) based on the protocol described by Kozasa (69) using His-tagged Gtβ1 and affinity purification of the 6xHis-Gtβγ on affinity Ni-NTA resin. His-tagged Gtβ1 baculovirus is a gift from Dr. T. Kozasa.

1. Prior to large-scale protein expression, titrate the β1:γ1 ratio to that of the native Gtβγ complex by adjusting the ratios of the β1 and γ1 viruses and by

monitoring protein expression by immunoblotting with antibodies against Gtγ1 and Gtβ1 subunits on a Western blot (26).

2. Make large scale protein preparations in 500 mL cultures at 2×10^6 of SF9 cells/mL in Grace's complete. On average, the protein production peaks at 36–48 h post infection.

3. Pellet infected cells at $3000 \times g$, 4°C for 15 min, decant, and resuspend in 30 mL of ice-cold membrane wash buffer-MWB, (50 mM Na-phosphate pH 8, 150 mM NaCl, 10 mM β-mercaptoethanol, 21 μg/mL PMSF, 0.01% TLCK, 0.01% TPCK).

4. Lyse the cells in a Parr bomb on ice at 1000 PSI for 30 min. Centrifuge cell lysate at $310,000 \times g$ in a Beckman ultracentrifuge at 4°C for 30 min, and resuspend the cell membrane pellet in 2 mL MWB. If membrane-binding studies are desirable, aliquots of the pellet, which represent the membrane fraction, and aliquots of the supernatant, containing the cytosolic fraction, are saved for immunoblotting analysis.

5. For protein purification, homogenize the pellet with a Potter-Elvenjem homogenizer, and measure the protein concentration using a Bradford assay (BioRad). Solubilize the protein with 1% octyl glucoside, and adjust the protein concentration to 5 mg/mL. Incubate on ice for 1 h with frequent homogenizing, then pellet in a 70Ti at $310,000 \times g$, 4°C for 30 min.

6. Affinity-purify recombinant 6xHis-Gtβγ on a 4 mL Ni-NTA Superflow agarose column (Qiagen) controlled by AKTA FPLC (Pharmacia). Pre-equilibrate the column with five column volumes of ice-cold buffer, 50 mM NaH_2PO_4, 300 mM Nacl, 10 mM imidazole, pH 8.0. Adjust the NaCl concentration in the extract to 0.3 M and the imidazole to 10 mM, and apply the extract to the column at a speed of 5 mL/h or overnight. Alternatively, this step can be incubated batchwise overnight on a lab rotator at 4°C, followed by the column-packing procedure.

7. Wash the column with 50 mL of buffer A containing 50 mM NaH_2PO_4, 300 mM NaCl, 20 mM imidazole, pH 8.0. This step eliminates the proteins bound to the resin nonspecifically.

8. Elute 6xHis-Gtβγ with 16 mL of buffer B containing 50 mM NaH_2PO_4, 300 mM NaCl, 500 mM imidazole, pH 8.0. Collect fractions of 0.5 mL. Analyze the fractions by silver staining (70) and immunoblotting with antibodies specific to transducin β- and γ-subunits as we have described (48) (figure 2.9). Dialyze the fractions that contain βγ-subunits against buffer 10 mM, Tris pH 7.4, 100 mM NaCl, 5 mM $MgCl_2$, 1 mM DTT, 0.1 mM PMSF, aliquot and store at −20°C in the presence of 50% glycerol. Normalize concentration of all mutants to that of a recombinant wild type to ensure accurate comparison of activities in functional assays.

9. Estimate the extent of posttranslational prenylation of the Gtγ subunit in Sf9 cells based on the differential electrophoretic mobilities of modified versus unmodified forms of Gtγ on urea-SDS-PAGE (71). In this gel system Gtγ-farnesyl has an apparent molecular weight of 6 kDa, whereas unfarnesylated Gtγ subunit has a slower mobility corresponding to the molecular weight of 15 kDa. Gel preparations are based on those by Laemmli, with

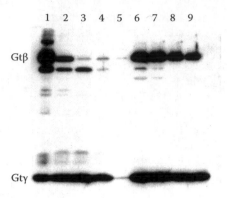

FIGURE 2.9 Western blot analysis of recombinant Gtβγ. Lane 1, Sf9 cell membranes infected with 6xHis-Gtβ and Gtγ baculoviruses. Lane 2, detergent extract. Lane 3, unbound fraction from the Ni-NTA Superflow agarose column. Lanes 4 and 5 are high-salt washes. Lanes 6–9 are fractions collected during elution of 6xHis-Gtβγ from the resin with high imidazole buffer.

6–8 M urea included in the 15% separating polyacrylamide gel. Analyze the relative amounts of the Gtγ subunit forms by quantitative densitometry of immunoblots probed with Gtγ-specific antibodies.

10. Estimate the ability of Gtβγ to interact with Gtα using ADP-ribosylation. Dissolve Pertussis Toxin (PT) (List Laboratories, Campbell, CA) and aliquot according to the company's instructions and store at −80°C. Before use, activate PT in the buffer containing 25 mM Tris-HCl, pH 7.5, 100 mM DTT, and 100 µM ATP at 30°C for 20 min. Final reaction is in 20 mM Tris-HCl, pH 7.8, 1 mM $MgCl_2$, 1 mM EDTA, 10 mM thymidine, 10 mM DTT, 100 µM GDP, 0.5% $C_{12}E_{10}$, 100 mM NaCl, and 0.1 mg/mL ovalbumin at 32°C for 45 min. The assay includes 5 µg of Gtα purified from bovine retinas and reconstituted with various amounts of purified Gtβγ complexes, and 5 µM [^{32}P]NAD$^+$ (New England Nuclear, Boston, MA) (10^6 cpm/sample). Terminate the reaction by addition of Laemmli sample buffer. Calculate the amount of [^{32}P]ADP-ribose transferred from [^{32}P]NAD$^+$ to Gtα from the autoradiograms of the dried SDS-gels by densitometry. Alternatively, slice the gels and determine the amount of [^{32}P]ADP-ribose incorporation by scintillation counting.

2.6 Gtβγ TRANSLOCATION

Light-dependent redistribution of Gt and other rod phototransduction proteins between rod outer and inner segments was discovered in the late 1980s (72–75), but only a few years ago a conclusive connection between Gt translocation and light adaptation was proved (76,77).

It is now firmly established that in response to strong illumination, Gtαβγ translocates from the rod outer segments to the rod inner segments, reducing overall concentration of Gt in the outer segment and attenuating phototransduction. The light-induced movement from the outer segments to the inner segments was shown to be energy independent, whereas the reverse process appears to require ATP or GTP (78).

The functional effect of Gt redistribution is enhanced by the reciprocal translocation of arrestin in the opposite direction from the inner to outer segments, facilitating more effective shutoff of the light-activated rhodopsin (78–81). Other rod proteins, such as recoverin, also show light-dependent translocation (82), so it appears that light adaptation involves concerted redistribution of many phototransduction proteins between the outer and inner segments.

Protocols 10, 11, and 12 describe mouse retina processing and staining for light microscopy and immunohistochemical analysis. For excellent cell biology reviews broadly covering photoreceptor and RPE cells, we would highly recommend the book *Photoreceptor Cell Biology and Inherited Retinal Degenerations* (83) (ed. Williams, D.S.). The in-depth topics related to the mechanisms of retinitis pigmentosa and age-related macular degeneration are compiled superbly in *Retinal Degenerative Diseases* (84) (eds. Hollyfield, J.G., Anderson, R.E., and LaVail, M.M.).

PROTOCOL 10: PROCESSING FOR LIGHT MICROSCOPY

1. Remove the eyes from the animal and make a slit posterior to the cornea.
2. Fix the eyes by immersion for 1 h at room temperature (RT) in 0.1 M sodium cacodylate buffer, pH 7.4, containing 2% paraformaldehyde, 2% glutaraldehyde, and 0.025% (w/v) $CaCl_2$.
3. Remove the lens and the inferior half of the cornea and incubate the eyes in fixative overnight at 4°C.
4. Wash eyecups twice for 10 min each in 0.1 M sodium cacodylate buffer, pH 7.4, containing 0.025% $CaCl_2$; postfix 1 h in 1% OsO_4 in 0.1 M sodium cacodylate, RT in the dark, and rinse twice for 10 min in sodium cacodylate buffer and once for 10 min in distilled water (DW).
5. Dehydrate fixed eyecups through a graded ethanol series (30–95%), 3 × 10 min each at RT, followed by 100% ethanol and propylene oxide (3 × 10 min each) and infiltrate with EtOH–Spurr mixtures and 100% Spurr resin over a 2 day period. Embedment is done in fresh Spurr in flat molds (to permit orientation) and polymerized at 70°C for 48 h.
6. Perform sectioning on a Reichert-Jung Ultracut E microtome using glass and diamond (Diatome) knives.
7. Stain semithin (0.75 µm) sections with 1% toluidine blue in 1% sodium borate; view and photograph with an Olympus BH-2 photomicroscope (figure 2.10).

PROTOCOL 11: PROCESSING IN LR WHITE FOR IMMUNOCYTOCHEMISTRY

1. After removal, fix eyes for 30 min at 4°C by immersion in freshly prepared 0.1 M phosphate buffer, pH 7.4, containing 4% paraformaldehyde and 0.1% glutaraldehyde.
2. After removing the lens and cornea, fix eyes for an additional hour, at 4°C, in the same fixative.

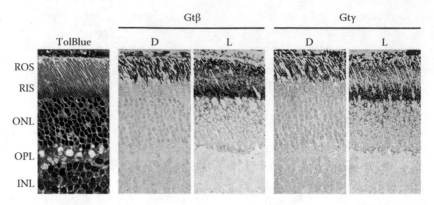

FIGURE 2.10 (See accompanying color CD.) Expression of Gtβγ complex in C57BL/6 mouse photoreceptors. TolBlue is toluidine blue staining; Gtβ and Gtγ panels are stained with specific anti-Gtβ and anti-Gtγ antibodies; D refers to the mouse dark adapted overnight; in L the mouse was exposed to bright laboratory light. (Kisselev, O.G. and Fliesler, S.J., unpublished.)

3. Wash fixed eyecups in phosphate buffer 3 × 10 min at 4°C, dehydrate through a graded ethanol series (30–100%) 1 × 10 min each at 4°C, then infiltrate with EtOH–LR White (LRW) mixtures and 100% LRW, at 4°C, on a rotator over a 2 day period.
4. Embed eyes in fresh LRW in gelatin capsules and polymerize under UV light at 4°C for 48–72 h.

PROTOCOL 12: IMMUNOLABELING

1. For light microscopic immunolabeling, dry semithin sections on glass slides, wash once for 5 min with 0.1 M phosphate buffer, pH 7.4, containing 0.9% NaCl (PBS), treat at RT for 15 min with 50 mM NH_4Cl in PBS, and wash in PBS 3 × 10 min.
2. Block sections for 30 min, RT, in PBS containing 1% (w/v) bovine serum albumin (BSA) and 5% (v/v) normal goat serum (NGS).
3. Dilute primary antibodies in antibody buffer (PBS containing 1% BSA, 1% NGS) and incubate sections overnight at 4°C in a humid chamber.
4. Wash thrice with PBS (15 min each) at room temperature in a Coplin jar.
5. Dilute secondary antibody (1 nm colloidal gold-conjugated) in antibody buffer containing 0.1% IgSS gelatin, apply to sections, and incubate for 2 h, RT, in a humid chamber.
6. Wash thrice with PBS (15 min each) and follow by fixation in 2% glutaraldehyde in PBS for 10 min, RT.
7. After two 10 min DW washes, proceed to silver intensification using an IntenSE™ M Silver Enhancement Kit (Amersham Biosciences), per kit directions.
8. Rinse sections with DW, counterstain with 0.1% toluidine blue, rinse again with DW, air-dry, and coverslip using Permount (Fisher Scientific). View and photograph with an Olympus BH-2 photomicroscope (figure 2.10).

ACKNOWLEDGMENTS

The author thanks Dr. Steven J. Fliesler for helpful discussions, Barbara Nagel for providing electron micrographs of retinal rod cells, and Loryn Rikimaru for help in manuscript preparation. We are indebted to Dr. Klaus Peter Horman for support and stimulating discussions. Supported in part by Research to Prevent Blindness and NIH GM63203.

REFERENCES

1. Ridge, K.D. and Palczewski, K. Visual rhodopsin sees the light: Structure and mechanism of G protein signaling. *J Biol Chem* 282, 9297–301 (2007).
2. Kisselev, O.G. Focus on molecules: Rhodopsin. *Exp Eye Res* 81, 366–7 (2005).
3. Filipek, S., Stenkamp, R.E., Teller, D.C., and Palczewski, K.G. Protein-coupled receptor rhodopsin: A prospectus. *Annu Rev Physiol* 65, 851–79 (2003).
4. Hubbell, W.L., Altenbach, C., Hubbell, C.M., and Khorana, H.G. Rhodopsin structure, dynamics, and activation: A perspective from crystallography, site-directed spin labeling, sulfhydryl reactivity, and disulfide cross-linking. *Adv Protein Chem* 63, 243–90 (2003).
5. Arshavsky, V.Y., Lamb, T.D., and Pugh, E.N., Jr. G proteins and phototransduction. *Annu Rev Physiol* 64, 153–87 (2002).
6. Hamm, H.E. How activated receptors couple to G proteins. *Proc Natl Acad Sci USA* 98, 4819–21 (2001).
7. Sakmar, T.P., Menon, S.T., Marin, E.P., and Awad, E.S. Rhodopsin: Insights from recent structural studies. *Annu Rev Biophys Biomol Struct* 31, 443–84 (2002).
8. Yeagle, P.L. and Albert, A.D. A conformational trigger for activation of a G protein by a G protein-coupled receptor. *Biochemistry* 42, 1365–8 (2003).
9. Abdulaev, N.G. Building a stage for interhelical play in rhodopsin. *Trends Biochem Sci* 28, 399–402 (2003).
10. Rao, V.R. and Oprian, D.D. Activating mutations of rhodopsin and other G protein-coupled receptors. *Annu. Rev. Biophys. Biomol. Struct.* 25, 287–314 (1996).
11. Schertler, G.F. Structure of rhodopsin and the metarhodopsin I photointermediate. *Curr Opin Struct Biol* 15, 408–15 (2005).
12. Okada, T., Ernst, O.P., Palczewski, K., and Hofmann, K.P. Activation of rhodopsin: New insights from structural and biochemical studies. *Trends Biochem Sci* 26, 318–24 (2001).
13. Hofmann, K.P. Signalling states of photoactivated rhodopsin. *Novartis Found Symp* 224, 158–75; discussion 175–80 (1999).
14. Bornancin, F., Pfister, C., and Chabre, M. The transitory complex between photoexcited rhodopsin and transducin. Reciprocal interaction between the retinal site in rhodopsin and the nucleotide site in transducin. *Eur J Biochem* 184, 687–98 (1989).
15. Fung, B.K., Hurley, J.B., and Stryer, L. Flow of information in the light-triggered cyclic nucleotide cascade of vision. *Proc Natl Acad Sci USA* 78, 152–6 (1981).
16. Maeda, T., Imanishi, Y., and Palczewski, K. Rhodopsin phosphorylation: 30 years later. *Prog Retin Eye Res* 22, 417–34 (2003).
17. Arshavsky, V.Y. Rhodopsin phosphorylation: From terminating single photon responses to photoreceptor dark adaptation. *Trends Neurosci* 25, 124–6 (2002).
18. Gurevich, V.V. and Gurevich, E.V. The molecular acrobatics of arrestin activation. *Trends Pharmacol Sci* 25, 105–11 (2004).
19. Calvert, P.D. et al. Phototransduction in transgenic mice after targeted deletion of the rod transducin alpha-subunit. *Proc Natl Acad Sci USA* 97, 13913–8 (2000).
20. Gautam, N., Downes, G.B., Yan, K., and Kisselev, O. The G-protein betagamma complex. *Cell Signalling* 10, 447–55 (1998).

52 Signal Transduction in the Retina

21. *Photoreceptor Cells.* ed. Hargrave, P.A. Academic Press, Inc., 1993.
22. *Signal Transduction in Photoreceptor Cells.* eds. Hargrave, P.A., Hofmann, K.P., and Kaupp, U.B. Springer-Verlag Berlin and Heidelberg GmbH, and Co. K, 1992.
23. Papermaster, D.S. and Dreyer, W.J. Rhodopsin content in the outer segment membranes of bovine and frog retinal rods. *Biochemistry* 13, 2438–44 (1974).
24. Yamazaki, A., Bartucca, F., Ting, A., and Bitensky, M.W. Reciprocal effects of an inhibitory factor on catalytic activity and noncatalytic cGMP binding sites of rod phosphodiesterase. *Proc Natl Acad Sci USA* 79, 3702–6 (1982).
25. Willardson, B.M., Pou, B., Yoshida, T., and Bitensky, M.W. Cooperative binding of the retinal rod G-protein, transducin, to light-activated rhodopsin. *J Biol Chem* 268, 6371–82 (1993).
26. Kisselev, O.G., Pronin, A.P., and Gautam, N. In *G-proteins: Tech Anal* (Ed. Manning, D.R.) 85–95 (CRC Press, 1999).
27. Hong, K. and Hubbell, W.L. Preparation and properties of phospholipid bilayers containing rhodopsin. *Proc Natl Acad Sci USA* 69, 2617–21 (1972).
28. Kisselev, O.G., Ermolaeva, M.V., and Gautam, N. A farnesylated domain in the G protein gamma subunit is a specific determinant of receptor coupling. *J Biol Chem* 269, 21399–402 (1994).
29. Kisselev, O.G., Meyer, C.K., Heck, M., Ernst, O.P., and Hofmann, K.P. Signal transfer from rhodopsin to the G-protein: Evidence for a two-site sequential fit mechanism. *Proc Natl Acad Sci USA* 96, 4898–4903 (1999).
30. Shinozawa, T. et al. Additional component required for activity and reconstitution of light-activated vertebrate photoreceptor GTPase. *Proc Natl Acad Sci USA* 77, 1408–11 (1980).
31. Kühn, H. Light- and GTP-regulated interaction of GTPase and other proteins with bovine photoreceptor membranes. *Nature* 283, 587–9 (1980).
32. Wensel, T.G. and Stryer, L. Reciprocal control of retinal rod cyclic GMP phosphodiesterase by its gamma subunit and transducin. *Proteins* 1, 90–9 (1986).
33. Bennett, N. and Clerc, A. Activation of cGMP phosphodiesterase in retinal rods: Mechanism of interaction with the GTP-binding protein (transducin). *Biochemistry* 28, 7418–24 (1989).
34. Arshavsky, V.Y., Dumke, C.L., and Bownds, M.D. Noncatalytic cGMP-binding sites of amphibian rod cGMP phosphodiesterase control interaction with its inhibitory gamma-subunits. A putative regulatory mechanism of the rod photoresponse. *J Biol Chem* 267, 24501–7 (1992).
35. Rarick, H.M., Artemyev, N.O., and Hamm, H.E. A site on rod G protein alpha subunit that mediates effector activation. *Science* 256, 1031–3 (1992).
36. Natochin, M. and Artemyev, N.O. A point mutation uncouples transducin-alpha from the photoreceptor RGS and effector proteins. *J Neurochem* 87, 1262–71 (2003).
37. Bockaert, J., Deterre, P., Pfister, C., Guillon, G., and Chabre, M. Inhibition of hormonally regulated adenylate cyclase by the beta gamma subunit of transducin. *EMBO J* 4, 1413–7 (1985).
38. Gaudet, R., Savage, J.R., McLaughlin, J.N., Willardson, B.M., and Sigler, P.B. A molecular mechanism for the phosphorylation-dependent regulation of heterotrimeric G proteins by phosducin. *Mol Cell* 3, 649–60 (1999).
39. Thulin, C.D. et al. Modulation of the G protein regulator phosducin by Ca^{2+}/calmodulin-dependent protein kinase II phosphorylation and 14–3–3 protein binding. *J Biol Chem* 276, 23805–15 (2001).
40. Ho, Y.K., Ting, T.D., and Lee, R.H. Phosducin down-regulation of G-protein coupling: Reconstitution of phosducin and transducin of cGMP cascade in bovine rod photoreceptor cells. *Methods Enzymol* 344, 126–39 (2002).

41. Wedegaertner, P.B., Wilson, P.T., and Bourne, H.R. Lipid modifications of trimeric G proteins. *J Biol Chem* 270, 503–6 (1995).
42. Bigay, J., Faurobert, E., Franco, M., and Chabre, M. Roles of lipid modifications of transducin subunits in their GDP-dependent association and membrane binding. *Biochemistry* 33, 14081–90 (1994).
43. Zhang, Z. et al. How a G protein binds a membrane. *J Biol Chem* 279, 33937–45 (2004).
44. Melia, T.J., Malinski, J.A., He, F., and Wensel, T.G. Enhancement of phototransduction protein interactions by lipid surfaces. *J Biol Chem* 275, 3535–42 (2000).
45. Matsuda, T. et al. Characterization of interactions between transducin alpha/beta gamma-subunits and lipid membranes. *J Biol Chem* 269, 30358–63 (1994).
46. Kassai, H. et al. Farnesylation of retinal transducin underlies its translocation during light adaptation. *Neuron* 47, 529–39 (2005).
47. Fukada, Y. et al. Effects of carboxyl methylation of photoreceptor G protein gamma-subunit in visual transduction. *J Biol Chem* 269, 5163–70 (1994).
48. Kisselev, O., Pronin, A., Ermolaeva, M., and Gautam, N. Receptor-G protein coupling is established by a potential conformational switch in the beta-gamma complex. *Proc Natl Acad Sci USA* 92, 9102–6 (1995).
49. Downs, M.A., Arimoto, R., Marshall, G.R., and Kisselev, O.G. G-protein alpha and beta-gamma subunits interact with conformationally distinct signaling states of rhodopsin. *Vision Res* 46, 4442–8 (2006).
50. Nikiforovich, G.V. and Marshall, G.R. Three-dimensional model for meta-II rhodopsin, an activated G-protein-coupled receptor. *Biochemistry* 42, 9110–20 (2003).
51. Nikiforovich, G.V. and Marshall, G.R. Modeling flexible loops in the dark-adapted and activated states of rhodopsin, a prototypical G-protein-coupled receptor. *Biophys J* 89, 3780–9 (2005).
52. Park, P.S., Filipek, S., Wells, J.W., and Palczewski, K. Oligomerization of G protein coupled receptors: Past, present, and future. *Biochemistry* 43, 15643–56 (2004).
53. Iiri, T., Farfel, Z., and Bourne, H.R. G-protein diseases furnish a model for the turn-on switch. *Nature* 394, 35–8 (1998).
54. Herrmann, R. et al. Sequence of interactions in receptor-G protein coupling. *J Biol Chem* 279, 24283–90 (2004).
55. Cherfils, J. and Chabre, M. Activation of G-protein Galpha subunits by receptors through Galpha-Gbeta and Galpha-Ggamma interactions. *Trends Biochem Sci* 28, 13–7 (2003).
56. Wang, Y., Jiang, Y., Meyering-Voss, M., Sprinzl, M., and Sigler, P.B. Crystal structure of the EF-Tu.EF-Ts complex from *Thermus thermophilus*. *Nat Struct Biol* 4, 650–6 (1997).
57. Fotiadis, D. et al. Atomic-force microscopy: Rhodopsin dimers in native disc membranes. *Nature* 421, 127–8 (2003).
58. Chabre, M. and le Maire, M. Monomeric G-protein-coupled receptor as a functional unit. *Biochemistry* 44, 9395–403 (2005).
59. Fotiadis, D. et al. Structure of the rhodopsin dimer: A working model for G-protein-coupled receptors. *Curr Opin Struct Biol* 16, 252–9 (2006).
60. Jastrzebska, B. et al. Functional and structural characterization of rhodopsin oligomers. *J Biol Chem* 281, 11917–22 (2006).
61. Nikiforovich, G.V., Taylor, C.M., and Marshall, G.R. Modeling of the complex between transducin and photoactivated rhodopsin, a prototypical g-protein-coupled receptor. *Biochemistry* (2007).
62. Dell'Orco, D., Seeber, M., and Fanelli, F. Monomeric dark rhodopsin holds the molecular determinants for transducin recognition: Insights from computational analysis. *FEBS Lett* 581, 944–8 (2007).
63. Bayburt, T.H., Leitz, A.J., Xie, G., Oprian, D.D., and Sligar, S.G. Transducin activation by nanoscale lipid bilayers containing one and two rhodopsins. *J Biol Chem* (2007).

64. Pulvermuller, A. et al. Functional differences in the interaction of arrestin and its splice variant, p44, with rhodopsin. *Biochemistry* 36, 9253–60 (1997).
65. Faurobert, E., Otto-Bruc, A., Chardin, P., and Chabre, M. Tryptophan W207 in transducin T alpha is the fluorescence sensor of the G protein activation switch and is involved in the effector binding. *EMBO J* 12, 4191–8 (1993).
66. *Molecular Mechanisms in Visual Transduction* 1st ed. eds. Stavenga, D.G., De Grip, W.J., and Pugh, E.N., Jr. North Holland, 2000.
67. *Vertebrate Phototransduction and the Visual Cycle.* eds. Abelson, J.N., Simon, M., and Palczewski, K. Academic Press, 2000.
68. Kozasa, T. Purification of G protein subunits from Sf9 insect cells using hexahistidine-tagged alpha and beta gamma subunits. *Methods Mol Biol* 237, 21–38 (2004).
69. Kozasa, T. and Gilman, A.G. Purification of recombinant G proteins from Sf9 cells by hexahistidine tagging of associated subunits. Characterization of alpha 12 and inhibition of adenylyl cyclase by alpha z. *J Biol Chem* 270, 1734–41 (1995).
70. Wray, W., Boulikas, T., Wray, V.P., and Hancock, R. Silver staining of proteins in polyacrylamide gels. *Anal Biochem* 118, 197–203 (1981).
71. Fukada, Y., Ohguro, H., Saito, T., Yoshizawa, T., and Akino, T. Beta gamma-subunit of bovine transducin composed of two components with distinctive gamma-subunits. *J Biol Chem* 264, 5937–43 (1989).
72. Brann, M.R. and Cohen, L.V. Diurnal expression of transducin mRNA and translocation of transducin in rods of rat retina. *Science* 235, 585–7 (1987).
73. Philp, N.J., Chang, W., and Long, K. Light-stimulated protein movement in rod photoreceptor cells of the rat retina. *FEBS Lett* 225, 127–32 (1987).
74. Whelan, J.P. and McGinnis, J.F. Light-dependent subcellular movement of photoreceptor proteins. *J Neurosci Res* 20, 263–70 (1988).
75. Organisciak, D.T. et al. Adaptive changes in visual cell transduction protein levels: Effect of light. *Exp Eye Res* 53, 773–9 (1991).
76. Sokolov, M. et al. Massive light-driven translocation of transducin between the two major compartments of rod cells: A novel mechanism of light adaptation. *Neuron* 34, 95–106 (2002).
77. Arshavsky, V.Y. Protein translocation in photoreceptor light adaptation: A common theme in vertebrate and invertebrate vision. *Sci STKE* 2003, PE43 (2003).
78. Nair, K.S. et al. Light-dependent redistribution of arrestin in vertebrate rods is an energy-independent process governed by protein-protein interactions. *Neuron* 46, 555–67 (2005).
79. Zhang, H. et al. Light-dependent redistribution of visual arrestins and transducin subunits in mice with defective phototransduction. *Mol Vis* 9, 231–7 (2003).
80. Strissel, K.J. and Arshavsky, V.Y. Myosin III illuminates the mechanism of arrestin translocation. *Neuron* 43, 2–4 (2004).
81. Lee, S.J. and Montell, C. Light-dependent translocation of visual arrestin regulated by the NINAC myosin III. *Neuron* 43, 95–103 (2004).
82. Strissel, K.J. et al. Recoverin undergoes light-dependent intracellular translocation in rod photoreceptors. *J Biol Chem* 280, 29250–5 (2005).
83. *Photoreceptor Cell Biology and Inherited Retinal Degenerations.* ed. Williams, D. S. World Scientific Publishing Company, 2004.
84. *Retinal Degenerative Diseases* 1st ed. eds. Hollyfield, J.G., Anderson, R.E., and LaVail, M.M. Springer, 2006.
85. Kisselev, O.G. and Downs, M.A. Rhodopsin controls a conformational switch on the transducin gamma subunit. *Structure (Camb)* 11, 367–73 (2003).
86. Kisselev, O.G. et al. Light-activated rhodopsin induces structural binding motif in G protein alpha subunit. *Proc Natl Acad Sci USA* 95, 4270–5 (1998).

3 How Rod Arrestin Achieved Perfection
Regulation of Its Availability and Binding Selectivity

Vsevolod V. Gurevich, Susan M. Hanson,
Eugenia V. Gurevich, and Sergey A. Vishnivetskiy

CONTENTS

3.1   INTRODUCTION

The visual amplification cascade initiated by light-activated rhodopsin has long served as a model of G-protein-coupled receptor (GPCR) signaling. The same activation, signal propagation, and termination mechanisms are used by the majority of the ~1000 mammalian GPCRs (1). However, in many ways, the rod photoreceptor cell has no equal: incredibly low noise in a cell that has 3 billion molecules of the light receptor rhodopsin, single-photon sensitivity, and a dynamic range of seven orders of magnitude of light intensity (2). This level of perfection is achieved through several unique features. Rods have a specialized signaling compartment, the outer segment, where all rhodopsin molecules are tightly packed in disks, separated from the inner segment, where soluble signaling proteins are stored until needed. Rhodopsin has a covalently attached inverse agonist, 11-*cis*-retinal, that is converted by light into a potent covalently attached agonist, all-*trans*-retinal, whereas other GPCRs are ligand-free in the resting state and demonstrate much higher constitutive activity. Rhodopsin kinase is prenylated and therefore constitutively localized to the membrane (3), whereas other G-protein-coupled receptor kinases are recruited to the membrane by other proteins (4). There is also the massive light-dependent translocation of three important signaling molecules that has no parallel in other GPCR-driven signaling systems. The visual G-protein transducin moves to the outer segment in the dark to amplify the signal, and in bright light moves away from the receptor to the inner segment (5). The key quenching protein, rod arrestin, moves in the opposite direction (6). Recoverin, a Ca^{2+}-binding protein that regulates the activity of rhodopsin kinase (7–9), moves away from the outer segment in the light (10). Here we describe the molecular mechanisms underlying the unique functional features of rod arrestin that contribute to the amazing sensitivity and enormous dynamic range of rod photoreceptors and the methods used to study these mechanisms.

3.2   ARRESTIN LOCALIZATION IN ROD PHOTORECEPTORS: JUST-IN-TIME DELIVERY

Prolonged illumination causes massive translocation of arrestin to the outer segment (OS) (figure 3.1), whereas in the dark arrestin returns to the inner segment (IS) and cell bodies. Transducin moves in the opposite direction. The movement of both proteins is believed to play a role in light and dark adaptation (11). The translocation of arrestin and transducin was described more than two decades ago (12–14). However, only the creation of various lines of genetically modified mice in the last few years provided suitable tools for the mechanistic studies of this phenomenon in vivo and yielded some definitive answers. Impaired synthesis of 11-*cis*-retinal (RPE65$^{-/-}$) prevents light-dependent arrestin movement to the OS, indicating that active rhodopsin is necessary for this process (15). The role of rhodopsin phosphorylation in arrestin translocation to the OS was addressed using two mouse models deficient in this process: rhodopsin knockout mice expressing mutant rhodopsin in which all of the phosphorylation sites at the C terminus were mutated to alanines (16) and rhodopsin kinase knockout mice (17). The results of these studies clearly show that rhodopsin phosphorylation is not required for arrestin movement to the OS (15,18), although the absence of rhodopsin phosphorylation changes the kinetics of arrestin translocation (6). Arrestin and transducin move independently: arrestin translocation is grossly normal in transducin knockout mice (15) and the movement of transducin is unchanged in arrestin knockout mice (18). However, the initial phase of arrestin translocation may be facilitated by transducin-mediated signaling, as evidenced by the enhancement of arrestin movement induced by relatively low light in RGS$^{-/-}$ mice in which transducin deactivation is impaired (19). Understanding the mechanism of light-dependent arrestin translocation requires the answers to three questions. First, does arrestin simply diffuse to its destination in light- and dark-adapted rods or is it actively transported in one or both directions? Second, how

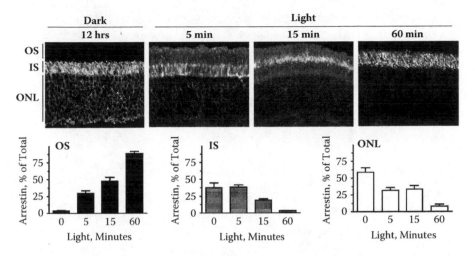

FIGURE 3.1 The time course of arrestin translocation in the light. After dark adaptation for 12 h, animals were kept in the dark or 2700 lux light for the indicated time. Arrestin was visualized with anti-arrestin antibody, and the proportion of arrestin localized in the outer segments (OSs), inner segments (ISs), and cell bodies (ONLs = outer nuclear layer) was determined by quantitative image analysis (protocols 1 and 3). Note that after 5 min a considerable proportion of arrestin has already moved to the OS, and this process is virtually complete after 60 min. At early stages, arrestin content in the cell body declines most, as arrestin "passes through" the IS to the OS.

is arrestin retained in the OS in the light? Third, what keeps arrestin in the inner segment in the dark? Considering the amount of arrestin present in rod photoreceptors (recent studies show that it is expressed at about 0.8:1 molar ratio to rhodopsin (19,20) and the huge energy costs of active transport, specific mechanisms retaining arrestin in the appropriate compartment are necessary regardless of its "mode of transportation." Otherwise, diffusion down a steep concentration gradient, apparent in both light- and dark-adapted rods (6,19), would rapidly destroy compartment-specific localization achieved by active transport.

Initial observations of the fairly rapid movement of arrestin and transducin in opposite directions upon the onset of light (12–14) almost inevitably led to the hypothesis that multiple active transport mechanisms are involved (14). The most complicated of these models proposes the participation of four transport mechanisms, one for the movement of each of these two proteins in each direction (21). This was based on the findings that upon exposure of dark-adapted rods to light, arrestin moves from the IS to the OS much faster than in the opposite direction during dark adaptation, and that transducin return to the OS in the dark takes almost 100 times longer than its movement to the IS upon illumination. Regardless of the number of mechanisms involved, the energy costs to actively transport the hundreds of millions of arrestin molecules present in each rod photoreceptor within minutes would be staggering. The distinguishing feature of active transport is its energy dependence. Therefore, the elucidation of energy dependence or independence of arrestin movement is the only definitive way to discriminate between active transport and diffusion. Although it cannot be done in a living animal, these experiments can be performed ex vivo in isolated eyecups (posterior eye containing most of the retina) that survive in cell culture medium for hours. Incubation of eyecups in glucose-free medium supplemented with deoxyglucose and KCN decreases their ATP content by two orders of magnitude, depleting it so thoroughly that even rhodopsin kinase, which among cellular ATPases has an unusually low Km for ATP, 2 µM (22), does not have enough to appreciably phosphorylate rhodopsin (6). In these severely ATP-depleted photoreceptors, arrestin was found to translocate in both directions normally, clearly demonstrating that it is an energy-independent process, i.e., diffusion (6).

PROTOCOL 1: STUDYING ARRESTIN TRANSLOCATION IN VIVO

Animal research was conducted in compliance with NIH Guide for the Care and Use of Laboratory Animals and approved by the Institutional Animal Care and Use Committee. We use pigmented C57Bl mice as the base strain because with regard to visual function these mice are much closer to wild type than any strains with defective pigmentation.

1. Dark-adapt mice for at least 12 h.
2. Expose the mouse to light of the desired intensity (we use 2700 lux without pupil dilation) for the selected period of time. Note that the operations described below take 1–2 min (depending on experience) and should be performed under the same light conditions. The illumination time should be counted as the time elapsed between turning the light on and immersion of the eyecup into fixing solution. This is especially important for shorter time points (<15 min) to ensure reproducibility.
3. Anesthetize the animal in a desiccator containing paper towels saturated with isoflurane. When the mouse is completely unconscious, sacrifice it by cervical dislocation and immediately enucleate the eyes. Cut off the front half of the eye (removing the cornea and lens) and immerse the eyecup into fixing solution (4% paraformaldehyde in 0.1 M cacodylate buffer, pH 7.2; this solution is best when made fresh; however, it can be stored for 1–2 weeks in the dark at 4°C). Note that fixing eyecups rather than whole eyes is important for correct estimates of the translocation rate by immunohistochemistry: arrestin movement stops only when paraformaldehyde reaches the photoreceptor cells, which occurs almost immediately in eyecups, but is delayed because of slow diffusion through the tissue when the whole eye is used. In the latter case the rate of protein translocation would be seriously overestimated, which likely explains the discrepancies between the published estimates of the rate of arrestin movement (compare (6,19) with (21).
4. After 4 h fixing at room temperature, rinse the eyecups thrice with 0.1 M cacodylate buffer, pH 7.2, cryoprotect overnight in 30% sucrose, and freeze at −80°C. The eyecups can be stored for months.

PROTOCOL 2: PREPARATION AND USE OF MOUSE EYECUPS TO STUDY PROTEIN TRANSLOCATION

The eyecups from dark-adapted animals survive for several hours in regular cell culture medium (e.g., DMEM) that can be supplemented with various pharmacological agents. With younger (4–6 weeks) animals, the retina is typically firmly attached to the RPE; with older animals the retina detaches more readily. Preparations with the retina attached to the RPE should be used, unless rhodopsin regeneration mimicking in vivo conditions is undesirable.

1. Remove the eyes from the dark-adapted anesthetized and sacrificed animal, as described earlier, and place them in a 5 mL dish with DMEM under a dissection microscope with 12.5× or 25× magnification.
2. Use one pair of forceps to hold the eye, and another pair to firmly grab the cornea (figure S1 in Reference 6). Using microsurgical scissors, carefully cut away the cornea. Remove the iris using two pairs of forceps; then gently pull out the lens and vitreous so that the remaining eyecup is not damaged.
3. To deplete ATP, incubate the eyecup in glucose-free DMEM supplemented with 2 mM deoxyglucose and 10 mM KCN for 1 h (6). Other pharmacological agents, such as channel and enzyme inhibitors and activators, etc., can be used for particular experiments, as necessary, making this a perfect ex vivo preparation that preserves the in vivo-like retina in contact with the RPE, yet is amenable to various manipulations, essentially as in cell culture.
4. Incubate the eyecup in the dark or under illumination of the appropriate intensity. The eyecups can be incubated at room temperature or at 37°C; incubation at 37°C dramatically accelerates the decay of light-activated rhodopsin, so that it completely disappears in about 5 min (6).
5. For immunohistochemical studies, fix and store the eyecups, as described in Protocol 1. Alternatively, the retina can be removed and used for biochemical studies.

PROTOCOL 3: IMMUNOHISTOCHEMICAL DETECTION OF ARRESTIN IN PHOTORECEPTORS AND QUANTITATIVE IMAGE ANALYSIS

In essence, there are two methods for quantitative studies of protein translocation: (1) immunohistochemical detection followed by image analysis (e.g., references 6 and 21); (2) serial tangential sectioning of the flattened retina followed by Western blot analysis of the protein content in the sections (e.g., references 5, 10, and 19). Each method has its advantages and disadvantages. The use of immunohistochemistry is based on two assumptions (which are, strictly speaking, virtually impossible to prove): that the antigen is equally available to the detecting antibody in all cellular compartments and that the signal increases linearly with the amount of protein. In contrast, Western blot can be truly quantitative (see Protocol 8). The main limitation of the serial sectioning, especially in small animals, is that the amount of tissue in individual sections is very small. As Western blot with any antibody has certain limits of sensitivity, it may be impossible to detect low concentrations of arrestin in specific compartments with the quantities of material available for the analysis. For example, arrestin is definitely present in the mouse OS in the dark (its presence in this compartment underlies a striking difference between single-photon responses in wild-type and arrestin knockout mice (23), but it was not detectable by the serial sectioning method (19). The serial sectioning procedure is also labor-intensive and requires special skills. The effort may be well justified, however, if quantification of arrestin in absolute units is desired, which is clearly impossible to do by immunohistochemistry. The immunohistochemical procedure, on the other hand, is relatively quick and uncomplicated. With proper controls and careful image analysis, immunohistochemistry is quantitative enough to provide an accurate picture of arrestin translocation in rod photoreceptors. Here we describe our procedure for detection and quantification of arrestin in the mouse retina by immunocytochemistry.

1. Cut fixed eyecups into 30 µm thick sections.
2. Block free-floating sections for 1 h in phosphate-buffered saline (PBS) with 0.3% Triton X-100, 3% BSA at room temperature. We prefer to use BSA for blocking, because the same buffer is suitable regardless of the type of secondary antibody. BSA could be substituted by 5% serum of the animal in which the secondary antibodies are made.
3. Incubate with an appropriate anti-arrestin antibody in PBS with 0.03% Triton X-100, 1% BSA (IC buffer). We found that the most quantitative staining is achieved after 48 h incubation of free-floating sections at 4°C, as judged by the relative intensity of staining of WT and arrestin hemizygous ($Arr^{+/-}$) mice. Retinas from arrestin knockout ($Arr^{-/-}$) mice should be used as a control to gauge the signal-to-noise ratio. We use rabbit polyclonal primary antibodies to avoid using antimouse secondary antibodies on mouse tissue.
4. After washing with IC buffer, incubate sections with goat antirabbit biotinylated antibody (Vector Laboratories, 1:200) followed by streptavidin conjugated with Alexa488 (Molecular Probes, 1:200) for 1 h each at room temperature. Alternatively, sequential incubation with chicken antirabbit-Alexa488 and goat antichicken-Alexa488 antibodies (Molecular Probes) at a dilution of 1:1000 for 1 h each at room temperature could be employed for signal amplification. The concentration of arrestin in rod photoreceptors is quite high. Therefore, arrestin can be detected simply with fluorescently labeled secondary antibody. However, we have found that using amplification cascades for detection, in addition to conserving primary anti-arrestin antibody, improves the signal-to-noise ratio (figure 3.1).
5. After the final wash with IC buffer, mount the sections and view using the 40× 1.3PlanNeoFluar objective lens of a Zeiss LSM 510 laser scanning confocal microscope. Detect immunofluorescence by excitation with a 488 nm laser line and an LP505 filter (figure 3.1). For subsequent quantitative analysis, it is imperative to use the same settings for the acquisition of all images

that will be compared. Parallel acquisition of a DIC image helps identify subcellular compartments (figure 3.1).

6. For quantitative image analysis, integrate intensity of the outer segments (I_{OS}), inner segments (I_{IS}), whole retina (I_R), and background in the extracellular regions (I_B) by multiplying the area of each compartment by the average intensity of that compartment. The percentage of arrestin in the outer segments ($\%I_{OS}$) and inner segments ($\%I_{IS}$) is then calculated using the following formula: $\%I_{OS} = (I_{OS} - I_B)/I_r \times 100$, and $\%I_{IS} = (I_{IS} - I_B)/I_r \times 100$, respectively.

7. For statistical analysis we use StatView software (SAS Institute, Cary, North Carolina). The data are analyzed by one-way ANOVA with time as the main factor, followed by Scheffe post hoc test where appropriate.

3.3 ARRESTIN INTERACTION WITH MICROTUBULES

As far as retaining arrestin in the outer segment in the light is concerned, its binding to light-activated rhodopsin is the most obvious solution. This was the first hypothesis proposed in 1985 (12), and now is supported by a wealth of additional data (6,15,18). Early studies also revealed that arrestin is not evenly distributed throughout the cytoplasm in dark-adapted photoreceptors: it appears to concentrate in the IS, perinuclear area, and synaptic terminals (12–14). Recent thorough quantitative analysis of arrestin distribution in *Xenopus* rods in the dark proved beyond reasonable doubt that it is at disequilibrium (24). The existence of an alternative nonmobile arrestin binding partner that must be enriched in the IS and cell bodies is the most straightforward way to rationalize these observations. This putative interaction partner was not known until a recent finding that arrestin directly binds microtubules (MTs) (25), which are particularly abundant in the IS (26). It should be noted that arrestin colocalization with MTs was reported as early as 1993 (27), but this phenomenon was usually interpreted as an indication of cytoskeleton participation in the active transport of arrestin (28) rather than direct arrestin–MT interaction. Recent studies clearly establish that a significant proportion of rod arrestin is bound to MTs in dark-adapted photoreceptors, whereas in the light the amount of cytoskeleton-associated arrestin is negligible (6). In vitro experiments showed that rhodopsin and MTs actually compete for arrestin, and that any form of phosphorylated or light-activated rhodopsin competent to bind arrestin wins in this competition (figure 3.2) (6). The finding that the MT-binding site on arrestin largely overlaps with the rhodopsin-binding site explained the mechanistic basis of the competition (29). The relatively low affinity of arrestin for MTs (Kd > 50 µM) (29) compared to its subnanomolar affinity for P-Rh* (30) explains why every form of rhodopsin, with the exception of dark unphosphorylated rhodopsin (prevalent in dark-adapted photoreceptors) and opsin, outcompetes microtubules (figure 3.2). This makes microtubules a perfect default binding partner for arrestin in dark-adapted photoreceptors, when more "attractive" partners, such as light-activated rhodopsin, are absent. Obviously, MT-bound and free arrestin are in equilibrium. The low affinity of the arrestin–microtubule interaction ensures that whenever free arrestin is "consumed" by rhodopsin binding, microtubules readily supply more without any appreciable delay, thus supporting arrestin movement to the OS while keeping the concentration of available free arrestin virtually constant.

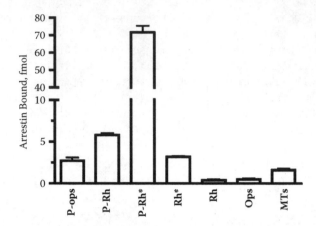

FIGURE 3.2 Comparative arrestin binding to different functional forms of rhodopsin and microtubules. The binding of translated radiolabeled arrestin to 0.6 µg of phosphoopsin (P-ops), phosphorylated dark (P-Rh) and light-activated rhodopsin (P-Rh*), unphosphorylated dark (Rh) and light-activated rhodopsin (Rh*), opsin (Ops), and microtubules (MTs) was determined as described in protocols 5 and 7. Note that equivalent amounts of every form of rhodopsin except dark Rh and opsin bind arrestin better than microtubules.

PROTOCOL 4: EXPRESSION AND PURIFICATION OF ROD ARRESTIN

Arrestins express very well in *E. coli* (strain BL21) using the pTrcHisB vector (Invitrogen). We introduced an extra codon (Ala) between the normal first and second codons to create an Nco I site and an optimal Kozak consensus for the initiator codon (the insertion of Ala between Met-1 and Lys-2 does not affect the functional characteristics of arrestin (31). The arrestin ORF was subcloned between Nco I and Hind III sites into pTrcHisB (note that this construct does not retain the His tag). We have previously described a number of modifications of this expression/purification protocol suitable for different members of the arrestin family and particular mutants (29,31–36). We take advantage of the ability of arrestins to bind heparin-Sepharose to purify all members of the family by conventional chromatography. This is particularly important because in our experience the addition of a His tag to either the N- or C-terminus of arrestin proteins significantly compromises their functional characteristics, making these tagged proteins unsuitable for functional studies.

1. Transform BL21 cells with an appropriate arrestin construct and plate. Use an individual colony to inoculate 3 mL of LB containing 0.1 µg/mL of ampicillin (LB/Amp) and grow at 250 rpm at 30°C to OD_{600} ~0.2–0.3. Add glycerol to 10% (v/v), aliquot cells, freeze, and store at −80°C until needed (these cells are stable for years).
2. Cell growth: Inoculate 160 mL LB/Amp in a 500 mL flask with a 20–50 µL aliquot of frozen cells. Grow at 30°C, 250 rpm, for 6–8 h (to OD_{600} ~0.1–0.4). Divide the culture between 4 × 1.1 L LB/Amp containing 20–30 µM IPTG in 3 L flasks, and grow at 30°C, 250 rpm, for 12–16 h. For arrestin proteins that do not express well or have in vivo proteolysis problems, grow

4×1.1 L cultures for 12–16 h without IPTG, then add 20–30 µM IPTG and grow for an additional 3–5 h.

3. Cell lysis: The following protocol applies to 4.5 L of culture. Pellet cells for 10 min, 5000 rpm at 4°C. Carefully discard supernatant, let the bottles stand for 1 min upside down, and wipe away all traces of supernatant using Kimwipes. Immediately add 37.5 mL ice-cold lysis buffer (50 mM Tris-HCl, pH 8.0, 5 mM EGTA, 1 mM DTT, 50 mM glucose, 2 mM benzamidine, protease inhibitor cocktail for bacterial cell extracts (Sigma), and 1 mM PMSF (from fresh 1 M stock in DMSO) to each bottle (150 mL total). Resuspend the cells thoroughly by pipetting, then add 1.5 mL of fresh 3 mg/mL lysozyme (dissolved in lysis buffer) to each bottle and incubate for 30–40 min on ice. Freeze the cells at −80°C. Cells can be kept frozen at this point for weeks. If you intend to proceed immediately, freeze the cells for 20 min, thaw at 4°C for 20–30 min, then add an additional 110 mL of lysis buffer (w/o glucose) per bottle, carefully resuspend, and sonicate for 2 min. Add 1.2 mL 1 M $MgCl_2$ (to final concentration of >7 mM) and 1 mL each of 3 mg/mL DNase I, DNase II, and RNase (Sigma) to each bottle, mix thoroughly, and incubate on ice for 40 min. Transfer the suspension to 250 mL bottles and pellet cell debris in a GSA rotor at 9000 rpm for 90 min. Transfer supernatant (~600 mL) to a 1 L beaker. Under continuous gentle stirring at 4°C, add 192 g (to 0.32 g/mL) of ammonium sulfate in 4–5 portions over a 20–40 min period. Let stirring continue until all ammonium sulfate is dissolved, and then pellet the precipitated protein in a GSA rotor at 9000 rpm for 90 min. Carefully remove supernatant and floating white material, and wipe the walls clean. Cover the tube with foil and store at −80°C until needed (stable for at least a month).

4. Heparin-Sepharose chromatography. Dissolve the protein in 200 mL of ice-cold column buffer (10 mM Tris-HCl, pH 7.5, 2 mM EDTA, 2 mM EGTA, 2 mM benzamidine, 1 mM PMSF) (CB). It takes 20–40 min of gentle shaking (avoid frothing). Pellet the insoluble material (GSA rotor, 9000 rpm, 90 min), and carefully determine the volume of the supernatant to estimate the volume of the initial pellet (it is usually 12–18 mL and contains 1200–1600 mg of total protein). Filter the supernatant through a 0.8 µm Millipore filter. Dilute the filtrate with CB to 65 times the volume of the initial pellet for wild-type visual arrestin during loading onto a 25 mL column of Heparin-Sepharose, equilibrated with CB/0.1 M NaCl using a gradient mixer and an appropriate ratio of filtrate and CB. Load at 1.5–2.5 mL/min, wash the column with ~200 mL of CB/0.1 M NaCl. Elute with a 400 mL linear gradient (CB/0.1 M NaCl -> CB/0.4 M NaCl) and collect 10 mL fractions. The main rod arrestin peak elutes in 8–12 fractions between 200–240 mM NaCl.

5. Q-Sepharose chromatography: Pool arrestin-containing fractions (total protein 15–30 mg, ~50% pure arrestin), concentrate to ~10 mL (Amicon, YM-30 membrane), and filter through a 0.8 µm Millipore filter. Calculate the concentration of NaCl in this pool, and load the sample at 1 mL/min onto a 10 mL Q-Sepharose column (equilibrated with CB) while diluting

the sample (using a gradient mixer) with CB to a final NaCl concentration of ~10 mM. It is important that the protein be diluted just before it is pumped onto the column because arrestins aggregate at low salt, but survive it while bound to the resin. Wash the column with 40–80 mL of CB, and then elute with a 400 mL linear gradient (CB -> CB/0.1 M NaCl). Arrestin (>95% purity) elutes in a peak at about 60 mM NaCl. Pool arrestin-containing fractions, add NaCl to 100 mM, and concentrate to 0.5–3 mg/mL (expect 2–5 mg/L of bacterial culture), filter through a 0.8 μm Millipore filter, aliquot, and freeze at −80°C. Frozen arrestins are stable for >> 2 years, and 2–3 freeze–thaw cycles do not appreciably reduce their activity.

PROTOCOL 5: DIRECT BINDING ASSAY FOR ARRESTIN–MICROTUBULE INTERACTIONS

We developed two alternative methods, using either radiolabeled in vitro translated arrestin or purified protein. Expression of radiolabeled arrestins in cell-free translation is simple and easy, allowing the production of up to 20 different mutants suitable for functional analysis in 1 day (35,37). However, the level of nonspecific "binding" in this case is relatively high, up to 30% of specific binding. The assay with purified arrestin gives an excellent signal-to-noise ratio, but requires prior protein purification and involves much more labor-intensive quantification by Western blot.

Microtubule preparation from purified tubulin (Cytoskeleton, Inc.) should be performed according to the manufacturer's instructions. Briefly, purified tubulin (>99%) is incubated at a concentration of 5 mg/mL in a buffer containing 80 mM PIPES, pH 6.9, 1 mM $MgCl_2$, 1 mM EGTA, 10% glycerol, and 1 mM GTP at 37°C for 40 min. Taxol (20 μM) is added, and the tubulin is incubated at room temperature for 10 min. The polymerized microtubules are then diluted to a final concentration of 1 mg/mL in 80 mM PIPES, pH 6.9, 1 mM $MgCl_2$, 1 mM EGTA, and 20 μM taxol.

Version 1—Using radiolabeled arrestin produced in cell-free translation (35,38). This assay is similar to the assay we use for direct receptor binding, which has been previously described in detail (35). The separation of microtubule-bound and free arrestin by centrifugation is based on the huge difference in size between polymerized MTs and free arrestin (29). It is important to keep the samples warm (>25°C) throughout the procedure to prevent depolymerization of tubulin.

1. Incubate 200 fmol of radiolabeled in vitro translated arrestin in 100 μL of 50 mM Tris-HCl, pH 7.4, 0.5 mM $MgCl_2$, 1.5 mM dithiothreitol, 1 mM EGTA, and 50 mM potassium acetate for 20 min at 25°C with 20 μg of prepolymerized tubulin.
2. Pellet microtubules along with bound arrestin by centrifugation at 90,000 rpm in a Beckman TLA 120.1 rotor for 10 min at 25°C. MT-arrestin pellets cannot be washed because of the low affinity of the interaction (i.e., high dissociation rate; see following text).

3. Dissolve the pellet in 0.1 mL of 1% SDS, 50 mM NaOH, transfer to scintillation vials, add 5 mL of water-miscible scintillation fluid (we use ScintiSafe Econo2, Fisher), and quantify bound arrestin by liquid scintillation counting.
4. Nonspecific "binding" (arrestin pelleted in the absence of microtubules) should be determined in the same experiment and subtracted.

Version 2—Using purified arrestin. We express wild-type rod arrestin and its various mutant forms in *E. coli* and purify it, as described in Protocol 4 and elsewhere (35), with yields ranging from 2 to 5 mg/L of bacterial culture. This method with minor modifications can be used to obtain most members of the arrestin family with purity suitable for crystallization (32,33,39).

1. Incubate 4 μg of purified arrestin in 50 μL of 50 mM Tris-HCl, pH 7.4, 100 mM NaCl, 2 mM EDTA, and 1 mM EGTA for 30 min at 30°C with 40 μg of prepolymerized tubulin.
2. Pellet microtubules along with bound arrestin by centrifugation at 90,000 rpm in a Beckman TLA 120.1 rotor for 10 min at 30°C. Parallel samples with the same amount of arrestin without microtubules serve as controls.
3. Dissolve pellet in SDS sample buffer. Because arrestin runs too close to tubulin on SDS-PAGE, bound arrestin must be quantified by Western blot (running different known aliquots of the same arrestin protein in parallel on each gel to construct a calibration curve, as described in detail in protocol 8).

3.4 ROD ARRESTIN SELF-ASSOCIATION: THE SHAPE AND FUNCTION OF THE TETRAMER

Curiously, self-association of visual arrestin was discovered many years ago, when this protein was commonly known as "S-antigen" and its function remained obscure (40). It was recognized as a special feature of the visual system after the finding that rod arrestin crystallizes as a tetramer (dimer of dimers) (33,41), in sharp contrast to arrestin2 (32,42) and cone arrestin (39). Subsequent studies using sedimentation equilibrium (34) and small-angle x-ray scattering (43,44) to investigate arrestin self-association in solution yielded contradictory results: arrestin was found to preferentially form either a dimer (34,44) or a tetramer (43). More importantly, these studies did not experimentally address the structure of the arrestin oligomer in solution and its function in rod photoreceptors. A recent study comprehensively reexamined the mechanism of rod arrestin self-association using multiangle laser light scattering to determine the weight-average molecular weight (MW_{av}) as a function of total monomer concentration (45). The advantages of this method include high resolution to within a few hundred Daltons, wide molecular mass range, relatively small sample size, and high sample throughput. Importantly, because the wavelength of light is large compared to the dimensions of proteins, the scattering is independent of molecular shape (46), making the interpretation of the data independent of speculative assumptions.

In this recent analysis, the working range of arrestin concentrations was significantly expanded (up to 100 μM), which resulted in experimental detection of molecular masses far in excess of that of arrestin dimer (90 kDa), proving the existence of tetramer in solution (45). These data confirmed the model: $2M = D\ (K_1)$, $2D = T\ (K_2)$, where M, D, and T are monomer, dimer, and tetramer, respectively (MDT model). The wide range of concentrations used (1–100 µM) greatly increased the precision of the measurements, allowing reliable determination of monomer–monomer (dimerization) and dimer–dimer (tetramer formation) association constants, K_1 and K_2. The corresponding dissociation constants ($K_{D1} = 1/K_1$; $K_{D2} = 1/K_2$; expression in molar units makes these more informative for a biologist) were found to be 37.2 and 7.4 µM, respectively (45). Thus, arrestin oligomerization is cooperative in the sense that the association constant $K_2 > K_1$ ($K_{D1} > K_{D2}$), i.e., dimer formation is rate limiting, so that as soon as the dimer is formed, it associates into a tetramer. Based on the values of K_1 and K_2, tetramer is expected to be the dominant form at physiological arrestin concentrations in rod photoreceptors (>> 200 µM, based on the arrestin-rhodopsin expression ratio of 0.8:1 in mouse rods that has been determined independently by two groups (19,20).

PROTOCOL 6: ANALYSIS OF ARRESTIN SELF-ASSOCIATION BY LIGHT SCATTERING

This is the most direct approach for studying the self-association of arrestin (or any other protein, for that matter). The main advantage is that the interpretation of light scattering data does not require any assumptions regarding the molecular shape of the complex. This turned out to be particularly advantageous for rod arrestin because the shape of the biologically relevant solution tetramer at physiological pH and ionic strength was found to be dramatically different from that of the crystal tetramer (45), even though essentially the same tetramer was observed in crystals grown under different conditions (33,41). This method is fairly fast (one measurement takes 10–15 min), allowing one to perform multiple experimental measurements within a few hours. The only drawback of this method is that for a comprehensive series of measurements one needs fairly large amounts (>5 mg) of highly purified arrestin (45), although still not as much as for analytical centrifugation (34). The following procedure described was adapted from Reference 45. The design and construction of a fully functional cysteine-less base mutant of visual arrestin that was necessary for complementary site-directed spin labeling and electron resonance studies has been described in sufficient detail previously, so the reader is referred to the original papers (29,36,45).

1. Thaw purified wild-type or mutant arrestin (protocol 4) and centrifuge at >100,000 × g for 60 min at 4°C to remove any aggregated material.
2. Make light scattering measurements with a DAWN EOS detector coupled to an Optilab refractometer (Wyatt Technologies) following gel filtration on a 7.8 mm (ID) × 15.0 cm (L) QC-PAK GFC 300 column (Tosoh Bioscience). Load the arrestin samples (100 µL) at different concentrations at 25°C, at a flow rate of 0.8 mL/min in 50 mM MOPS, 100 mM NaCl, pH 7.2. Note that this column does not resolve oligomeric species, but simply acts as a filter to

remove highly scattering particulates. Measure light scattering at 18 angles (15°–160°), absorbance at 280 nm, and refractive index (at 690 nm) for each sample for a narrow slice at the peak of the elution profile (47).

3. Calculations: The relationships describing the arrestin equilibria according to the MDT model are:

$$K_1 = \frac{D}{M^2} \qquad K_2 = \frac{T}{D^2}$$

where: M, D, and T are the monomer, dimer, and tetramer concentrations, respectively, and C is the total protein concentration. From these relationships and particular values for K_1, K_2, and C, the concentrations of monomer, dimer, and tetramer may be obtained as a solution to equations (3.1)–(3.3):

$$M + 2 \cdot K_1 \cdot M^2 + 4 \cdot K_1^2 \cdot K_2 \cdot M^4 - C = 0 \qquad (3.1)$$

$$D = K_1 \cdot M^2 \qquad (3.2)$$

$$T = D^2 \cdot K_2 = K_1^2 \cdot K_2 \cdot M^4 \qquad (3.3)$$

Using the values of M, D, and T, defined earlier, compute MW_{av} (average molecular weight measured by light scattering) as:

$$MW_{av} = M_m \left((M + 4D + 16T) / (M + 2D + 4T) \right) \qquad (3.4)$$

where M_m is the monomer molecular weight (45,000). Determine the value for C in the light scattering cell at the point where data were collected from the refractive index increment (0.184 g/mL) and A_{280} (both of which should yield the same value within experimental error). The experimental M_{av} values are then obtained from the concentration and light scattering data using Astra for Windows 4.90 software (Wyatt Technologies). Fit the experimental data for MW_{av} as a function of C to equations 3.1–3.4 with only K_1 and K_2 as adjustable parameters using a least-squares method. Collecting the data at more than 10 experimental points ensures high precision of determination of arrestin self-association constants.

In all published crystals of rod arrestin (33,41) grown in different conditions, the asymmetric unit contains four arrestin molecules arranged such that one pair is related by a local twofold rotation axis to a second pair, i.e., essentially the same tetramer (dimer of dimers). Therefore, all subsequent studies of arrestin self-association were based on the assumption that the solution tetramer had the same shape and structure (34,43,44). From a functional point of view, the main problem with this assumption was that based on the accessibility of two of the four monomers, it was impossible to explain why arrestin does not form "infinite" chains consisting of these tetramers, i.e., why self-association stops at the tetramer level. A recent study using arrestins spin-labeled in different positions finally solved this problem presenting several lines of compelling evidence that the shape of the physiologically relevant solution tetramer is quite different (45). First, numerous mutations in the "crystal" interfaces did not affect arrestin self-association, whereas others far from any "crystal" intersubunit interfaces severely interfered with it. Second, the pattern of spin label immobilization due to tetramer formation did not match the changes that would be expected if the crystal interfaces were involved. Finally, the great majority of long-range distance measurements (using double electron–electron resonance) made between the spin labels in different monomers within the tetramer were completely at odds with the distances expected in the crystal tetramer (45). This study established that in solution visual arrestin forms "closed" diamond-shaped tetramers, which explains both the cooperativity of arrestin self-association and the absence of higher-order oligomers (45).

Most important from the biological viewpoint, long-range distance measurements that report the self-association status of visual arrestin also made testing the functional capabilities of the tetramer possible for the first time. It was shown that the addition of light-activated phosphorhodopsin (in an amount sufficient to bind the arrestin present in the assay) resulted in complete disappearance of all distances characteristic of arrestin oligomers, unambiguously demonstrating that only arrestin monomer can bind rhodopsin (45). In fact, in the solution tetramer a large part of the receptor-binding surface of arrestin that was previously mapped using a variety of methods (36,48–50), is engaged by "sister" monomers, thus providing a simple mechanistic explanation for its inability to bind rhodopsin. These data support the idea that the arrestin oligomer is a "storage" form, supplying active binding-competent monomer, as needed. Unexpectedly, the addition of microtubules at a concentration high enough to bind most of the arrestin did not affect the distances, indicating that arrestin dimer and tetramer are capable of interacting with microtubules (45). Because microtubules engage essentially the same surface in arrestin monomer as the receptor (29), the interaction of the tetramer with MTs must be mechanistically different. It is tempting to speculate that the ability of arrestin tetramer to bind microtubules serves to increase their arrestin-binding capacity to ensure virtually complete arrestin translocation to microtubule-rich compartments (26) of the rod photoreceptor in the dark (6,19,21). Thus, arrestin tetramer not only serves as a "storage" form, but it also directly binds microtubules to facilitate the removal of arrestin away from the outer segment in dark-adapted photoreceptors. This ensures maximum gain of photoresponse by preventing premature signal termination by the extremely abundant arrestin. It is highly likely that this mechanism enables dark-adapted rods to generate a response to a single photon, while keeping enough stored arrestin to bind virtually every rhodopsin molecule in bright light.

3.5 HOW RHODOPSIN PHOSPHORYLATION LEVEL REGULATES ARRESTIN BINDING

It was established more than 20 years ago that for high-affinity binding, arrestin requires that rhodopsin be both activated and phosphorylated (51). However, the effect of increasing the level of rhodopsin phosphorylation on arrestin binding remains controversial. Every conceivable model has been proposed: that arrestin affinity gradually increases with the level of rhodopsin phosphorylation (52,53) or that tight arrestin binding is an all-or-nothing event requiring a certain number of receptor-attached phosphates. This number was reported to be one (54,55), two (38,48,56), or three (16,57,58), and in some cases conflicting reports came from the same group (54,57).

The use of a sensitive direct binding assay (described in the following text) revealed that arrestin independently recognizes the activation and phosphorylation state of rhodopsin, but its binding to light-activated phosphorhodopsin (P-Rh*) is at least 10 times higher than to dark P-Rh or unphosphorylated Rh* (figure 3.3). This remarkable arrestin selectivity for P-Rh* is explained by the current model of sequential multisite interaction (reviewed in references 59 to 61). This model, proposed in 1993 (38), is based on the hypothesis that arrestin has two *primary binding sites*: an *activation sensor*, which interacts with parts of rhodopsin that change conformation

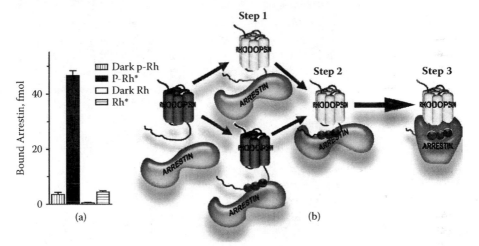

FIGURE 3.3 Exquisite arrestin selectivity for P-Rh* (a) is explained by the model of sequential multisite arrestin–rhodopsin interaction (b). Weak binding to any form of rhodopsin enables arrestin to "probe" its functional state. When arrestin encounters phosphorylated or activated rhodopsin, it binds either to the phosphorylated C-terminus or to the elements that change conformation upon activation via the arrestin phosphorylation-recognition or activation-recognition site, respectively (Step 1). If rhodopsin is only activated or phosphorylated, arrestin dissociates rapidly. When rhodopsin is phosphorylated and activated, both arrestin sites are engaged (Step 2). The simultaneous binding of the phosphorylation- and activation-recognition sites (termed *primary binding sites*) to respective parts of rhodopsin triggers a substantial conformational change in arrestin, which mobilizes additional potent binding sites for the interaction (Step 3). This conformation is termed the high-affinity receptor-binding state, in contrast to its basal low-affinity state. This complex multistep mechanism, which is shared by all arrestin proteins, makes arrestin binding to P-Rh* semiirreversible, so that arrestin dissociates only after Meta II decays to opsin, providing for high fidelity of signal termination. The binding in panel A was determined as described in Protocol 7 using 0.3 μg of the indicated form of rhodopsin and 50 fmol of radiolabeled arrestin. In panel B rhodopsin activation is indicated by lighter shading, and phosphorylation is indicated by three spheres on the rhodopsin C-terminus. Global conformational change in arrestin upon its binding to P-Rh* is also shown (Step 3).

upon light activation, and a *phosphate sensor*, binding to rhodopsin-attached phosphates. Arrestin works as a coincidence detector, i.e., when both primary sites are simultaneously engaged (which can happen only when arrestin encounters P-Rh*), arrestin undergoes a transition into a high-affinity receptor-binding state involving a global conformational change (62) that results in the mobilization of additional binding sites (38). This model has been validated by the discovery of the main (63) and auxiliary (56) phosphate sensors and a convincing description of the mechanism of their function in the context of the arrestin crystal structure (33,64). Thus, in the context of this model the question can be restated, as follows: How many rhodopsin-attached phosphates are necessary to activate the phosphate sensor in arrestin? A surprisingly large number of the exposed positively charged residues in rod arrestin have been shown to bind rhodopsin-attached phosphates (39,50,56,63,65), so that this issue cannot be unambiguously settled based solely on structural considerations.

Obviously, direct comparison of arrestin binding to preparations of rhodopsin with different precisely determined phosphorylation levels is necessary to resolve this controversy. To this end, we used previously described procedures to phosphorylate rhodopsin in bovine rod outer segments by endogenous rhodopsin kinase (66), to purify differentially phosphorylated rhodopsin species, and to reconstitute them into liposomes (67,68). The actual phosphorylation level of rhodopsin in each fraction was carefully quantified by mass spectrometry, as described earlier (69,70). We ascertained that there is no appreciable difference in the orientation of reconstituted differentially phosphorylated rhodopsin by comparing the binding of all fractions to 4D2 antibody that recognizes the rhodopsin N-terminus (71) (which binds rhodopsin in the inside-out orientation). Finally, we used our extremely sensitive direct binding assay with radiolabeled arrestin to compare its interaction with rhodopsin carrying from 0 to 7 phosphates. The data reveal an interesting multithreshold mechanism: one rhodopsin-attached phosphate does not increase arrestin binding above the level observed with unphosphorylated Rh*. Two phosphates are the minimum number sufficient to facilitate arrestin interaction, three apparently activate arrestin fully, although four phosphates slightly but statistically significantly increase the binding, whereas additional phosphates do not appreciably enhance the binding further (figure 3.4) (70).

These data indicate that in the context of the rod photoreceptor cell, which expresses only 2–3 rhodopsin kinase molecules (72,73) and up to 800 arrestin molecules (19,20) per 1000 rhodopsins, arrestin is likely to bind light-activated rhodopsin as soon as two to three phosphates are attached. Indeed, in vivo studies revealed the presence of rhodopsin species carrying one, two, and three phosphates, with negligible amounts of higher levels of phosphorylation (69). It has also been shown in transgenic mice expressing mutant rhodopsin with different numbers of phosphorylation sites that three are required to observe arrestin-dependent acceleration of photoresponse shutoff (16). Collectively, these data suggest that rhodopsin signaling is turned off in just 3–5 steps (2–4 phosphorylation events followed by arrestin binding), favoring the models that explain limited variability of the single-photon response by a small number of steps of rhodopsin inactivation (e.g., reference 74). However, the C-terminus of mammalian rhodopsin carries 6–7 phosphorylation sites, suggesting that their abundance is biologically important. Moreover, the elimination of even

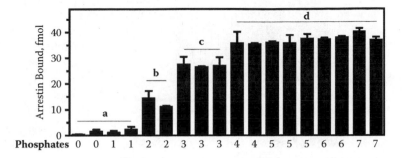

FIGURE 3.4 The effect of rhodopsin phosphorylation level on arrestin binding. Purified reconstituted phosphorhodopsin (0.15 μg) with the indicated level of phosphorylation was incubated with 50 fmol of radiolabeled arrestin in 50 μL for 5 min at 37°C in the light, and then bound arrestin was separated from free and quantified (Protocol 7). Binding data were analyzed using one-way ANOVA with phosphorylation level as a main factor, followed by a Bonferroni/Dunn post hoc test with correction for multiple comparisons. The binding levels were grouped ("a" through "d") so that they were not significantly different within the group and significantly different from other groups on the same panel ($p < 0.0001$ in all cases).

one or two out of six sites in mouse rhodopsin clearly affects photoresponse kinetics, and the elimination of three substantially slows down rhodopsin inactivation (16). It is tempting to speculate that the large number of sites is necessary to increase the probability that transient relatively low-affinity binding of rhodopsin kinase to light-activated rhodopsin is productive, i.e., results in the phosphorylation of at least one of the sites. This would become increasingly important with the progression of phosphorylation, e.g., after two sites are already phosphorylated, wild-type rhodopsin would retain four accessible targets, whereas the mutant carrying three sites would have only one left. In vivo comparison of the phosphorylation kinetics of wild-type and mutant rhodopsin is necessary to test this hypothesis.

PROTOCOL 7: DIRECT BINDING ASSAY WITH RADIOLABELED ARRESTIN

This assay takes advantage of the expression of arrestins with high specific activity in cell-free translation (described in detail in references 35 and 75), which provides sufficient sensitivity to study high-affinity arrestin binding to P-Rh*, as well as low-affinity interactions with other forms of rhodopsin. As an added bonus, in vitro translated visual arrestin can be used without further purification, because it is the only radiolabeled protein in the translation mix. This relatively high-throughput assay is perfectly suited for the functional characterization of multiple mutant forms of arrestin (38,48,63,65), as well as for the direct comparison of wild-type arrestin binding to multiple rhodopsin preparations (figures 3.2 and 3.4) (70).

1. Store translated radiolabeled arrestin at –80°C, where it is stable for weeks. It survives several freeze–thaw cycles very well, much better than prolonged incubation at 4°C, so that it is advisable to thaw the sample "just in time" and then refreeze it immediately after taking the aliquot for the binding experiment.

2. Dilute arrestin (50–100 fmol per assay, for 1–2 nM final concentration) with 50 mM Tris-HCl, pH 7.5, 50 mM potassium acetate, 0.5 mM MgCl$_2$ (RB buffer) containing 0.5 mM DTT to 2–4 fmol/μL, and add 25 μL of this solution per tube.

3. Under dim red light, dilute Rh and P-Rh with the same buffer to 12 μg/mL and distribute between tubes (25 μL/assay, i.e., 0.3 μg or ~7.5 pmol per tube).

4. Incubate arrestin with the appropriate form of rhodopsin in the dark or with illumination at 37°C for 5 min. Precise timing is crucial, because of the rapid decay of light-activated rhodopsin.

5. Immediately cool the samples on ice. Separate bound and free arrestin by gel filtration at 4°C (under dim red light for dark rhodopsin) on a 2 mL Sepharose CL-2B column equilibrated with 10 mM Tris-HCl, pH 7.5, 100 mM NaCl (100/10). The membrane-containing fraction elutes in the void volume between 0.5 and 1.1 mL and is counted in a liquid scintillation counter. This is accomplished by having columns in a 10 × 10 rack that can be directly placed over a standard 10 × 10 box of scintillation vials. Samples are loaded on the columns and allowed to soak in. The column is then washed with 100 μL of 100/10 followed by 400 μL of 100/10 after the first portion enters the matrix, and then rhodopsin with bound arrestin is eluted into scintillation vials with 600 μL of 100/10. It is crucial that the columns for dark samples are run under dim red light. Determine nonspecific arrestin binding (using 0.3 μg of liposomes) and subtract. The columns for this assay can be reused many times, provided that after each experiment the columns are immediately washed with 4 × 3 mL of 100/10, and stored capped at 4°C in 100/10 buffer. If columns are not used for an extended period of time, the addition of 3 M NaCl to the storage buffer is advised to prevent microbial growth.

3.6 STOICHIOMETRY OF THE ARRESTIN–RHODOPSIN INTERACTION

Based on the fact that rods faithfully respond to a single photon (73) and that a photon can only activate one rhodopsin (1), it was widely believed for a long time that arrestin interacts with an individual rhodopsin molecule (1,59,76). All arrestins in their basal state are elongated molecules consisting of two cuplike domains, each with a diameter of about 35 Å (32,33,39,41,42). Recent discoveries suggest that rhodopsin (77) and many other G-protein-coupled receptors (78) form dimers under certain circumstances. Dark (inactive) rhodopsin is the only receptor for which the crystal structure is known (79–81). Its relatively small cytoplasmic tip measures about 40 Å across. On the basis of this rather provocative geometry, an alternative model in which a single arrestin binds to two receptors in a dimer was proposed (82). The biological implications of these two models are profoundly different, so it is very important to resolve this issue experimentally.

Luckily, rod photoreceptors provide a unique model in which the stoichiometry of the arrestin–receptor interaction can be determined in vivo because they express comparable amounts of rhodopsin and arrestin at very high levels unparalleled in any other cell type (5,19,83,84). As discussed earlier, light induces arrestin translocation from the inner to the outer segment, where it is retained by virtue of its binding to light-activated rhodopsin (6). It has been established that the expression of rhodopsin and arrestin can be genetically manipulated independently: hemizygous rhodopsin (Rh$^{+/-}$) and arrestin (Arr$^{+/-}$) mice express about half of these respective proteins as compared to wild-type animals (23,85). Thus, the extent of arrestin translocation in these genetic backgrounds can be used to determine the stoichiometry of the arrestin–rhodopsin interaction using exclusively wild-type proteins for this purpose. To this end, we bred Rh$^{+/-}$, Arr$^{+/-}$, and Rh$^{+/-}$Arr$^{+/-}$ animals and compared the content of rhodopsin and arrestin in their retinas with that of wild-type mice (table 3.1). Both proteins were measured by quantitative Western blot in the homogenates of whole eyecups using the corresponding purified proteins to construct calibration curves. The results confirmed that the elimination of one allele reduces the expression by ~50%, so that the arrestin/rhodopsin ratio in wild-type and Rh$^{+/-}$Arr$^{+/-}$ animals is similar (0.8:1 and 0.94:1, respectively), whereas in Arr$^{+/-}$ and Rh$^{+/-}$ mice, it is significantly shifted in the expected direction, to 0.38:1 and 1.74:1, respectively (table 3.1).

If two rhodopsins were necessary to bind one arrestin, virtually complete arrestin translocation to the OS in bright light would only be expected in Arr$^{+/-}$ animals, but not in wild-type mice. In contrast, in the case of a one-to-one interaction, partial translocation would be expected only in Rh$^{+/-}$ animals that express more arrestin than rhodopsin. In agreement with previous reports (6,12–14,19,21,28), we invariably observed virtually complete arrestin translocation in the light-adapted retinas of wild-type mice, as well as other lines expressing more rhodopsin than arrestin (figure 3.5). Quantitative image analysis shows that in wild-type mice and the other two lines that express excess rhodopsin, 81–89% of arrestin translocates to the OS in the light (table 3.1) (20). In contrast, in Rh$^{+/-}$ mice, only about half of that amount of arrestin moves to the OS (table 3.1, figure 3.5). These data are consistent with one-to-one binding and cannot be reconciled with the model of rhodopsin dimer interacting with just one arrestin molecule. Based on the extent of arrestin translocation and the absolute expression levels of both proteins, one can calculate that in the light-adapted retinas of these mice, 65–83% of rhodopsin is occupied at steady state by bound arrestin (table 3.1), strongly supporting the one-to-one binding model. Notably, the rhodopsin occupancy in both Rh$^{+/-}$ lines (75–83%) exceeds the theoretical maximum for the one-to-two model (50%) even more than that in Rh$^{+/+}$ animals. Interestingly, the reduction of rhodopsin concentration in the disks in Rh$^{+/-}$ mice has been shown to accelerate the kinetics of photoresponse, possibly by increasing the lateral diffusion of rhodopsin that facilitates its interactions with transducin (20). Higher rhodopsin occupancy by arrestin in Rh$^{+/-}$ lines than in the Rh$^{+/+}$ lines suggests that the reduction of rhodopsin concentration in the disk membrane favors arrestin binding as well; i.e., the relief of rhodopsin "overcrowding" facilitates its interactions with all signaling proteins.

Thus, arrestin translocation in every genetic background is only consistent with the one-to-one model of arrestin–rhodopsin binding in vivo. However, one can argue that in the photoreceptor cell other proteins could participate in arrestin retention in the OS, so that these in vivo data are not necessarily conclusive. To ascertain that other proteins present in live photoreceptors do not affect the translocation of arrestin to the OS, we reproduced similar arrestin/rhodopsin ratios (0.5, 1.1, 2.2, and

TABLE 3.1

Light-Dependent Translocation of Arrestin to the Outer Segment as a Function of Arrestin and Rhodopsin Expression

Genotype	Rh Content (nmol/retina)	Arrestin Content (nmol/retina)	Arr/Rh Molar Ratio	Percentage Arr in the OS (dark)	Percentage Arr in the OS (light)	Arr in the OS (light) nmol	Rh Occupied by Arr (%)
WT	0.40 ± 0.05 (7)	0.32 ± 0.04 (7)	0.80	1.8 ± 0.8	81.4 ± 1.3	0.26	65
A$^{+/-}$	0.40 ± 0.03 (5)	0.15 ± 0.02 (5)	0.38	3.1 ± 1.1	88.8 ± 0.5	0.13	33
A$^{+/-}$Rh$^{+/-}$	0.18 ± 0.02 (4)	0.17 ± 0.02 (4)	0.94	1.4 ± 0.5	89.1 ± 1.0	0.15	83
Rh$^{+/-}$	0.19 ± 0.01 (4)	0.33 ± 0.02 (4)	1.74	1.1 ± 0.4	43.5 ± 6.5	0.14	75

Note: Mice with the indicated genotypes were dark-adapted overnight (dark) or exposed to 2700 lux for 1 h (light). One eyecup from each mouse was fixed and processed for immunohistochemistry (Protocol 3), whereas the other was homogenized for rhodopsin and arrestin quantification by Western blot (Protocol 8). The proportion of arrestin localized in the OS was quantified by the intensity of arrestin immunostaining in 10 images per animal from 3 to 5 animals per genotype per light condition. Means ± SD are shown. The data were analyzed by one-way ANOVA with genotype as a main factor. The rhodopsin content of A$^{+/-}$Rh$^{+/-}$ and Rh$^{+/-}$ and the arrestin content of A$^{+/-}$ and A$^{+/-}$Rh$^{+/-}$ were statistically different from all other genotypes ($p < 0.0001$) but were not different from each other. The amount of arrestin in the OS (nmol) was calculated by multiplying the arrestin content by the percentage of arrestin in the OS in the light. The percentage of rhodopsin occupied was determined by dividing this value by the total rhodopsin content.

3.3) in vitro with purified proteins carefully quantified by amino acid analysis. In these experiments, recombinant purified bovine arrestin was mixed in the dark with bovine rhodopsin in native disk membranes that was phosphorylated with endogenous rhodopsin kinase (66) and fully regenerated with 11-*cis*-retinal. The samples were illuminated for 5 min at 37°C, and then rhodopsin along with bound arrestin was pelleted by centrifugation. Samples containing the same amounts of arrestin and no rhodopsin were used as controls. Equal aliquots of the original samples, pellets, and supernatants were resolved by SDS-PAGE and stained with Coomassie blue, revealing clearly saturable arrestin binding (figure 3.6). To achieve high precision,

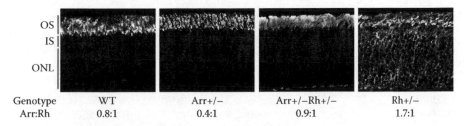

Genotype	WT	Arr+/−	Arr+/−Rh+/−	Rh+/−
Arr:Rh	0.8:1	0.4:1	0.9:1	1.7:1

FIGURE 3.5 Complete arrestin translocation to the OS requires the presence of equivalent number of rhodopsin molecules. Mice with the indicated genotypes were dark-adapted overnight and exposed to 2700 lux for 1 h. Arrestin localization was visualized and quantified as described in Protocol 3. The positions of the outer (OSs) and inner segments (ISs) and of the outer nuclear layer (ONL) are indicated. Arrestin/rhodopsin expression ratios in each genotype are shown. Note that arrestin translocation is incomplete only when arrestin is expressed in excess over rhodopsin.

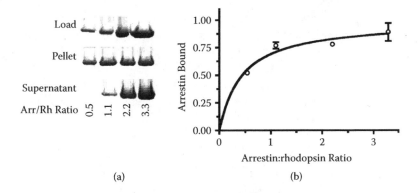

(a)   (b)

FIGURE 3.6 The stoichiometry of arrestin–rhodopsin interaction. (a) Purified bovine arrestin (142, 286, 572, and 858 pmol) was incubated for 5 min in the light with 10 μg (261 pmol) of phosphorhodopsin in 25 μL of 50 mM MOPS-Na, pH 7.2, 100 mM NaCl. Rhodopsin with bound arrestin was pelleted through a 50 μL cushion of the same buffer with 0.2 M sucrose. Equal aliquots of the sample before centrifugation (Load) (1/14), Pellet (1/4), and supernatant (Sup) (1/4) were subjected to SDS-PAGE. Arrestin was visualized by Coomassie blue staining. Arrestin/rhodopsin ratios are shown under the corresponding lanes. (b) The absolute amount of arrestin and rhodopsin in the pellet was measured by quantitative Western blot (Protocol 8). Arrestin binding is plotted as a function of the arrestin/rhodopsin molar ratio in the sample. Means + SD of two experiments are shown. The analysis of the binding data (using GraphPad Prizm) yields Bmax (saturation) at 0.99 ± 0.08 mol/mol.

we measured the amounts of rhodopsin and arrestin in the original samples, pellets, and supernatants by quantitative Western blot with appropriate standards (Protocol 8). Rhodopsin was found to pellet quantitatively; the amount of arrestin pelleted in the absence of rhodopsin did not exceed 4.5%. Under these conditions, we observed clear saturation of rhodopsin by increasing concentrations of arrestin, reaching up to 0.9 mol of specifically bound arrestin per mol of rhodopsin (figure 3.6). These results are in perfect agreement with the in vivo data (figure 3.5). Because only arrestin monomer can bind rhodopsin (45), these data definitively establish that one arrestin molecule binds one rhodopsin molecule.

Interestingly, even at a very high arrestin/rhodopsin expression ratio in $Arr^{+/+}Rh^{+/-}$ mice, the proportion of arrestin-occupied rhodopsin "saturates" at ~80%, which almost exactly corresponds to the relative expression of these proteins in wild-type animals (table 3.1 and references 19 and 20). Apparently, mice express just enough arrestin to occupy all rhodopsin that is binding competent (i.e., light-activated, light-activated phosphorylated, dark phosphorylated, and phosphoopsin (38,70) (figure 3.2)) at equilibrium in fully light-adapted rods. Some of the rhodopsin in light-adapted OS exists as unphosphorylated opsin that does not bind arrestin and does not support arrestin translocation when the equilibrium is shifted toward this form by hydroxylamine treatment (6). This actually reduces the amount of rhodopsin competent to bind arrestin, further strengthening the conclusion that only the binding of arrestin by each individual rhodopsin can account for the observed extent of arrestin translocation in vivo.

PROTOCOL 8: QUANTIFICATION OF THE CONTENT OF INDIVIDUAL PROTEIN IN TISSUE SAMPLES BY WESTERN BLOT

Sometimes it is important to actually measure the content of a particular protein in a tissue sample that can be expressed in absolute (weight or molar) units. This task becomes even more complicated when the absolute content of different proteins and their expression ratios must be established (20,86,87). This can be accomplished by Western blot only when a rigorous set of conditions is met. Here we describe these conditions, necessary reagents, and quality control procedures that ensure reliable quantitative results.

Protein standards

1. First, one needs the pure proteins of interest, recombinant or purified from the tissue. The amounts of purified protein (>0.2 mg) must be sufficient to perform quality control experiments and have enough left for the actual work. The purity of each protein in preparations used to prepare the standards must be determined by SDS-PAGE followed by Coomassie blue staining. To achieve this, it is important to run different amounts of protein: "overloaded" lanes with 3–10 µg to detect even minor impurities, as well as lanes with 0.2–1 µg to ascertain that proteolytic products of a size similar to the full-length protein do not contaminate the sample. Purity below 95% is unacceptable.
2. Initial protein quantification by any standard method (e.g., Bradford), which is usually based on the comparison of the protein of interest with BSA,

γ-globulin, or some other protein, must be checked by one of the "absolute" methods: sequence-calibrated UV absorbance or amino acid analysis.

3. For the sake of stability many proteins are kept in rather complex buffers with a variety of components, some of which absorb UV. Therefore, for spectroscopic measurements it is imperative that the spectrum of exactly the same buffer in which the protein is dissolved is also taken and subtracted. The best way to achieve this is to dialyze the protein extensively (>2 h for relatively small 0.1–0.5 mL samples) against the buffer, and then use the dialysis buffer in the beaker as a control.

4. Quantification by amino acid analysis requires the elimination of all buffer components from the sample, and in the case of membrane proteins (e.g., rhodopsin), the elimination of the lipid. The simplest way to accomplish this is to precipitate the protein by the addition of nine volumes of methanol and pellet it by centrifugation in tubes suitable for subsequent acid hydrolysis (e.g., 10 min at 13,000 rpm in 6 × 50 mm borosilicate tubes with appropriate holders in a standard tabletop microcentrifuge). The pellets should be washed with 90% methanol (0.2–0.5 mL) by vortexing and centrifugation, dried, and sent to a commercial or in-house protein chemistry facility. The proteins will be hydrolyzed with HCl (vapor phase, 150°C, 75 min) along with a standard amino acid mix (to control for the loss of amino acids under these conditions). Samples will then be dissolved in an appropriate buffer, derivatized, and aliquots of each sample will be analyzed by HPLC. The quantification of amino acids stable under the conditions of acid hydrolysis (Ala, Val, Leu, Phe, and Lys) should be used to calculate the amount of each protein in the sample based on its known amino acid composition. The numbers based on each of these amino acids must agree to within 5–7%. If they do not, the sample likely contains too many impurities to be used to prepare standards.

5. The standards should be prepared by serial dilution of the purified protein, which should be kept in aliquots at −80°C. In most cases the protein can be diluted in SDS sample buffer and stored frozen. Some proteins in SDS sample buffer, especially in very diluted solutions, "go bad" after a few freeze–thaw cycles, as manifested by a significant decrease in signal intensity, so that it becomes necessary to make fresh dilutions for each experiment. Therefore, it is advisable to compare fresh and refrozen standards of the protein of interest early on to test whether this problem exists.

6. The signal on the blot from the same amount of protein sometimes decreases in the presence of other proteins in the sample (possibly because of "shielding" of the antigen by proteins with the same electrophoretic mobility transferred to the membrane). Therefore, whenever possible, the standards should be mixed with the same amount of protein from the same tissue of corresponding knockout animals that are going to be present in the experimental samples. For example, retinal protein from arrestin and rhodopsin knockout mice was added to arrestin and rhodopsin standards, respectively, in the study described earlier (84). However, many proteins cannot be knocked out because in their absence mice are not viable, and for

other species there are very few knockouts available. In this case the protein in each experimental sample should be run at 2–3 different concentrations to ascertain that the signal increases linearly with the load.

Quantitative Western blot

1. After standard disk-electrophoresis, we transfer the proteins to Immobilon-P (Millipore) membrane and block with 5% nonfat dry milk in TBS containing 0.1% Tween-20 (TBST) for 30–60 min at 37°C or for 60 min at room temperature with gentle rocking.

2. Several pilot experiments are usually required to establish conditions under which the blot is actually quantitative. First, it is necessary to establish the linear range of the signal by running 7–10 standards containing from 10 pg to 10 ng of the protein. Usually, this has to be done more than once (adjusting the range as you go) with different dilutions of primary antibody to optimize this parameter at the same time. In many cases the signal is linear only in a fairly narrow range and only at certain antibody concentration, both of which may be very different depending on the antibody, (e.g., in our hands rhodopsin signal is linear between 0.1 and 1.5 ng/lane at 1:3,000 dilution of 4D2 antibody (71), whereas arrestin signal is linear between 10 and 100 pg/lane at 1:10,000 dilution of F4C1 antibody (40). In some cases incubation with the primary antibody overnight at 4°C works better than 1 h at room temperature, but in other cases the opposite is true. Changes in the dilution or incubation time of the secondary antibody cannot compensate for the nonlinearity of primary antibody binding. Therefore secondaries should be used at a standard concentration, (e.g., we use 1:12,000 dilution of HRP-conjugated antirabbit and antimouse antibodies from Jackson ImmunoResearch) and incubate for 1 h at room temperature.

3. The great majority of commercial antibodies marketed as "specific" on the strength of a blot of the material from cells overexpressing the protein in question label more than one band in real-life tissue samples. Sometimes, only the comparison with the standard allows the identification of the specific band among others. It is often worthwhile to adjust the acrylamide concentration in the gel to ensure optimal separation of the band of interest from the nearest nonspecific bands. This will greatly facilitate subsequent quantification and improve its precision. The pattern of nonspecific bands may be very different in different tissues or cell lines.

4. Working solutions of many (but not all) primary antibodies can be frozen at −20°C and reused several times. This is only possible if the antibody is diluted in TBST supplemented with 2% BSA. As a rule, freezing antibodies in TBST with nonfat dry milk kills them. If a working solution of the primary antibody is to be reused, the milk from the blocking solution should be thoroughly washed away with TBST before the addition of the antibody. It is advisable to test whether the antibody survives freezing by comparing the signal on two identical blots incubated with fresh and thawed antibody solution.

5. After incubation with HRP-conjugated secondary antibody we develop the blot with SuperSignal West Pico reagent (Pierce) for 1–1.5 min, and then expose it using SuperRX x-ray film (Fujifilm). For quantification purposes, it is imperative to obtain exposures where the bands in the standards and samples are not saturated ("pretty" blots with thick black bands that look good in papers are not quantitative). Often, it is useful to get two or more different exposures in which the bands to be used for quantification are "gray."

6. A series of standards should be run on every blot. For quantification we use VersaDoc with QuantityOne software (Bio-Rad). Background (signal in an equal "empty" area, which is never zero even after background subtraction) should be subtracted from the integrated signal in the band (volume = intensity $\times$ mm^2). The specific signal from standards should be plotted as a function of protein amount. This calibration curve (or at least the part of it that will be used) should be linear. Bands in experimental samples with a signal that falls within the linear range of the calibration curve can be used for quantification. It is very important to quantify each protein in each sample on at least two independent blots. These numbers should agree within 10%.

3.7 RELEASE OF BOUND ARRESTIN

To ensure high fidelity of quenching, arrestin has to stay bound as long as rhodopsin is in a functional state that has activity toward transducin. Indeed, arrestin was found to be the most slowly released protein among all that bind rod outer segment membranes in a light-dependent manner (88). The half-life of the complex at 0°C exceeds 2 h (38,70), suggesting that even at physiological temperature it would last for minutes, i.e., long enough for Meta II decay. Ultimately, to return the system to the ground state, arrestin must be released, and rhodopsin has to be dephosphorylated and regenerated with 11-*cis*-retinal. Because arrestin precludes rhodopsin dephosphorylation (89), it apparently has to dissociate first. The model of sequential multisite arrestin binding (figure 3.3) suggests that rhodopsin deactivation (decay of Meta II accompanied by the corresponding conformational changes in opsin) decreases arrestin affinity, thereby facilitating its release. Although in live photoreceptors arrestin translocation back to the inner segment in the dark occurs at the same rate as rhodopsin dephosphorylation (6), this correlation does not tell which event is the cause and which is the effect. However, dramatic acceleration of arrestin release by the addition of hydroxylamine (that facilitates Meta II decay) to the eyecups (6) indicates that the release of retinal from Meta II promotes arrestin dissociation. Interestingly, the release of retinal and bound arrestin may be interdependent events (70,90), and this area certainly requires further investigation.

3.8 PUTTING THE PIECES TOGETHER: THE FUNCTIONAL CYCLE OF ARRESTIN IN RODS

To better understand what the affinities of arrestin for rhodopsin, microtubules, and itself actually tell us, we need to look at arrestin interactions with various partners from a thermodynamic point of view. The interaction of two proteins is a chemical reaction. The Gibbs standard free energy change, $\Delta G°$, for a reaction at constant temperature and pressure is given by:

$$\Delta G° = \Delta H° - T\Delta S°$$

where $\Delta H°$ is the enthalpy difference between reactants (free proteins in this case) and products (protein–protein complex) in their standard states, i.e., at concentrations of 1 M and at a specified temperature (usually 25°C); $\Delta S°$ is the entropy difference between reactants and products (in their standard states); and T is the temperature (in degrees Kelvin).

The association of two proteins, similar to the association of a ligand with a receptor, is a second-order reaction (A + B = AB). The reaction proceeds when the product, complex AB, has less free energy than the reactants (free proteins A and B), so that a negative value of $\Delta G°$ is thermodynamically favored. The enthalpy term $\Delta H°$ shows whether heat is being released ($\Delta H° < 0$) or absorbed ($\Delta H° > 0$) during the reaction. The changes in enthalpy usually reflect the formation of intermolecular bonds (ion pairs, hydrogen bonds, dipole–dipole interactions). The entropy term $\Delta S°$ is intuitively perceived as an expression of a degree of randomness of the system. Strictly speaking, it represents the number of equivalent energy states available to a particular molecular species, and the changes in entropy often reflect the participation of hydrophobic interactions (because the removal of hydrophobic side chains from water environment yields favorable changes of entropy).

For a second-order reaction, the free energy for association, $\Delta G°$, is related to the equilibrium association constant as:

$$\Delta G° = -RT \, ln K_A,$$

where R is the gas constant (1.99 cal/mol × degree); T is the temperature (in degrees Kelvin); and K_A is the equilibrium association constant (M^{-1}); $K_A = 1/K_D$.

Because a decrease in $\Delta G°$ drives the reaction forward, high-affinity interactions (large K_A, i.e., small K_D) are always favored over lower-affinity interactions. Thus, arrestin binding to rhodopsin ($K_D < 1$ nM) is much more favorable than its binding to microtubules ($K_D > 50$ μM) or self-association (dimerization [K_{D1}] and tetramerization [K_{D2}] constants 37.2 and 7.4 μM, respectively). Thus, arrestin will self-associate and bind MTs only when more "attractive" partners (P-Rh*, Rh*, dark P-Rh, P-opsin; cf. figure 3.2) are absent, i.e., in the dark. By the same token, self-association of arrestin monomers is somewhat more favorable than microtubule binding, so that arrestin is more likely to tetramerize first and bind MTs later. The ability of the tetramer to bind MTs without dissociating (45) greatly favors this scenario.

The kinetics of arrestin binding to itself and other partners and the rate of its dissociation from these complexes are of utmost biological importance. This information can also be derived from the affinity of these interactions, because $K_D = k_{-1}/k_1$, where

K_D is the equilibrium dissociation constant, and k_1 and k_{-1} are kinetic association and dissociation constants. The association rate constant is proportional to the rate of diffusion, and because both rhodopsin and microtubules do not move between the compartments of the photoreceptor cell (at least on the subsecond time scale of photoresponse), we only need to take into account the rate of arrestin diffusion. The diffusion-limited association rate constant for ~45 kDa arrestin monomer can be estimated as $1 \times 10^6 M^{-1}sec^{-1}$ (91). The estimates of the amount of arrestin present in the outer segment in the dark vary from 1 to 3% (6,20) to <7% of the total (19). The estimated rhodopsin concentration in the OS is ~3 mM (73), and arrestin is expressed at 0.8:1 ratio to rhodopsin (19,20). So if 1, 2, 3, or 7% of arrestin is present in the OS in the dark, it translates into 24, 48, 72, or 168 μM concentrations, which at first glance look very different. However, based on the self-association constants, one can calculate that at any of these concentrations, a significant proportion of arrestin would be in the form of dimer and tetramer with ~ 45, 30, 22, and 14%, respectively, being a monomer (45), which yields the concentrations of the active monomer in dark-adapted OS at 11, 14, 16, 23 μM, respectively. Thus, because of self-association alone, a sevenfold change in the total arrestin in the OS results in just a twofold change in the concentration of the active monomer. This probably explains why Arr$^{+/-}$ animals that express half of the WT amount of arrestin (table 3.1) demonstrate perfectly normal photoresponse kinetics (23). An important corollary of the calculations of the concentration of free monomer in the OS when from 1 to 7% of total arrestin is present in this compartment is that a fairly large proportion of arrestin must be quasi-irreversibly "consumed" by rhodopsin binding before the concentration of the free monomer changes significantly.

Together with k_1 of $1 \times 10^6 M^{-1}s^{-1}$ monomer concentration of 10–20 μM (i.e., 10–$20 \times 10^{-6} M$) gives an on-rate ($k_1 \times$ [arrestin concentration]) of 10–20 s^{-1}, which means that in dark-adapted OS arrestin "checks" the functional state of each rhodopsin 10–20 times per second. This estimate is in good agreement with the finding that the time course of a single photon response in Arr$^{-/-}$ mice begins to deviate from that in WT mice at ~0.16 s (23) and with a recent estimate of ~80 ms half-life of active rhodopsin in vivo (92). The affinity of arrestin tetramer for microtubules cannot be measured directly because arrestin monomers, dimers, and tetramers are always at equilibrium. The fact that microtubule binding does not promote tetramer dissociation (45) indicates that tetramer has higher affinity for the MTs than monomer; i.e., the K_D of this interaction is likely <50 μM. The ability of P-Rh* and Rh* to remove arrestin from MTs in vivo and in vitro (6,29) suggests that even Rh* has substantially higher affinity for arrestin than MTs. Arrestin affinity for Rh* has not been measured, but based on P-Rh* affinity and relative arrestin binding levels to these two forms (figure 3.2), it can be estimated at about 1 μM. This gives a range of 10–40 μM for the K_D of arrestin tetramer–MTs interaction. The lack of effect of MT binding on the DEER signal reporting intersubunit distances (45) suggests that arrestin dimer also binds MTs.

Remembering that $K_D = k_{-1}/k_1$ along with the estimate of k_1 given earlier, we can calculate k_{-1} and thereby half-life ($t_{1/2} = 0.693/k_{-1}$) of these complexes. Arrestin affinity for its preferred partner P-Rh* (figure 3.2), is in the nanomolar range (30,62,93), which yields a half-life of several minutes. This is reasonably close to the half-time of Meta II decay, as well as to the rate of arrestin return from the OS (where it is held by rhodopsin) after the lights are turned off (6,21). It should be noted that the three reported values of the K_D yield rather different half-lives: 1 nM (30), 12 min; 20 nM (93), ~0.6 min; and 50 nM (62), ~0.2 min. Only the first of these three values would ensure that arrestin stays bound until Meta II decays, which is the most likely scenario in vivo. The long half-life of the arrestin–P-Rh* complex means that arrestin bound to P-Rh* is definitely out of the game for the duration of the flash response. Because in real life light comes continuously rather than in brief flashes, more arrestin monomer must be supplied to keep up with demand. Calculations show that arrestin oligomers and microtubules are in a good position to fulfill this function: the half-life of arrestin dimer ($K_D = 37.2$ µM) is ~60 ms, the half-time of arrestin monomer release from MTs is ~45 ms. Interestingly, about 20% of arrestin exists as a dimer in a wide range of concentrations from 25 to >200 µM (45), ensuring that a large supply of arrestin is available at very short notice. The generation of active monomer from the tetramer ($K_D = 7.4$ µM; half-life ~430 ms; dissociation would release dimers with a half-life of 60 ms) takes a lot longer, especially if the tetramer is MT bound, so this process is unlikely to contribute much to a single flash response, but is still fast enough to generate "fresh" monomer within a half second.

With these numbers in mind, we can deduce the following functional cycle of arrestin in rods (figure 3.7). In the dark, the great majority (93–99%) of arrestin is in the inner segment and cell body. Although the cytoplasmic volume of these compartments is greater than that of the OS, this still translates into very high (up to 1 mM) concentration. Thus, the great majority of arrestin in the IS and other compartments is in the form of tetramer, most of which is bound to microtubules (1 mM of monomer produces almost 250 µM of tetramer, which is many times higher than the most conservative estimate of the K_D of its MT binding). This explains why arrestin is concentrated in MT-rich areas, such as IS, perinuclear space, and synaptic terminals in the dark (figure 3.7). About 10% of this arrestin is present as a dimer (the proportion of dimer gradually decreases with the increase of arrestin concentration), and a small fraction exists as a monomer. All three forms of arrestin are in equilibrium with one another, and each form is also in equilibrium between a free and MT-bound state. Note that the concentrations of free monomer, dimer, and tetramer in the OS and the rest of the cell are equalized by diffusion, so the absolute concentration of free monomer must be ~10–20 µM. Considering that some of the monomer may be bound with very low affinity to dark rhodopsin (this form being in equilibrium with free and MT-bound monomer), the overall proportion of monomeric arrestin in the OS may be higher than in other compartments of the photoreceptor cell. Free arrestin monomer in the OS constantly diffuses around, "probing" the functional state of each rhodopsin via random encounters many times per second. The extremely low arrestin affinity for dark Rh present at 3 mM concentration ensures that this interaction does not significantly impede arrestin diffusion (e.g., at 1 mM K_D the half-life of the complex would be ~2 ms). The activation of a small proportion of rhodopsin

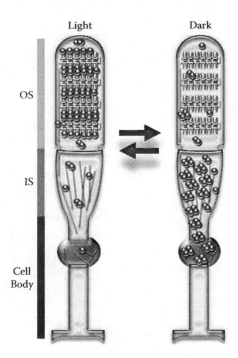

FIGURE 3.7 The functional cycle of arrestin in rods. In the dark, arrestin (shown as spheres) is localized in the inner segment (IS) and cell body (where the great majority of it is bound to microtubules shown as thin rods), with a small fraction of the total localized in the outer segment (OS). High arrestin concentration in the rod ensures that in all these compartments most of the arrestin exists as a tetramer (for spheres together) and dimer (two joint spheres). Because of self-association and microtubule binding, the concentration of free monomer is fairly low and the same in all compartments. In the light, activated rhodopsin binds arrestin, "consuming" active free monomer and thereby shifting the equilibrium toward arrestin release from microtubules and the dissociation of dimers and tetramers. After prolonged illumination, most of the arrestin localizes to the OS, where it stays as long as rhodopsin is in one of the functional forms (P-Rh*, dark P-Rh, Rh*, phosphoopsin) that has higher affinity for arrestin than microtubules and other arrestin molecules. As active rhodopsin decays and is dephosphorylated and regenerated with 11-*cis*-retinal (generating opsin and dark Rh with very low affinity for arrestin), other arrestin molecules and microtubules become competitive again. Because of this, arrestin reassociates into dimers and tetramers and relocalizes to the compartments particularly rich in microtubules, such as the inner segment, perinuclear space, and synaptic terminals, thus leaving the outer segment.

by a dim flash does not change the situation: bound free monomer is immediately replenished by the dissociation of the dimer, with minimal shift of the overall equilibrium and without appreciable arrestin redistribution between the compartments. Calculations suggest that there is enough arrestin monomer to bind ~0.5% of rhodopsin. In response to the "consumption" of a significant proportion of the monomer in the OS, it is replenished by the release from MTs and dissociation of arrestin oligomers in the OS, as well as by the diffusion of arrestin from the IS that equalizes the concentrations of free monomers, dimers, and tetramers across cellular compartments. The ensuing reduction of the concentrations of free arrestin in the IS promotes the release of MT-bound forms to restore the equilibrium. Prolonged bright illumination induces the accumulation of Rh*, P-Rh*, and P-opsin, all of which bind arrestin, so that a considerable proportion of the arrestin content is bound up by these rhodopsin species. This promotes massive diffusion of arrestin from the IS to the OS, causing the dissociation of the bulk of bound arrestin from MTs to replenish the pool of free forms (figure 3.7). These events are observed as light-dependent arrestin translocation to the OS (figure 3.1). Apparently, during the day when the light intensity far exceeds the working range of rods, a significant proportion of arrestin stays in the OS in complex with several binding-competent functional forms of rhodopsin (figure 3.2), possibly helping the rods survive through this period of signal overload. Recent findings that the initial phase of arrestin movement may be signal dependent and leads to the relocalization to the OS of a greater number of arrestin molecules than the number of light-activated rhodopsins (19) cannot be rationalized in the context of this "minimalist" model, suggesting that at least some arrestin interactions are regulated. Importantly, the model suggests the most likely points that can be subject to such a regulation: any biochemical event that inhibits arrestin self-association and/or weakens tetramer binding to the microtubules would greatly increase the concentration of free arrestin in the IS, thereby inducing its translocation to the OS driven by diffusion, equalizing the concentrations of free arrestin species. These hypothetical mechanisms remain to be elucidated.

3.9   END OF THE STORY? ARRESTIN IS FULL OF SURPRISES!

Until very recently the prevailing belief was that the only biological function of rod arrestin was to limit the duration of rhodopsin signaling. Arrestin self-association (34,43–45) and its interaction with microtubules (6,25,29) were regarded exclusively as mechanisms that regulate arrestin availability for rhodopsin binding in the photoreceptor cell. Several recent findings challenged this view, suggesting that arrestin may have other functions. Rod arrestin was shown to directly bind c-Jun N-terminal kinase (JNK3) and ubiquitin ligase Mdm2 and redistribute these proteins from the nucleus to the cytoplasm (94). Because these two proteins are intimately involved in "life-or-death" decisions in the cell, these data suggest that arrestin-mediated signaling may be involved in photoreceptor death associated with the formation of stable arrestin–rhodopsin complexes that has been recently reported in flies (95,96) and mice (97). Recent findings also indicate that microtubule binding does not simply remove rod arrestin until it is needed to quench rhodopsin. Microtubule-bound rod arrestin was found to recruit extracellular signal-regulated protein kinase (ERK1/2) and Mdm2 with very different consequences (98). Arrestin dramatically reduces the overall level of ERK1/2 activation in the cell and increases Mdm2-mediated ubiquitination of microtubule-associated proteins (98). ERK1/2 signaling promotes cell survival and proliferation, whereas microtubules play an important role in a variety of cellular functions. Thus, these "unconventional" arrestin-mediated processes may also contribute to pathological phenotypes in arrestin-associated diseases. Both rod and cone arrestins were also found to interact with the most ubiquitous calcium-dependent regulator in eukaryotic cells, calmodulin, in a strictly calcium-dependent fashion (99). Calmodulin is abundant in photoreceptors and regulates cGMP sensitivity of crucial cation channels (100,101); thus, visual arrestins may play an unexpected additional role in light adaptation.

Considering that its ability to interact with JNK3, ERK1/2, Mdm2, and calmodulin were all discovered within the last year, rod arrestin likely has quite a few additional surprises in store. Retinal degeneration in mice and in patients with compromised arrestin function was traditionally interpreted as a consequence of excessive rhodopsin signaling. Emerging new data suggest that the role of rod arrestin in photoreceptor survival is likely much more sophisticated and may involve its direct interactions with a number of key players in pro-survival and/or pro-apoptotic signaling.

ACKNOWLEDGMENTS

The authors are grateful to numerous colleagues in collaboration with whom many of these studies were performed, especially to the laboratories of Paul B. Sigler and Javier Navarro (crystallography), Wayne L. Hubbell and Candice S. Klug (site-directed spin labeling-EPR and DEER), Jeannie Chen (genetically modified mice), James B. Hurley (quantification of rhodopsin phosphorylation), and Vladlen Z. Slepak (arrestin translocation). We are also grateful to Drs. Rosalie K. Crouch, Narsing Rao, Robert S. Molday, Cheryl M. Craft, and Toshimichi Shinohara for 11-*cis*-retinal, rabbit polyclonal anti-arrestin antibody, monoclonal anti-rhodopsin antibody 4D2, mouse and bovine arrestin cDNAs, respectively. This work was supported in part by NIH grants EY11500 (VVG) and NS45117 (EVG). SMH was a recipient of predoctoral NIH training grant GM07628.

REFERENCES

1. Gurevich, V.V. and Gurevich, E.V. (2006) *Pharm Ther* 110, 465–502.
2. Burns, M.E. and Baylor, D.A. (2001) *Annu Rev Neurosci* 24, 779–805.
3. Lorenz, W., Inglese, J., Palczewski, K., Onorato, J.J., Caron, M.G., and Lefkowitz, R.J. (1991) *Proc Natl Acad Sci USA* 88, 8715–8719.
4. Kohout, T.A. and Lefkowitz, R.J. (2003) *Mol Pharmacol* 63, 9–18.
5. Sokolov, M., Lyubarsky, A.L., Strissel, K.J., Savchenko, A.B., Govardovskii, V.I., Pugh, E.N., Jr., and Arshavsky, V.Y. (2002) *Neuron* 34, 95–106.
6. Nair, K.S., Hanson, S.M., Mendez, A., Gurevich, E.V., Kennedy, M.J., Shestopalov, V.I., Vishnivetskiy, S.A., Chen, J., Hurley, J.B., Gurevich, V.V., and Slepak, V.Z. (2005) *Neuron* 46, 555–567.
7. Chen, C.K., Inglese, J., Lefkowitz, R.J., and Hurley, J.B. (1995) *J Biol Chem* 270, 18060–18066.
8. Calvert, P.D., Klenchin, V.A., and Bownds, M.D. (1995) *J Biol Chem* 270, 24127–24129.
9. Higgins, M.K., Oprian, D.D., and Schertler, G.F. (2006) *J Biol Chem* 281, 19426–19432.
10. Strissel, K.J., Lishko, P.V., Trieu, L.H., Kennedy, M.J., Hurley, J.B., and Arshavsky, V.Y. (2005) *J Biol Chem* 280, 29250–29255.
11. Arshavsky, V.Y. (2003) *Sci STKE* 204, PE43.

12. Broekhuyse, R.M., Tolhuizen, E.F., Janssen, A.P., and Winkens, H.J. (1985) *Curr Eye Res* 4, 613–618.
13. Philp, N.J., Chang, W., and Long, K. (1987) *FEBS Lett* 225, 127–132.
14. Whelan, J.P. and McGinnis, J.F. (1988) *J Neurosci Res* 20, 263–270.
15. Mendez, A., Lem, J., Simon, M., and Chen, J. (2003) *J Neurosci* 23, 3124–3129.
16. Mendez, A., Burns, M.E., Roca, A., Lem, J., Wu, L.W., Simon, M.I., Baylor, D.A., and Chen, J. (2000) *Neuron* 28, 153–164.
17. Chen, C.K., Burns, M.E., Spencer, M., Niemi, G.A., Chen, J., Hurley, J.B., Baylor, D.A., and Simon, M.I. (1999) *Proc Natl Acad Sci USA* 96, 3718–3722.
18. Zhang, H., Huang, W., Zhu, X., Craft, C.M., Baehr, W., and Chen, C.K. (2003) *Mol Vis* 9, 231–237.
19. Strissel, K.J., Sokolov, M., Trieu, L.H., and Arshavsky, V.Y. (2006) *J Neurosci* 26, 1146–1153.
20. Hanson, S.M., Gurevich, E.V., Vishnivetskiy, S.A., Ahmed, M.R., Song, X., and Gurevich, V.V. (2007) *Proc Natl Acad Sci USA* 104, 3125–3128.
21. Elias, R.V., Sezate, S.S., Cao, W., and McGinnis, J.F. (2004) *Mol Vis* 10, 672–681.
22. Palczewski, K., McDowell, J.H., and Hargrave, P.A. (1988) *J Biol Chem* 263, 14067–14073.
23. Xu, J., Dodd, R.L., Makino, C.L., Simon, M.I., Baylor, D.A., and Chen, J. (1997) *Nature* 389, 505–509.
24. Peet, J.A., Bragin, A., Calvert, P.D., Nikonov, S.S., Mani, S., Zhao, X., Besharse, J.C., Pierce, E.A., Knox, B.E., and Pugh, E.N., Jr. (2004) *J Cell Sci* 117, 3049–3059.
25. Nair, K.S., Hanson, S.M., Kennedy, M.J., Hurley, J.B., Gurevich, V.V., and Slepak, V.Z. (2004) *J Biol Chem* 279, 41240–41248.
26. Eckmiller, M.S. (2000) *Vis Neurosci* 17, 711–722.
27. Nir, I. and Ransom, N. (1993) *Exp Eye Res* 57, 307–318.
28. McGinnis, J.F., Matsumoto, B., Whelan, J.P., and Cao, W. (2002) *J Neurosci Res* 67, 290–297.
29. Hanson, S.M., Francis, D.J., Vishnivetskiy, S.A., Klug, C.S., and Gurevich, V.V. (2006) *J Biol Chem* 281, 9765–9772.
30. Osawa, S., Raman, D., and Weiss, E.R. (2000) *Methods Enzymol* 315, 411–422.
31. Gray–Keller, M.P., Detwiler, P.B., Benovic, J.L., and Gurevich, V.V. (1997) *Biochemistry* 36, 7058–7063.
32. Han, M., Gurevich, V.V., Vishnivetskiy, S.A., Sigler, P.B., and Schubert, C. (2001) *Structure* 9, 869–880.
33. Hirsch, J.A., Schubert, C., Gurevich, V.V., and Sigler, P.B. (1999) *Cell* 97, 257–269.
34. Schubert, C., Hirsch, J.A., Gurevich, V.V., Engelman, D.M., Sigler, P.B., and Fleming, K.G. (1999) *J Biol Chem* 274, 21186–21190.
35. Gurevich, V.V. and Benovic, J.L. (2000) *Methods Enzymol* 315, 422–437.
36. Hanson, S.M., Francis, D.J., Vishnivetskiy, S.A., Kolobova, E.A., Hubbell, W.L., Klug, C.S., and Gurevich, V.V. (2006) *Proc Natl Acad Sci USA* 103, 4900–4905.
37. Gurevich, V.V. and Benovic, J.L. (1992) *J Biol Chem* 267, 21919–21923.
38. Gurevich, V.V. and Benovic, J.L. (1993) *J Biol Chem* 268, 11628–11638.
39. Sutton, R.B., Vishnivetskiy, S.A., Robert, J., Hanson, S.M., Raman, D., Knox, B.E., Kono, M., Navarro, J., and Gurevich, V.V. (2005) *J Mol Biol* 354, 1069–1080.
40. Wacker, W.B., Donoso, L.A., Kalsow, C.M., Yankeelov, J.A., Jr., and Organisciak, D.T. (1977) *J Immunol* 119, 1949–1958.
41. Granzin, J., Wilden, U., Choe, H.W., Labahn, J., Krafft, B., and Buldt, G. (1998) *Nature* 391, 918–921.
42. Milano, S.K., Pace, H.C., Kim, Y.M., Brenner, C., and Benovic, J.L. (2002) *Biochemistry* 41, 3321–3328.

43. Imamoto, Y., Tamura, C., Kamikubo, H., and Kataoka, M. (2003) *Biophys J* 85, 1186–1195.
44. Shilton, B.H., McDowell, J.H., Smith, W.C., and Hargrave, P.A. (2002) *Eur J Biochem* 269, 3801–3809.
45. Hanson, S.M., Van Eps, N., Francis, D.J., Altenbach, C., Vishnivetskiy, S.A., Arshavsky, V.Y., Klug, C.S., Hubbell, W.L., and Gurevich, V.V. (2007) *EMBO J* 26, 1726–1736.
46. Mogridge, J. (2004) *Methods Mol Biol* 261, 113–118.
47. Woodbury, R.L., Hardy, S.J., and Randall, L.L. (2002) *Protein Sci* 11, 875–882.
48. Gurevich, V.V., Dion, S.B., Onorato, J.J., Ptasienski, J., Kim, C.M., Sterne-Marr, R., Hosey, M.M., and Benovic, J.L. (1995) *J Biol Chem* 270, 720–731.
49. Vishnivetskiy, S.A., Hosey, M.M., Benovic, J.L., and Gurevich, V.V. (2004) *J Biol Chem* 279, 1262–1268.
50. Hanson, S.M. and Gurevich, V.V. (2006) *J Biol Chem* 281, 3458–3462.
51. Kuhn, H., Hall, S.W., and Wilden, U. (1984) *FEBS Lett* 176, 473–478.
52. Wilden, U. (1995) *Biochemistry* 34, 1446–1454.
53. Gibson, S.K., Parkes, J.H., and Liebman, P.A. (2000) *Biochemistry* 39, 5738–5749.
54. Ohguro, H., Van Hooser, J.P., Milam, A.H., and Palczewski, K. (1995) *J Biol Chem* 270, 14259–14262.
55. Krupnick, J.G., Gurevich, V.V., and Benovic, J.L. (1997) *J Biol Chem* 272, 18125–18131.
56. Vishnivetskiy, S.A., Schubert, C., Climaco, G.C., Gurevich, Y.V., Velez, M.-G., and Gurevich, V.V. (2000) *J Biol Chem* 275, 41049–41057.
57. Ohguro, H., Johnson, R.S., Ericsson, L.H., Walsh, K.A., and Palczewski, K. (1994) *Biochemistry* 33, 1023–1028.
58. Mendez, A., Burns, M.E., Roca, A., Lem, J., Wu, L. W., Simon, M.I., Baylor, D.A., and Chen, J. (2000) *Neuron* 28, 153–164.
59. Gurevich, V.V. and Gurevich, E.V. (2004) *TIPS* 25, 59–112.
60. Gurevich, V.V. and Gurevich, E.V. (2006) *Pharmacol Ther* 110, 465–502.
61. Gurevich, E.V. and Gurevich, V.V. (2006) *Genome Biol* 7, 236.
62. Schleicher, A., Kuhn, H., and Hofmann, K.P. (1989) *Biochemistry* 28, 1770–1775.
63. Gurevich, V.V. and Benovic, J.L. (1995) *J Biol Chem* 270, 6010–6016.
64. Vishnivetskiy, S.A., Paz, C.L., Schubert, C., Hirsch, J.A., Sigler, P.B., and Gurevich, V.V. (1999) *J Biol Chem* 274, 11451–11454.
65. Gurevich, V.V. and Benovic, J.L. (1997) *Mol Pharmacol* 51, 161–169.
66. Wilden, U. and Kuhn, H. (1982) *Biochemistry* 21, 3014–3022.
67. McDowell, J.H., Nawrocki, J.P., and Hargrave, P.A. (2000) *Methods Enzymol* 315.
68. Niu, L., Kim, J.M., and Khorana, H.G. (2002) *Proc Natl Acad Sci USA* 99, 13409–13412.
69. Kennedy, M.J., Lee, K.A., Niemi, G.A., Craven, K.B., Garwin, G.G., Saari, J.C., and Hurley, J.B. (2001) *Neuron* 31, 87–101.
70. Vishnivetskiy, S.A., Raman, D., Wei, J., Kennedy, M.J., Hurley, J.B., and Gurevich, V.V. (2007) *J Biol Chem* 282, in press.
71. Hicks, D. and Molday, R.S. (1986) *Exp Eye Res* 42, 55–71.
72. Sitaramayya, A. (1986) *Biochemistry* 25, 5460–5468.
73. Pugh, E.N., Jr. and Lamb, T.D. (2000) in *Handbook of biological physics. Molecular mechanisms in visual transduction* (Stavenga, D.G., DeGrip, W.J., and Pugh, E.N., Jr. Eds.), Elsevier, Amsterdam, pp. 183–255.
74. Whitlock, G.G. and Lamb, T.D. (1999) *Neuron* 23, 337–351.
75. Gurevich, V.V., Orsini, M.J., and Benovic, J.L. (1999) *Receptor Biochem Methodol* 4, 157–178.
76. Lefkowitz, R.J. and Shenoy, S.K. (2005) *Science* 308, 512–517.

77. Fotiadis, D., Liang, Y., Filipek, S., Saperstein, D.A., Engel, A., and Palczewski, K. (2003) *Nature* 421, 127–128.
78. Angers, S., Salahpour, A., and Bouvier, M. (2002) *Annu Rev Pharmacol Toxicol* 42, 409–435.
79. Palczewski, K., Kumasaka, T., Hori, T., Behnke, C.A., Motoshima, H., Fox, B.A., LeTrong, I., Teller, D.C., Okada, T., Stenkamp, R.E., Yamamoto, M., and Miyano, M. (2000) *Science* 289, 739–745.
80. Okada, T., Fujiyoshi, Y., Silow, M., Navarro, J., Landau, E.M., and Shichida, Y. (2002) *Proc Natl Acad Sci USA* 99, 5982–5987.
81. Li, J., Edwards, P.C., Burghammer, M., Villa, C., and Schertler, G.F. (2004) *J Mol Biol* 343, 1409–1438.
82. Liang, Y., Fotiadis, D., Filipek, S., Saperstein, D.A., Palczewski, K., and Engel, A. (2003) *J Biol Chem* 278, 21655–21662.
83. Wenzel, A., Oberhauser, V., Pugh, E.N., Jr., Lamb, T.D., Grimm, C., Samardzija, M., Fahl, E., Seeliger, M.W., Reme, C.E., and Lintig, V.J. (2005) *J Biol Chem* 280, 29874–29884.
84. Hanson, S.M., Gurevich, E.V., Vishnivetskiy, S.A., Ahmed, M.R., Song, X., and Gurevich, V.V. (2007) *Proc Natl Acad Sci USA* 104, 3125–3128.
85. Calvert, P.D., Govardovskii, V.I., Krasnoperova, N., Anderson, R.E., Lem, J., and Makino, C.L. (2001) *Nature* 411, 90–94.
86. Gurevich, E.V., Benovic, J.L., and Gurevich, V.V. (2004) *J Neurochem* 91, 1404–1416.
87. Gurevich, E.V., Benovic, J.L., and Gurevich, V.V. (2002) *Neuroscience* 109, 421–436.
88. Kuhn, H. (1978) *Biochemistry* 17, 4389–4395.
89. Palczewski, K., McDowell, H., Jakes, S., Ingebritsen, T.S., and Hargrave, P.A. (1989) *J Biol Chem* 264, 15770–15773.
90. Sommer, M.E., Smith, W.C., and Farrens, D.L. (2005) *J Biol Chem* 280, 6861–6871.
91. Northrup, S.H. and Erickson, H.P. (1992) *Proc Natl Acad Sci USA* 89, 3338–3342.
92. Krispel, C.M., Chen, D., Melling, N., Chen, Y.J., Martemyanov, K.A., Quillinan, N., Arshavsky, V.Y., Wensel, T.G., Chen, C.K., and Burns, M.E. (2006) *Neuron* 51, 409–416.
93. Pulvermuller, A., Maretzki, D., Rudnicka–Nawrot, M., Smith, W.C., Palczewski, K., and Hofmann, K.P. (1997) *Biochemistry* 36, 9253–9260.
94. Song, X., Raman, D., Gurevich, E.V., Vishnivetskiy, S.A., and Gurevich, V.V. (2006) *J Biol Chem* 281, 21491–21499.
95. Alloway, P.G., Howard, L., and Dolph, P.J. (2000) *Neuron* 28, 129–138.
96. Kiselev, A., Socolich, M., Vinos, J., Hardy, R.W., Zuker, C.S., and Ranganathan, R. (2000) *Neuron* 28, 139–152.
97. Chen, J., Shi, G., Concepcion, F.A., Xie, G., Oprian, D., and Chen, J. (2006) *J Neurosci* 26, 11929–11937.
98. Hanson, S.M., Cleghorn, W.M., Francis, D.J., Vishnivetskiy, S.A., Raman, D., Song, X., Nair, K.S., Slepak, V.Z., Klug, C.S., and Gurevich, V.V. (2007) *J Mol Biol* 368, 375–387.
99. Wu, N., Hanson, S.M., Francis, D.J., Vishnivetskiy, S.A., Thibonnier, M., Klug, C.S., Shoham, M., and Gurevich, V.V. (2006) *J Mol Biol* 364, 955–963.
100. Haynes, L.W. and Stotz, S.C. (1997) *Vis Neurosci* 14, 233–239.
101. Rebrik, T.I. and Korenbrot, J.I. (2004) *J Gen Physiol* 123, 63–75.

4 Function and Regulation of PDE6

*Michael Natochin, Hakim Muradov,
and Nikolai O. Artemyev[1]*

CONTENTS

4.1 INTRODUCTION

Members of the cGMP phosphodiesterase (PDE6) family are the key photoreceptor proteins regulating the concentration of the second messenger, cGMP, in response to light in cones and rods. Rod PDE6 consists of two catalytic subunits PDE6α and PDE6β, which form a heterodimer. The cone-specific PDE6, PDE6α', forms a homodimer in cones. Under dark conditions, both PDE6s are inhibited by two copies of rod- and cone-specific subunits, Pγ, respectively. The domain structure of the catalytic subunits of PDE6 is very similar to another member of the PDE family, PDE5 (1–4). However, the unique functional features of PDE6 that distinguish it from PDE5 and all other members of the PDE family are the presence of the tightly bound inhibitory Pγ subunits and a very high k_{cat} value approaching the diffusion barrier. The inhibition is relieved by interaction of the Pγ subunits with transducin-α (Tα)-GTP to allow robust PDE activation and rapid decrease in the concentration of cGMP, which in turn leads to the closure of cGMP-gated Na/Ca channels and hyperpolarization of the photoreceptor cells. Here, we describe procedures routinely used in our laboratory to study the function and regulation of PDE6.

PROTOCOL 1: PDE6 PURIFICATION FROM THE BOVINE RETINAS

PDE6 can be easily purified from rod outer segment (ROS) membranes with **a** yield of ~1 mg holoenzyme per 100 bovine retinas. Because PDE6 is a peripheral membrane-associated protein and does not dissociate from the ROS membranes in isotonic buffers but dissociates under hypotonic conditions, it is of vital importance to isolate ROS membranes that are not contaminated by other membrane fractions from the retina. Purification of PDE6 has been described elsewhere (5). This procedure is outlined in the following text:

1. Suspend 500 retinas in 450 mL of 10 mM Tris-HCl buffer, pH 7.5, containing 150 mM NaCl, 2 mM MgSO$_4$, 1 mM PMSF, 0.5 mM DTT (buffer A), and 45% sucrose.
2. Vigorously shake retinas for 3 min to break ROS. Centrifuge suspension for 5 min at 3000 × g.
3. Filter supernatant through four single layers of gauze and dilute 1:2 with buffer A.
4. Centrifuge supernatants for 7 min at 6400 × g. Collect pellets.
5. Prepare a step sucrose gradient using buffer A with the following sucrose concentrations (w/v): 38.9, 34.2, and 28.7%.
6. Resuspend pellets in 200 mL of buffer A with 25% sucrose using a potter homogenizer. Load the suspensions on top of the tubes and centrifuge them for 1 h at 25,000 × g.
7. Collect ROS membranes from the 28.7 and 34.2% interface and adjust their volume to 900 mL with isotonic buffer: 20 mM Tris-HCl, pH 7.5, 150 mM NaCl, 5 mM MgSO$_4$, and 0.5 mM DTT (buffer B).
8. Bleach ROS membranes for 30 min under ambient light and centrifuge them for 20 min at 40,000 × g four times. At this stage, the pellets can be stored at −80°C.
9. Homogenize pellets in hypotonic buffer: 20 mM Tris-HCl, pH 7.5, 5 mM MgSO$_4$, and 0.5 mM DTT (buffer C).
10. Centrifuge for 30 min at 40,000 × g, resuspend pellets in buffer C, and repeat centrifugation. Combine the supernatants containing crude PDE6.
11. Concentrate the supernatant by loading it onto 5 mL of DEAE-Sepharose. Elute the protein with a linear gradient of 500 mM NaCl in buffer C.
12. Concentrate the eluate on YM-30 membrane (Amicon) and load onto a Superose 12 gel-filtration column equilibrated with buffer B. Elute PDE6 with a rate of 0.5 mL/min. PDE6-containing fractions are combined and dialyzed against buffer B with 50% glycerin and stored at −20°C.

4.2 CHARACTERIZATION OF NATIVE PDE6

Protocol 2: PDE6 Activity Assay

Variations of the isotope-based PDE6 activity assay are described in a number of reports (6,7). We suggest the following protocol, which we routinely employ in the laboratory:

1. Mix, on ice, PDE samples in 10–50 μL of buffer B in the absence or presence of inhibitors or activating agents (trypsin, transducin, etc.).

2. Start the reaction by the addition of a mixture of 200 µM [8-³H]cGMP (100,000 cpm) and 0.1 u of bacterial alkaline phosphatase. The optimal reaction time should be determined experimentally.
3. Stop reaction with the addition of 500 µL of AG1 anion exchange resin (Bio-Rad) suspended in water (1:5). Incubate the slurry for 10 min at room temperature, mixing every 3 min.
4. Centrifuge the resin for 2 min at 10,000 × g and count supernatants in a liquid scintillation counter using the cocktail 3a70B.

PROTOCOL 3: ACTIVATION OF PDE6 BY TRYPSIN

The PDE6 holoenzyme can be potently activated by a brief incubation with trypsin (5,7), followed by trypsin inhibition with soybean trypsin inhibitor (SBTI). The cause of activation is the extremely high susceptibility of the Pγ subunit to trypsin. The trypsin-activated PDE6 (taPDE) hydrolyzes cGMP at a rate of ~2,000–3,000 mol cGMP/s·mol PDE6. We typically utilize the following protocol to obtain trypsin-activated PDE6 from holoPDE6:

1. Incubate holoPDE6 (10 mg) with 40 µg TPCK-treated trypsin for 2 h at room temperature.
2. Stop the trypsin treatment with the addition of 200 µg SBTI for 10 min at room temperature.
3. Purify taPDE on a MonoQ 5/50 column using a gradient of NaCl (0–500 mM) in buffer C.
4. Dialyze taPDE against buffer B with 50% glycerin and store at −20°C.
5. The activity of taPDE and its inhibition by Pγ or small-molecule drug PDE6 inhibitors can be determined as described in Protocol 2.

PROTOCOL 4: ACTIVATION OF PDE6 BY TRANSDUCIN

In the phototransduction cascade, holoPDE6 is activated by GTP-bound transducin-α (4,8). This activation can be mimicked in vitro by the addition of the nonhydrolyzable GTP analog, GTPγS, to bleached ROS membranes containing rhodopsin and holotransducin, Gtαβγ. The following protocol describes a method of generating GtαGTPγS and the GtαGTPγS-dependent activation of PDE6:

1. Following the addition of 50 µM GTPγS, incubate ROS membranes (protocol 1, step 8) in buffer B containing 100 µM rhodopsin for 30 min at 4°C under ambient light.
2. Remove ROS membranes by centrifugating for 30 min at 100,000 × g and load the supernatant containing GtαGTPγS and Gtβγ onto a MonoQ 5/50 column equilibrated with buffer C. To elute proteins, apply a linear gradient of NaCl (0–800 mM) at 0.5 mL/min.
3. Combined fractions containing GtαGTPγS are then dialyzed against buffer B containing 50% glycerin and stored at −20°C.
4. Activate holoPDE6 by incubation of the enzyme with a 5- to 10-fold molar excess of GtαGTPγS and measure the activity as described in Protocol 2.

4.3 CHARACTERIZATION OF CHIMERIC PDE5/PDE6

Although native holoPDE6 can be readily obtained from ROS membranes, an expression system for PDE6 is lacking, thus hindering the progress of understanding the structure–functional relationship of PDE6 (9–11). In contrast, similarly organized PDE5 has been readily expressed in various expression systems (12,13). PDE5 is not inhibited by Pγ and its k_{cat} for cGMP hydrolysis is at least 100-fold lower than that of PDE6. Still, PDE5 and PDE6 are quite similar in terms of domain organization (figure 4.1). Similar to PDE5, PDE6 has two N-terminal GAF domains (termed for their presence in cGMP-regulated PDE, *Anabaena* adenylyl cyclases, and the *Escherichia coli* protein Fh1A) and one catalytic domain (cd) located in the C-terminal part of the molecules. In addition, PDE5 and PDE6 share a relatively high homology of the catalytic domains, specificity to cGMP, and sensitivity to common catalytic-site inhibitors, such as sildenafil (Viagra), zaprinast, and IBMX (3-isobutyl-1-methylxanthine). Therefore, an approach utilizing PDE5/PDE6 chimeras has been a relatively successful way to overcome the limitations of PDE6 functional expression (14,15). We adhere to the following protocols for cloning and expressing PDE5/PDE6 chimeras using the baculovirus/sf9 cell system.

PROTOCOL 5: CLONING OF PDE5/PDE6 CHIMERAS FOR EXPRESSION USING THE BACULOVIRUS-SF9 CELL SYSTEM

1. Design and order synthetic PCR primers carrying the appropriate restriction sites or sequences coding the chimeric junctions.
2. Perform a PCR reaction according to standard protocols using proofreading Pfu DNA polymerase.
3. If appropriate restriction sites are not available, extend the first PCR product to the closest unique restriction site with a second-round PCR.

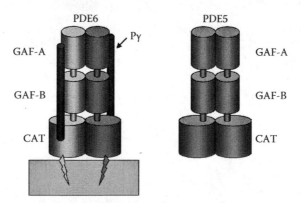

FIGURE 4.1   (See accompanying color CD.) Schematic representation of the domain structures of rod PDE6 and PDE5. The rod PDE6 catalytic dimer is composed of homologous PDE6α and PDE6β subunits, each containing the N-terminal GAF-A and GAF-B domains and the C-terminal catalytic domain. The C-termini of PDE6α and PDE6β are isoprenylated, thereby providing the enzyme attachment to the ROS disk membranes. PDE5 is a catalytic homodimer with a similar subunit domain structure.

4. Purify the digested PCR product using the Qiagen Gel Extraction kit and ligate it to the appropriately digested pFastBac vector with T4 DNA ligase using an insert to vector molar ratio of 3:1.
5. Transform DH5α *E. coli* cells, and select colonies for propagation and sequence analysis.

PROTOCOL 6: EXPRESSION AND PURIFICATION OF PDE5/PDE6 CHIMERAS USING THE BACULOVIRUS-SF9 CELL SYSTEM

Baculovirus expression systems have several important advantages for the expression of PDE5/PDE6 chimeras. They often provide high levels of heterologous gene expression and some posttranslational modifications. Moreover, simultaneous infection of insect cells with two or more viruses allows for the expression of hetero-oligomeric protein complexes. The Bac-to-Bac™ baculovirus expression system offers additional advantages allowing for rapid generation of recombinant viruses. Genes of interest, such as PDE5/PDE6 chimeric cDNAs, are inserted into the multiple cloning site of a pFastBac vector downstream of the baculovirus-specific promoter. The plasmid represents a convenient template for mutagenesis of PDE5 and PDE6 cDNAs. The standard *E. coli* strain DH5α can be used for all manipulations during the construction of PDE chimeras or mutants. Once the chimeric construct is prepared, the vector is used to transform DH10Bac *E. coli* cells that carry and propagate a baculovirus shuttle vector (bacmid). Site-specific transposition of the pFastBac vector into the bacmid produces a recombinant bacmid that allows for a white/blue selection of colonies. Isolated recombinant bacmids are used to infect insect cells to generate recombinant viruses:

1. Generate recombinant bacmids, transfect Sf9 cells, and carry out viral amplifications according to the manufacturer's recommendations (Life Technologies, Inc.).
2. Infect Sf9 cells with individual viruses or with a desired combination of viruses at a MOI of 3 to 10.
3. Pellet Sf9 cells 48 h post transfection. Wash the pellet with 20 mM Tris-HCl buffer, pH 8.0, containing 130 mM NaCl and 2 mM $MgSO_4$. Store the cell pellet at −80°C or process immediately.
4. Resuspend cells from 100 mL of culture in 10 mL of 20 mM Tris-HCl buffer, pH 8.0, containing 2 mM $MgSO_4$ and 1 tablet of EDTA-free™ Mini protease inhibitor cocktail (Roche).
5. Disrupt the cells by sonicating with two 10 s pulses using a microtip attached to a 550 Sonic Dismembrator (Fisher Scientific). Centrifuge the cell lysate at $100,000 \times g$ for 90 min at 4°C.
6. Load cleared lysate onto $NiSO_4$-charged His·bind resin (Novagen).
7. Elute the bound proteins with 20 mM Tris-HCl buffer, pH 8.0, containing 500 mM NaCl and 200 mM imidazole.
8. Dialyze the eluted protein against 20 mM Tris-HCl buffer, pH 7.5, containing 150 mM NaCl, 2 mM DTT, 2 mM $MgSO_4$, and 50% glycerin. Store at −20°C.

PROTOCOL 7: CONSTRUCTION, EXPRESSION, AND PURIFICATION OF CHIMERIC PDE5/PDE6 CATALYTIC DOMAINS IN *E. COLI*

Recent studies have demonstrated that the catalytic domain of PDE5 (PDE5cd) can be easily expressed in *E. coli* with high yields, permitting a solution of its crystal structure (12,13). Unfortunately, PDE6cd expression did not produce active soluble protein. The atomic structures of the PDE5cd facilitated the design of chimeric PDE5/PDE6cd constructs (figure 4.2) (16):

1. Primers designed to overlap junctions (15–20 nt) on each side of the PDE5 and PDE6 cDNA fragments to be joined should be paired with primers aligned to the MCS's *Nde*I and *Xho*I sites. The desired chimeras will be obtained with two rounds of PCR as described in Protocol 5.
2. Clone all constructs into the pET15b vector (Novagen) using *Nde*I and *Xho*I sites. Transform plasmids containing the chimeric constructs into BL21-codon plus competent cells (Stratagene).
3. Pick a single colony and transfer it into ampicillin-containing LB medium and grow overnight at 37°C.
4. Dilute the overnight cultures 1:100 with 2xTY medium containing 100 µg/mL ampicillin. At the cell density of OD_{600} ~0.8, adjust the temperature to 16°C, and induce the cultures at OD_{600} ~1.0 with 100 µM IPTG for 16–18 h.
5. Resuspend cell pellets in 20 mM Tris-HCl buffer, pH 8.0, containing 500 mM NaCl, 20 mM imidazole, add Complete Mini protease inhibitor cocktail (Roche) and sonicate with six 20 s pulses using a flat tip attached to a 550 Sonic Dismembrator (Fisher Scientific).
6. Clear cell lysates by centrifugation (100,000 × g, 1 h, 4°C) and load onto $NiSO_4$-charged His·bind resin (Novagen).
7. Elute the bound proteins with 20 mM Tris-HCl buffer, pH 8.0, containing 500 mM NaCl and 200 mM imidazole.

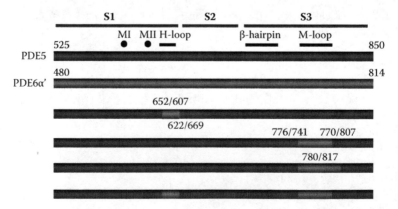

FIGURE 4.2   (See accompanying color CD.) Schematic representation of active PDE5/PDE6 chimeric catalytic domains. The borders of the subdomains S1, S2, S3 (12), and the sequences corresponding to the metal-binding sites MI and MII, the H- and M-loops, and the β-hairpin region are indicated.

8. Cleave the His$_6$ tag with thrombin (1 u per 10 mg of protein) for 18–20 h on ice and dialyze the protein against 20 mM Tris-HCl buffer, pH 7.5, containing 500 mM NaCl. Load the dialyzed protein onto a NiSO$_4$-charged His·bind column.
9. Elute the protein with 50 mM imidazole, and store at –80°C or dialyze against buffer B (Protocol 1, step 7) with 50% glycerin and store at –20°C. Typical yields are more than 20 mg of 95% pure protein from 1 L of culture.

PROTOCOL 8: EXPRESSION OF Pγ CONSTRUCTION AND EXPRESSION OF Pγ MUTANTS

Pγ can be purified from ROS membranes (17). A much more efficient procedure to obtain Pγ is to express the protein in bacteria.

The pET system for expression of Pγ

Synthetic Pγ cDNA (18) represents a convenient template for site-directed mutagenesis of Pγ. For protein expression, this synthetic gene has been PCR-amplified with primers containing *Nde*I and *Bam*HI sites and subcloned into the pET-11a or pET-15b vectors (Novagen). The resulting pET-Pγ vectors have been widely used for Pγ mutagenesis (19–21).

Cloning of Pγ mutants

The short size of Pγ cDNA makes PCR-based mutagenesis the method of choice to introduce mutations into virtually any part of the gene. A ~30 nt primer carrying the desired mutation in the middle is paired with either the primer containing the *Nde*I or *Bam*HI site. The PCR product is used as a primer in the second PCR reaction and is paired with the primer containing *Nde*I or *Bam*HI sites, whichever primer was not used in the first-round PCR. The second-round PCR will produce mutant Pγ cDNA, which can be cloned into pET vectors. We use the following protocol to express and purify Pγ mutants in our laboratory:

1. Transform BL21(DE3) *E. coli* cells with the wild-type or mutant pET-11a-Pγ plasmid using standard methods.
2. Grow BL21(DE3) cells carrying the pET-Pγ plasmid or mutant plasmid at 37°C overnight in LB medium containing 100 µg/mL ampicillin.
3. Dilute the overnight culture (1:100) with 2xTY medium containing 50 µg/mL ampicillin and continue to grow the cells at 37°C until an OD$_{600}$ of ~0.8.
4. Reduce the incubation temperature to 30°C and induce Pγ expression by the addition of 0.5 mM IPTG. Allow the induction to proceed for 3–4 h.
5. Following the induction, pellet the cells and wash the pellets twice with 50 mM Tris-HCl buffer, pH 7.5. Resuspend the pellet from 1 L of culture in 60 ml of 50 mM Tris-HCl buffer, pH 7.5, containing 20 mM NaCl, 5 mM EDTA, 1 mM DTT, and the protease inhibitors PMSF (1 mM) and pepstatin A (20 µg/mL).

6. Disrupt the cells by sonication with 30 s pulses for a total sonication time of 3 min. Centrifuge the cell lysate at 100,000 × g for 30 min and collect the supernatant.

7. Prepare a 5 mL column with SP Fast Flow Sepharose (Pharmacia) and equilibrate it with 50 mM Tris-HCl buffer, pH 7.5, containing 20 mM NaCl, 5 mM EDTA, and 1 mM DTT.

8. Load the supernatant from step 6 onto the column and elute the bound proteins at a flow rate of 0.5 mL/min using 100 mL of a 20–400 mM NaCl gradient. Pγ elutes at ~250 mM NaCl.

9. Collect fractions and analyze them for the presence of Pγ using the taPDE6 inhibition assay or SDS-PAGE and Western blotting using standard procedures.

10. Apply Pγ-containing fractions onto a reversed-phase protein C-4 HPLC column. Elute bound proteins at a flow rate of 0.5 mL/min with a gradient 0–80% of acetonitrile in 0.1% trifluoroacetic acid. Pγ elutes at ~45% acetonitrile.

11. Lyophilize HPLC fractions containing purified Pγ using a Speed Vac SC100 (Savant). Dissolve lyophilized Pγ protein in 20 mM HEPES buffer, pH 7.5, and store in aliquots at −80°C. This procedure typically yields up to 20 mg of >95% pure Pγ or mutant per liter of culture.

NOTE

1. This work was supported by National Institutes of Health Grant EY-10843.

REFERENCES

1. Beavo, J.A. Cyclic nucleotide phosphodiesterases: functional implications of multiple isoforms, *Physiol Rev*, 75, 725, 1995.

2. Francis, S.H., Turko, I.V., and Corbin, J.D., Cyclic nucleotide phosphodiesterases: relating structure and function, *Prog Nucl Acid Res Mol Biol*, 65, 1, 2001.

3. Arshavsky, V.Y., Lamb, T.D., and Pugh, Jr., E.N., G proteins and phototransduction, *Annu Rev Physiol*, 264, 153, 2002.

4. Chabre, M. and Deterre, P., Molecular mechanism of visual transduction, *Eur J Biochem,* 179, 255, 1989.

5. Muradov, K.G., Granovsky, A.E., Schey, K.L., and Artemyev, N.O., Direct interaction of the inhibitory γ-subunit of rod cGMP phosphodiesterase (PDE6) with the PDE6 GAF-A domains, *Biochemistry*, 41, 3884, 2002.

6. Thompson, J. and Appleman, M.M., Multiple cyclic nucleotide phosphodiesterase activities from rat brain, *Biochemistry*, 10, 311, 1971.

7. Hurley, J.B. and Stryer, L., Purification and characterization of the gamma regulatory subunit of the cyclic GMP phosphodiesterase from retinal rod outer segments, *J Biol Chem*, 257, 11094, 1982.

8. Yarfitz, S. and Hurley, J.B., Transduction mechanisms of vertebrate and invertebrate photoreceptors, *J Biol Chem*, 269, 14329, 1994.

9. Qin, N., Pittler, S.J., and Baehr, W., In vitro isoprenylation and membrane association of mouse rod photoreceptor cGMP phosphodiesterase α and β subunits expressed in bacteria, *J Biol Chem*, 267, 8458, 1992.

10. Qin, N. and Baehr, W., Expression and mutagenesis of mouse rod photoreceptor cGMP phosphodiesterase, *J Biol Chem*, 269, 3265, 1994.

11. Piriev, N.I., Yamashita, C., Samuel, G., and Farber, D.B., Rod photoreceptor cGMP-phosphodiesterase: analysis of α and β subunits expressed in human kidney cells, *Proc Natl Acad Sci USA.*, 90, 9340, 1993.

12. Sung, B.J., Hwang, K.Y., Jeon, Y.H., Lee, J.I., Heo, Y.S., Kim, J.H., Moon, J., Yoon, J.M., Hyun, Y.L., Kim, E., Eum, S.J., Park, S.Y., Lee, J.O., Lee, T.G., Ro, S., and Cho, J.M., Structure of the catalytic domain of human phosphodiesterase 5 with bound drug molecules, *Nature*, 425, 98, 2003.

13. Huai, Q., Liu, Y., Francis, S.H., Corbin, J.D., and Ke, H. Crystal structures of phosphodiesterases 4 and 5 in complex with inhibitor 3-isobutyl-1-methylxanthine suggest a conformation determinant of inhibitor selectivity, *J Biol Chem*, 279, 13095, 2004.

14. Granovsky, A.E., Natochin, M., McEntaffer, R.L., Haik, T.L. Francis, S.H., Corbin, J.D., and Artemyev, N.O., Probing domain functions of chimeric PDE6α/PDE5 cGMP-phosphodiesterase, *J Biol Chem*, 279, 24485, 1998.

15. Muradov, K.G., Boyd, K.K., Martinez, S.E, Beavo, J.A., and Artemyev, N.O. The GAF-A domains of rod cGMP-phosphodiesterase 6 determine the selectivity of the enzyme dimerization, *J Biol Chem*, 278, 10594, 2003.

16. Muradov, K.G., Boyd, K.K., and Artemyev N.O., Analysis of PDE6 function using chimeric PDE5/6 catalytic domains, *Vision Res*, 46, 860, 2006.

17. Ovchinnikov, Yu. A., Lipkin, V.M., Kumarev, V.P., Gubanov, V.V., Khramtsov N.V., Akhmedov, N.B., Zagranichny, V.E., and Muradov, K.G., Cyclic GMP phosphodiesterase from cattle retina. Amino acid sequence of the γ-subunit and nucleotide sequence of the corresponding cDNA, *FEBS Lett*, 204, 288, 1986.

18. Brown, R.L. and Stryer, L., Expression in bacteria of functional inhibitory subunit of retinal rod cGMP phosphodiesterase, *Proc Natl Acad Sci USA.*, 86, 4922, 1989.

19. Skiba, N.P., Artemyev, N.O., and Hamm, H.E., The carboxyl terminus of the γ-subunit of rod cGMP phosphodiesterase contains distinct sites of interaction with the enzyme catalytic subunits and the α-subunit of transducin, *J Biol Chem*, 278, 10594, 1995.

20. Slepak, V.Z., Artemyev, N.O., Zhu,Y., Dumke, C.L., Sabacan, L., Sondek, J., Hamm, H.E., Bownds, M.D., and Arshavsky, V.Y., An effector site that stimulates G-protein GTPase in photoreceptors, *J Biol Chem*, 278, 14319, 1995.

21. Artemyev, N.O., Natochin, M., Busman, M., Schey, K.L., and Hamm, H.E., Mechanism of photoreceptor cGMP phosphodiesterase inhibition by its γ-subunits, *Proc Natl Acad Sci USA.*, 93, 5407, 1996.

5 Biochemical Characterization of Phototransduction RGS9-1–GAP Complex

Qiong Wang and Theodore G. Wensel

CONTENTS

5.1 INTRODUCTION

A major player in the regulation of timing and sensitivity of photoresponses is the GTPase accelerating protein, RGS9-1 (1–4). RGS9-1 binds to the α subunits of the rod and cone G-proteins, $G_{\alpha t1}$ and $G_{\alpha t2}$, when they are in their activated GTP-bound conformations, and speeds up the rates at which they hydrolyze GTP, thereupon returning to the inactive GDP-bound conformations. This conceptually simple function is complex in its biochemical details and regulation. In addition to RGS9-1, two other subunits, $G_{\beta5L}$ (5–9) and R9AP (RGS9-1 anchor protein) (10–12), are required for the function and stability of the GTPase accelerating protein (GAP) complex, and the inhibitory subunit of the photoreceptor effector enzyme, PDE6γ, dramatically enhances the affinity of this complex for $G_{\alpha t}$-GTP. The activity of the complex and the time resolution of vision depend on its concentration in the cells (13), and the concentrations are very different in rods and cones (1,14). The $G_{\beta5}$ gene is subject to alternative splicing, and although only one splice variant, $G_{\beta5L}$, is found in rods (15), both $G_{\beta5L}$ and $G_{\beta5S}$ are found associated with RGS9-1 in cones (14). RGS9-1 is also subject to Ca^{2+}-regulated phosphorylation by protein kinase C (16,17), and may be further regulated by phosphoinositides. Levels of this complex have been shown to be essential for the timely recovery of photoresponse by loss-of-function studies in mice (18–21) and humans (22), and those proteins also control the rate-limiting step in vision confirmed by a recent discovery that overexpression of the whole complex speeds up the response (13). This chapter will focus on biochemical techniques used to characterize the molecular mechanisms of RGS9-1–GAP complex in phototransduction kinetics and their regulation.

5.2 EXPRESSION AND PURIFICATION OF RECOMBINANT COMPONENTS, FRAGMENTS, AND MUTANTS OF THE RGS9-1 COMPLEX

Characterization of the RGS9-1–$G_{\beta5}$–R9AP complex requires expression and purification of recombinant proteins, with either wild-type or reengineered sequences. Three major heterologous expression systems have been previously tested, and each has its own advantages. The choice of expression system depends on the requirement of these proteins for different applications: in vitro biochemical assay, generation of antibodies, or functional studies of domains and specific sites. There are some special considerations that must be taken into account when working with these proteins. One is that any RGS9-1 construct containing the GGL domain must be coexpressed with $G_{\beta5}$, if soluble and functional protein is to be obtained, and, conversely, no success has been achieved in expression of functional $G_{\beta5}$ without coexpression of a GGL-domain-containing construct. R9AP is a transmembrane protein, in contrast to the RGS9-1–$G_{\beta5}$ complex, so conditions for its expression and purification are necessarily different. Although R9AP is required for stability of both RGS9-1 and the long isoform of $G_{\beta5}$ in vivo in photoreceptors, in vitro expression systems do not require coexpression. All three proteins have been successfully expressed in both insect cells and mammalian tissue culture cells, and they appear to form functional complexes. The RGS9-1–$G_{\beta5}$ complex can be expressed and purified without R9AP, and then recombined either with full-length R9AP reconstituted into lipid vesicles, or with the cytoplasmic domain of R9AP, which binds tightly to RGS9-1 without the need for lipid or detergent.

5.2.1 EXPRESSION OF RGS9-1 FRAGMENTS IN *E. COLI*

The following procedure describes the expression and purification of either full-length RGS9-1 or fragments for antibody generation and functional assays (2,23,24). Both His_6 tags (pET vectors, Novagen) and glutathione-S-transferase (GST, pGEX vectors, GE) tags have been used successfully. Neither full-length RGS9-1 nor the RGS domain-containing fragments (i.e., containing residues 291–418) of RGS9-1 are initially soluble when overexpressed in *E. coli*. They are found primarily in inclusion bodies, so simply lysing the cells and washing the inclusion bodies gives a good first step of purification. Subsequently the proteins can be extracted under denaturing conditions. His_6-tagged and GST-tagged proteins both renature well into active form upon dialysis to remove denaturants. The His-tagged protein is purified by metal ion affinity chromatography under denaturing conditions, and then renatured by step dialysis, whereas the GST-tagged protein is renatured by dialysis prior to affinity purification using glutathione beads. In contrast, full-length RGS9-1 does not fold properly without $G_{\beta 5}$, so for this protein, *E. coli* expression is useful only to produce denatured protein for antibody production or for use as a standard in gel-based (Coomassie-blue staining or Western blotting) quantification procedures.

PROTOCOL 1: BACTERIAL EXPRESSION AND PURIFICATION OF TAGGED RGS9-1 FRAGMENTS

1. Cells (BL21 (DE3) pLysS) are freshly transformed with plasmid using standard procedures, and incubated overnight at 37°C on LB-ampicillin/ chloramphenicol plates. The next day multiple colonies are collected and used to inoculate the LB medium. A typical volume is 1 L. pLysS strains, expressing T7 lysozyme, are useful for efficient disruption of cell walls. However, when a Microfluidizer® (Microfluidizer Corp.) is used to break open the cells, BL21 (DE3) without pLysS can be used, and no chloramphenicol is needed.

2. For vectors with the T7-lac promoter (e.g., pET-14B–derived expression plasmids), 1 mM IPTG (isopropyl-beta-D-thiogalactopyranoside) is used to induce expression when the cells have reached an OD_{600} of 0.6, and growth is allowed to continue at 37°C if the intent is to extract protein from inclusion bodies, or at 25°C or even lower temperatures if the intent is to optimize the amount of initially soluble protein. For vectors with the T7 promoter (e.g., pGEX-2TK), 0.1 mM IPTG is used to induce expression. In general, it is a good idea to remove samples for SDS-PAGE assessment of protein expression at different time points.

3. Cells are harvested by 4000 × g centrifugation (20 min, 4°C), and pellets are sonicated and washed twice with lysis buffer: 50 mM sodium phosphate, pH 7.4, 300 mM NaCl, 1 mM dithiothreitol (DTT). The supernatants from these washes can then be used for affinity purification on immobilized

metal or glutathione beads, following the manufacturer's instructions. However, for most RGS9-1 fragments, a much higher yield of soluble protein is obtained by extracting and renaturing proteins from inclusion bodies.

4. The inclusion bodies (pellets from previous washes) are solubilized in 35 mL/L culture of guanidinium HCl buffer: 6 M guanidinium HCl, 100 mM sodium phosphate, 10 mM Tris, pH 8.0. The samples are rotated at room temperature for 2 h.

5. For His-tagged proteins, the solubilized protein is then applied to nickel nitrilotriacetic (Ni^{2+}–NTA) beads (Qiagen) and purified under denaturing conditions according to the manufacturer's instructions. The purified protein is then diluted to a concentration of 0.1 mg/mL with urea buffer: 8 M urea, 100 mM sodium phosphate, 10 mM Tris, pH 8.0. The protein is then dialyzed overnight at 4°C against a 10,000-fold larger volume of renaturing buffer: 50 mM sodium phosphate, pH 7.4, 300 mM NaCl, 10% glycerol, 0.1% 2-mercaptoethanol. Insoluble protein is removed by centrifugation at $20,000 \times g$ (20 min 4°C), and the protein in the supernatant is concentrated by centrifugal ultrafiltration (Centricon, Millipore). Most RGS9-1 fragments show a tendency to aggregate at concentrations above 10 μM, so there is not much point in trying to exceed this concentration.

6. For GST-tagged proteins, the crude solubilized inclusion bodies are renatured as described earlier to renature the GST moiety, prior to their being loaded onto glutathione beads and purified according to the manufacturer's instructions.

7. Full-length RGS9-1 does not renature, but instead forms precipitates when denaturants are dialyzed away. The denatured protein can be stored at −20°C after adding glycerol to 40% (v/v) and used as a standard for quantifying recombinant proteins expressed by other means using Coomassie staining on gels or quantitative immunoblotting.

5.2.2 R9AP Expression and Purification in *E. coli*

R9AP is a transmembrane protein, and thus the full-length form, containing its single C-terminal transmembrane helix, is expressed in insoluble form in bacteria. Fortunately, however, it is one of the few eukaryotic membrane proteins that can be solubilized from bacterial membrane pellets using nondenaturing detergents, purified, and reconstituted in functional form into lipid vesicles. Moreover, the cytoplasmic domain of R9AP can be expressed as a soluble fragment in bacteria. In both cases, the protein has been expressed in pET14b and purified using N-terminal His tags. Immobilized metal-ion affinity chromatography is not sufficient to obtain pure protein, so an additional step of ion exchange chromatography is used.

Protocol 2: Expression of R9AP and its Soluble Fragments in *E. coli* and Purification by Affinity and Ion Exchange Chromatography

1. Transformation of BL21 (DE3) pLysS cells is carried out as described earlier for RGS9-1 fragments. Cells are induced when $OD_{600} = 0.6$–0.8 with 0.3 mM IPTG, and allowed to grow for 4–5 h at 30°C.

2. Cells are harvested by centrifugation at $4000 \times g$ for 20 min at 4°C, and the pellets sonicated with lysis buffer (25 mM Tris, pH 8.0, 300 mM NaCl, 20 mM imidazole, 2 mM DTT, and ~20 mg/L PMSF, phenylmethylsulfonyl fluoride).

3. His-tagged bovine R9AP fragments lacking the C-terminal transmembrane helix (His-R9AP-ΔC, amino acids 1–212 or His-R9AP-ΔC2, amino acids 1–191) are largely soluble because they do not contain the C-terminal transmembrane helix. In these cases, cell debris and insoluble materials are removed by centrifugation at $24{,}000 \times g$ for 30 min, and the soluble fragments in the supernatant are purified under native conditions with immobilized metal ion beads following the manufacturer's protocol, and eluted with 200 mM imidazole in the lysis buffer. The purest fractions are combined and dialyzed against dialysis buffer (10 mM Tris-HCl, 1 mM DTT, pH 8.0). The protein is not exceptionally pure at this point, so it is applied onto a strong anion exchange column (e.g., POROS HQ HPLC anion exchange column, Applied Biosystems), washed with 20 mM NaCl in the dialysis buffer, and eluted with a linear gradient from 20 to 200 mM NaCl in the same buffer (at a flow rate of 2–3 mL/min for the POROS HPLC column). About 10 mg of protein can be obtained from 1 L of culture.

4. His-tagged mouse full-length R9AP is insoluble and is found in pellets after centrifugation of the cell lysates as described earlier. The proteins are extracted with 4% sodium cholate in lysis buffer for 0.5–1 h at 4°C with gentle agitation, followed by centrifugation; this procedure is repeated 3–4 times, yielding >70% of total His-mR9AP extracted in a soluble form. The pooled detergent-solubilized R9AP is purified using Ni^{2+}-NTA beads in 4% sodium cholate in the lysis buffer according to the manufacturer's protocol. Routinely, at least 5 mg R9AP with 90–95% purity can be obtained from 2 L *E. coli* cultures. Reconstitution procedures are described in section 5.3.

5.2.3 EXPRESSION OF THE RGS9-1 COMPLEX WITH $G_{\beta 5}$ IN INSECT CELLS USING BACULOVIRUS

The most useful expression system so far for production of functional full-length RGS9-1 and its partner $G_{\beta 5}$ is baculovirus-directed expression in insect cells. Attempts to express each separately have produced low yields of most insoluble proteins, for which no activity has been detected; however, coexpression using simultaneous infection with two viruses produces useful amounts of soluble, active protein complex. It is generally necessary to test different ratios of RGS9-1 virus to $G_{\beta 5}$ virus to determine the one that produces the highest yield of the complex; this ratio will vary over time if the titers of the virus stocks decline at different rates, so it is advisable to retiter the virus from time to time. The following procedure (protocol 3) uses adherent cells growing in plates. Alternatively, cells can be grown in suspension as described (25).

There are two known splice variants of $G_{\beta 5}$, $G_{\beta 5L}$ and $G_{\beta 5S}$. In insect cells, viruses encoding the long-variant $G_{\beta 5L}$ produce a mixture of $G_{\beta 5L}$ and $G_{\beta 5S}$, with the cells apparently using the first methionine residue of $G_{\beta 5S}$ as an alternative translation start site.

PROTOCOL 3: INSECT CELL EXPRESSION AND
PURIFICATION OF RGS9-1/G$_{\beta 5}$

1. If it is necessary to produce new virus, recombinant baculoviruses are isolated after cotransfection of the linearized BaculoGold viral DNA (PharMingen) and the recombinant transfer vector pVL1392 (PharMingen) with the proper insert into Sf9 cells following the manufacturer's protocol.
2. The cells are grown as monolayers in 150 mm culture dishes in Insect-Xpress medium (Bio Whittaker) supplemented with 8% fetal bovine serum and 10 µg/mL gentamycin. They are coinfected with two types of recombinant viruses containing RGS9-1 or G$_{\beta 5L}$ (the MOI [multiplicity of infection]) for both viruses is approximately 1:1) at 80% confluency and the cells are harvested 48 h later.
3. Cells are removed from the plates, collected by centrifugation, and the cell pellet is resuspended in ice-cold lysis buffer (50 mM Tris-HCl, pH 8.0, 500 mM NaCl, 1 mM DTT, and 1% Nonidet P-40) with freshly added protease inhibitors (0.03 mg/mL leupeptin, 0.017 mg/mL pepstatin A, 0.005 mg/mL aprotinin, 0.03 mg/mL lima bean trypsin inhibitor, and ~20 mg/L solid PMSF) and mixed by rocking for 1 h at 4°C. The cell suspensions are sonicated on ice, centrifuged at 20,000 × g for 30 min at 4°C, and the supernatants are collected.
4. For His-tagged proteins, the supernatants are supplemented with 20 mM imidazole and loaded onto a Ni^{2+}-NTA agarose column, washed with lysis buffer, and then with GAPN-H buffer (10 mM HEPES, pH 7.4, 100 mM NaCl, 2 mM MgCl$_2$, 1 mM DTT, and ~20 mg/L PMSF) containing 20 mM imidazole, and then eluted with 250 mM imidazole in GAPN-H buffer.
5. For GST-tagged proteins, the supernatants are loaded to a glutathione-sepharose 4B column, washed with lysis buffer followed by GAPN-H buffer (10 mM HEPES, pH 7.0, 100 mM NaCl, 2 mM MgCl$_2$, 1 mM DTT, and ~20 mg/L PMSF) and eluted with 40 mM glutathione in GAPN-H buffer.
6. Untagged G$_{\beta 5}$ expressed in the cells consistently copurify with RGS9-1 after affinity chromatography (figure 5.1).

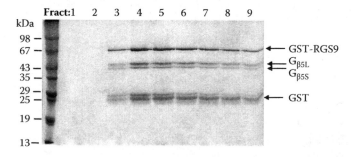

FIGURE 5.1   Coomassie-blue-stained SDS-PAGE gel of glutathione elution fractions from GST-RGS9-1-G$_{\beta 5}$ expressed in insect cells. After the column had been thoroughly washed with buffer, as described in protocol 3, glutathione was added to the buffer, and fractions collected, beginning with Fraction 1.

5.3 RECONSTITUTION OF FULL-LENGTH R9AP INTO LIPID VESICLES

R9AP appears to have several roles in regulating RGS9-1 function. These include ensuring the stability of the RGS9-1–$G_{\beta 5L}$ complex (20), controlling its localization to the outer segments (26), anchoring it to the disk membrane (10,11) where it interacts with $G_{\alpha t}$-GTP and PDE6, and enhancing the catalytic activity of RGS9-1 (27). To study its membrane-dependent functions using purified proteins, it is necessary to reconstitute it into lipid vesicles. This is conveniently carried out with protein purified from *E. coli* in nondenaturing detergents, as described earlier, and well-defined phospholipid mixtures. This procedure can be carried out with or without the simultaneous presence of detergent-purified rhodopsin, and remarkably homogeneous vesicles as determined by cryoelectron microscopy are obtained (11).

5.3.1 RECONSTITUTION OF PURIFIED R9AP INTO VESICLES CONTAINING PHOSPHOLIPIDS ONLY OR PHOSPHOLIPIDS AND RHODOPSIN

PROTOCOL 4: RHODOPSIN PURIFICATION

1. Buffers used are lysis buffer (300 mM NaCl, 25 mM Tris, pH 8.0); GAPN-H buffer (100 mM NaCl, 2 mM $MgCl_2$, 10 mM HEPES pH 7.4); ConA buffer (300 mM NaCl, 50 mM Tris-HCl, pH 7.0, 1 mM $CaCl_2$, 1 mM $MgCl_2$, 1 mM $MnCl_2$); high-salt buffer (1 M NH_4Cl, 10 mM HEPES, pH 7.4, 2 mM $MgCl_2$). DTT, to 1 mM, and solid PMSF, to 20 mg/L, were added to each buffer just before use.

2. Purification of rhodopsin uses a modified version of a published procedure (28), with all procedures carried out either under dim red light or using near-infrared illumination and infrared image-converting goggles. Sepharose beads containing immobilized concanavalin A (Con A) are stabilized before use to prevent bleeding off of Con A during elution by treatment with 0.05% glutaraldehyde in 250 mM $NaHCO_3$ and prepared as described (28). Rod outer segments are prepared using a standard discontinuous sucrose gradient procedure (29), and extracted twice with high-salt buffer at a concentration of 15 μM rhodopsin or lower. They are then washed twice at the same concentration with Con A buffer, and the pellets are solubilized in Con A buffer containing 4% (w/v) sodium cholate to the same concentration. The supernatant is loaded onto the column, which is washed first with 10 column volumes of Con A buffer containing 4% sodium cholate, and then with 3 column volumes of the same buffer supplemented with 300 mM α-methyl mannoside. The eluant is concentrated by ultrafiltration to obtain a final rhodopsin concentration (determined by absorbance at 500 nM) of 2–3 mg/mL.

PROTOCOL 5: RECONSTITUTION OF R9AP WITH OR WITHOUT RHODOPSIN IN LIPID VESICLES

1. A phospholipid solution is prepared by first mixing chloroform solutions of lipids from Avanti Polar Lipids or Molecular Probes at a mass ratio of

phosphatidylcholine:phosphatidylethanolamine:phosphatidylserine:rhoda-
mine-labeled phosphatidylethanolamine = 50:35:1:0.43, drying it under a
stream of argon, and then dissolving it in sufficient lysis buffer containing 4% sodium cholate to achieve a final lipid concentration of 20 mg/mL.
The mixture is sonicated as necessary under argon and on ice to achieve a
homogeneous solution. Then, cholate solutions of either rhodopsin, or His-
tagged R9AP, or both are added to the lipid solution to achieve a lipid-to-
protein mass ratio between 20:1 and 40:1.

2. The same procedure without added proteins produces protein-free vesicles
   to use in control experiments.
3. Typical yields of protein incorporated into vesicles are 42 molecules of
   rhodopsin per vesicle, and 12 molecules of R9AP, determined by 500 nm
   absorbance for rhodopsin, or by densitometry of Coomassie-stained gels
   and comparison to a standard for R9AP.
4. The amount of accessible (i.e., cytoplasmic domain facing outward) rhodopsin and R9AP can be determined by titration with either transducin,
   $G_{\alpha\beta\gamma}$-GDP, for rhodopsin, or with RGS9-1–$G_{\beta5}$, for R9AP, in centrifugation-
   based binding assays.
5. The unilammelar character of the vesicles and their size distribution can be
   readily assessed by cryoelectron microscopy.

5.4 SINGLE-TURNOVER GAP ASSAY FOR PURIFIED RECOMBINANT OR RETINAL PROTEINS

To detect effects of different factors on GAP activity, two kinetic approaches have
been commonly used (described in detail in Reference 30). The multiple-turnover
method detects the steady-state GTP hydrolysis rate during cycles of GTPase activation and inactivation, whereas the single-turnover approach described here has been
employed to avoid any possible interference caused by the G-protein recycling. A limitation of the latter approach is that it must be carried out at substrate (i.e., G_{α}-GTP)
concentrations well below saturating, and thus cannot be used by itself to determine
values of the Michaelis-Menten constant, Km. Rather, it yields a single-exponential
rate constant, k_{inact}, which when divided by the RGS9-1 concentration yields k_{cat}/Km,
the catalytic efficiency. This value when extrapolated to in vivo conditions allows a
comparison with the time constant for photoresponse recovery (figure 5.2).

PROTOCOL 6: PREPARATION OF UREA-WASHED ROS MEMBRANES

1. Rod outer segments are prepared in the dark from fresh or frozen bovine
   retinas (obtained from a local abattoir, or from Schenk Packing, Seattle,
   Washington, or Wanda Lawson Packing, Lincoln, Nebraska) using a standard sucrose gradient procedure (29,31).
2. Each wash is carried out using the following procedure in the dark:
   The membrane pellet is resuspended with a 22½ gauge needle and 16 mL
   of wash buffer (listed in the following text). Four aliquots of 4 mL each are
   removed and diluted to 55 mL in a Potter-Elvehjem homogenizer (with Teflon

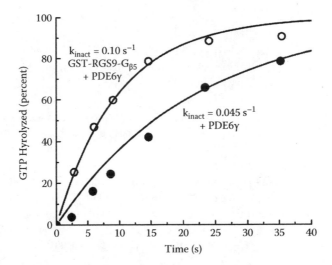

FIGURE 5.2   GAP assay. Enhancement of GTP hydrolysis in rod outer segment membranes by added recombinant RGS9-1-$G_{\beta 5}$. The release of [^{32}P]Pi from [γ–^{32}P]GTP was monitored by scintillation counting of the supernatant from a charcoal suspension in samples quenched by acid at the indicated times after addition of GTP plus the indicated proteins at time zero, as described in protocol 7.

pestle) and homogenized with 10 slow strokes, care being taken not to add air bubbles to the suspension. The homogenized membranes are poured into Ti-45 centrifuge tubes (55 mL per tube), and centrifuged for 30 min at 42,000 rpm, 4°C, in a Type 45-Ti rotor (Beckman). All washes are carried out at a final concentration of 15 μM rhodopsin or below. All buffers are supplemented with 1 mM DTT and solid PMSF just before use.

3. Wash once with 1 × GAPN-Tris buffer (10 mM Tris-HCl, pH 7.4, 100 mM NaCl, 2 mM $MgCl_2$), twice with low-salt buffer (5 mM Tris-HCl, pH 7.4, 0.5 mM $MgCl_2$), twice with high-salt buffer (5 mM Tris-HCl, pH 7.4, 0.5 mM $MgCl_2$, 1 M NaCl), twice with urea wash buffer (5 mM Tris-HCl, pH 7.4, 0.5 mM $MgCl_2$, 4 M urea—urea deionized with mixed-bed ion exchange resin). The final pellet is washed once in the final assay buffer, e.g., GAPN buffer (protocol 3) for the following GAP assay (protocol 7), and after resuspension in this buffer at the desired concentration is divided into 100–200 μL aliquots wrapped in foil and stored at −80°C.

PROTOCOL 7: SINGLE-TURNOVER GAP ASSAY

1. Urea-stripped ROS membranes are mixed with holo-transducin, $G_{\alpha\beta\gamma}$ purified from bovine retinas (see Reference 32 for purification procedure) at concentrations of 15 μM rhodopsin and 1 μM Gt in GAPN buffer (protocol 3).
2. Expose the sample to room light immediately before the assay and continuously vortex the reaction mechanically during the assay.
3. Initiate the assay by adding 50 nM [γ-^{32}P]GTP (GTP $\ll$ $G_{\alpha t}$) premixed with any acceleration factors to be assayed (e.g., recombinant RGS9 or PDEγ,

alone or together) with the tape recorder starting running and marking the first verbal note ($t = 0$), then stop the reaction at various times up to 2 min by adding trichloroacetic acid (TCA) to 10% (w/v) and recording another verbal note upon quenching. For early time points, it is useful to have a pipette preloaded with TCA, and to have a second person either hold the sample on the vortexer or handle the second pipette.

4. Incubate the stopped reaction with 5% activated charcoal in phosphate buffer to determine released [^{32}P]Pi by charcoal binding and scintillation counting. GTP remains bound to the charcoal, while the released [^{32}P]Pi is collected in the supernatant.

5. The following control reactions are needed: To determine the background of [^{32}P]Pi released not due to hydrolysis by $G_{\alpha t}$, GTPγS in excess of total $G_{\alpha t}$ is added prior to addition of GTP. To determine the total amount of releasable [^{32}P]Pi, one sample is incubated with 10-fold higher transducin for 10 min. For assaying effects of accelerating factors, one set of control samples should not have these factors.

6. Determine the GTP hydrolysis rate using single exponential curve fitting to: [^{32}P]Pi(t) − [^{32}P]Pi(t, GTPγS) = ([^{32}P]Pi(10 min, 10 × Gt) − [^{32}P]Pi(t, 10 × Gt, GTPγS))(1 − exp[−$k_{inact}t$]). The time values are determined accurately by using a stopwatch while playing back the tape recording of the assay.

5.5 QUANTIFICATION OF ROS PROTEINS INVOLVED IN GTPASE REGULATION BY QUANTITATIVE IMMUNOBLOTTING

A key method to study endogenous ROS proteins is to quantify protein concentrations in purified ROS or total retina using quantitative immunoblotting. The major ROS marker rhodopsin with known concentration in ROS is a commonly used standard. The method described in the following text uses chemiluminescence and x-ray film to detect horseradish peroxidase-labeled secondary antibodies. This method is very sensitive and convenient, but is well known to have a limited dynamic range for a given set of conditions, and to have a nonlinear response throughout much of its dynamic range. In addition to the nonlinearity inherent in the chemiluminescence detection method, quantitative immunoblotting is also subject to uncertainties due to unevenness of transfer from SDS-PAGE gel to nitrocellulose membrane, and possible nonlinearities in antibody binding as a function of immobilized antigen. At the upper end of the dynamic range of this method, it is easy to have too much antigen in a band, so that the peroxidase substrate is rapidly depleted, or excessive immobilized protein blocks antibody binding. In this case, there is actually a decrease in signal with increasing antigen, often seen as a negative image on x-ray film. For all these reasons, it is absolutely critical that every quantitative immunoblot be analyzed in replicates on the same blot as a series of samples for a standard curve prepared with varying known amounts of the antigen. In addition, multiple exposure times should be tested to increase the chances of the sample intensity falling within the linear range of the standard curve. The best way to determine the amount of protein in the standard sample is to use absorbance spectrophotometry of purified protein and the molar extinction coefficient. For most proteins, the molar extinction coefficient at 280 nm can be accurately calculated from the sequence and the protein's absorbance measured in 6 N guanidinium hydrochloride using the method of Gill and Von Hippel (33). For rhodopsin the known molar extinction coefficient at 500 nm of 42,700 M^{-1}-cm^{-1} can be used. Once the molar ratio of a given antigen to rhodopsin in a sample of purified ROS is determined using these methods, that sample can then be used in place of purified protein to establish a standard curve, if the purified protein is present in limited supply. Note that in most cases samples derived from rod outer segment or other retinal membranes should not be boiled before use in SDS-PAGE, as such treatment tends to cause formation of large protein aggregates that do not enter the gel (figure 5.3).

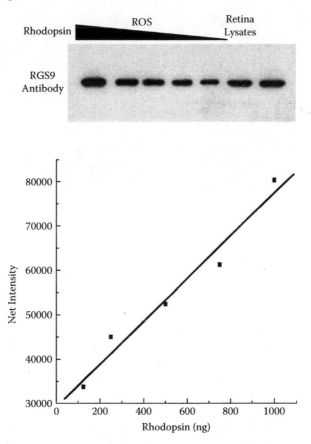

FIGURE 5.3 Quantitative immunoblots of RGS9-1. Top panel: x-ray film showing chemiluminescence signal from immunoblots using RGS9-1-specific antibody. Varying amounts of purified bovine rod outer segments were used to generate a standard curve (lanes 1–5) for comparison to a duplicate sample of retinal lysate (lanes 6 and 7). Lower panel: plot of the integrated intensity on the film as a function of amount of rhodopsin, determined by 500 nm absorbance.

**PROTOCOL 8: ISOLATION OF OSMOTICALLY INTACT RETINAL ROD
OUTER SEGMENTS USING ISO-OSMOTIC DENSITY GRADIENTS**

The following procedure is modified from Reference 34 for a small-scale ROS
preparation:

1. A 40% (w/v) iodixanol working solution is prepared by diluting 8 mL of
   OptiPrep™ (60% solution of iodixanol in water, density = 1.32 g/mL; Axis-
   Shield) to 12 mL Ringer's buffer (10 mM HEPES, pH7.4, 130 mM NaCl,
   3.6 mM KCl, 2.4 mM $MgCl_2$, 1.2 mM $CaCl_2$, 0.02 mM EDTA), and is
   diluted to a series of 4 mL iodixanol solution (4.8, 6, 10, 14, and 18%) with
   Ringer's buffer.
2. A gradient is formed by layering 300 μL of the iodixanol solutions from 6
   to 18% in polyallomer centrifuge tubes (11 × 34 mm, 2.2 mL; Beckman)
   using a syringe attached with a 26½ gauge needle. A continuous gradient is
   formed by gently rotating the tubes to a horizontal position and them allowing them to stand 45–60 min at room temperature, or by allowing iodixanol
   to diffuse at 4°C until the sharp lines between layers are no longer visible.
   For a reproducible gradient, the same procedure should be followed each
   time.
3. All the following procedures are performed under dim red light or using
   near-infrared illumination and infrared image-converting goggles. Four
   mouse retinas are placed in 120 μL of 4.8% iodixanol solution, vortexed for
   1 min, and then centrifuged at 200 × g for 1 min. The supernatant containing the ROS is carefully removed without disturbing the pellet. The pellet is
   resuspended in 120 μL of 4.8% iodixanol solution, vortexed, and centrifuged
   again. This procedure of vortexing and centrifugation is repeated once.
4. The collected supernatants (approximately 350 μL) are combined and
   placed on top of a 6–18% continuous gradient of iodixanol solution and
   centrifuged at 26,500 × g for 1 h in a TLS 55 swing bucket rotor (Beckman)
   using the slowest acceleration and deceleration modes.
5. The band containing ROS (about two-thirds of the way from the top of the
   gradient) is collected as a second band. Rhodopsin concentration is determined by measuring the absorbance at 500 nm.

**PROTOCOL 9: QUANTITATIVE IMMUNOBLOTTING
PROTOCOL FOR ROS PROTEINS**

1. To prepare mouse retina lysates, six whole retinas are collected in a
   Microfuge polyallomer tube (9.5 × 38 mm, 1.5 mL; Beckman), sonicated in
   500 μL SDS-PAGE sample application buffer (5% SDS (w/v), 15% sucrose
   (w/v), 50 mM Na_2CO3, small amounts of bromphenol blue, and 50 mM
   DTT). Insoluble debris is removed by centrifugation at 100,000 × g for 30
   min at room temperature in a TLA 100.3 rotor (Beckman).
2. RGS9-1 quantification serves as an example in quantification of ROS proteins in retina lysates. Whole-retina lysates are applied 5 μL per lane to
   a 12% polyacrylamide bis-tris gel. The bovine ROS containing rhodopsin

125, 250, 500, 750, and 1000 ng are loaded in equal volume on the same gel; these amounts of proteins could be a starting point to establish a standard curve for other ROS proteins with an unknown quantity but a determined molar ratio to rhodopsin.

3. The gel is electrophoresed at 90 V until the dye front passes the stacking gel and continued at 120 V for approximately 1 h. The proteins are wet-electroblotted to supported nitrocellulose membranes (NitroPure, Osmotics, Inc.) in transfer buffer (25 mM Tris, 192 mM glycine, 20% methanol (v/v), 0.1% SDS, pH 8.3) for 90 min at 350 mA at 4°C.

4. Blots are blocked with 5% (w/v) nonfat dry milk in TBST buffer (20 mM Tris, pH 7.6, 137 mM NaCl, 0.1% Tween 20) for 1 h at room temperature and then incubated overnight at 4°C with polyclonal anti-RGS9-1 antibody R4432at a dilution of 1:1000 in 0.5% nonfat dry milk/TBST buffer. After being washed thrice for 10 min each with TBST buffer, blots are incubated 40 min at room temperature with horseradish peroxidase-conjugated anti-rabbit antibody (Promega) at a dilution of 1:10,000.

5. Signals on the blot are developed on an x-ray film using Supersignal West Pico Chemiluminescent Substrate (Pierce), and exposure time is varied from 5 min to 1 s, depending on the signal intensity.

6. The x-ray film is scanned, and protein band densities are measured using densitometric analysis software such as UN-SCAN-IT or ImageJ.

7. A standard curve is generated by plotting band intensities on the blot against the known rhodopsin amount and fitted with a linear function. If the band intensities of the sample series fall within the linear range of this standard curve, the amount of RGS9-1 in retina lysates or ROS is predicted by the linear function or interpolating on the curve and the documented molar ratio of RGS9-1 to rhodopsin.

5.6 IMMUNOPRECIPITATION OF THE GAP COMPLEX

Immunoprecipitation is one of the most useful techniques to detect protein interactions in cells to understand the functional regulation of the proteins. Immunodepletion experiments were originally used to demonstrate that the PDE6γ-sensitive GAP activity in extracts of rod outer segment membranes was primarily due to RGS9-1 (1). Through RGS9-1 immunoprecipitation followed by mass spectrometry, a novel 25 kDa protein was identified and later characterized to be the membrane anchor and GAP activator (10). Immunoprecipitation was also essential in identifying the carboxy-terminal domain of RGS9-1 as a major site of Ca^{2+}-dependent phosphorylation by protein kinase C (16,17). Protocol 11 is for RGS9-1 immunoprecipitation. $G_{\beta 5L}$ and R9AP immunoprecipitation procedures are similar except the elution using peptides for making $G_{\beta 5L}$ antibody or SDS-PAGE sample buffer for R9AP. Because the RGS9-1 GAP complex is tightly membrane associated because of anchoring by the transmembrane protein R9AP, detergent must be used to extract the proteins in soluble form before binding to immobilized antibodies. A persistent problem in immunoprecipitation is nonspecific binding by proteins that are not associated directly with the antigen of interest. In general, better results are obtained when the antibodies

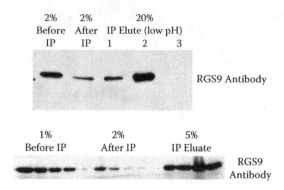

FIGURE 5.4 Immunoprecipitation of RGS9-1 from retinal extract. Detergent-solubilized bovine ROS proteins were immunoprecipitated as described in protocol 10, separated by SDS-PAGE, and immunoblotted with RGS9-1 antibody. (Above) In one immunoprecipitation trial, RGS9-1 was eluted by thrice the column volume of 0.1 M glycine, pH 2.5. (Below) In four trials, different amounts of starting ROS (containing 250, 125, 75, and 50 µg rhodopsin from left to right in lane 1–4) were used and compared in "after IP" flowthrough and "IP eluate" with the same order to optimize immunoprecipitation conditions.

used are more specific, higher affinity, and purer. The cleanest results are obtained when elution from the immunoaffinity matrix is carried out by competition with a peptide containing the epitope recognized by the antibody. This is generally feasible for antibodies raised against and affinity-purified with peptides, or with monoclonal antibodies with known epitopes. It is not generally feasible with polyclonal antibodies raised against proteins of 5 kDa size or larger. In this case, elution can be carried out by high pH, low pH, high concentrations of monovalent or divalent salts, etc. The most complete elution, but also the one giving the highest level of nonspecifically bound proteins, is to elute with the strongly denaturing detergent, SDS. Another common problem with immunoprecipitation is bleeding of antibodies off the affinity matrix; these often interfere with immunoblotting procedures that are frequently carried out on immunoprecipitated proteins. One partial solution is to cross-link the antibodies to the matrix more securely using glutaraldehyde. This procedure usually has the drawback of decreasing somewhat the capacity of the matrix for antigen. Another approach is simply to wash the affinity matrix very extensively with the same solutions used for washing and elution (e.g., low pH solution) while testing the eluent for antibodies until no more are detected. This procedure also decreases the antigen-binding capacity somewhat. In general, whenever the immobilized antibody is prepared, it is essential to have a control matrix with preimmune antibody (preferably from the same animal) identically prepared (unless the specific antibody was prepared by an antigen-affinity procedure) and coupled at the same density to the matrix (figure 5.4).

PROTOCOL 10: PREPARATION OF IMMOBILIZED ANTIBODY MATRIX

The procedure begins with purification of the antibody. At a minimum, protein A or protein G can be used to purify the IgG fraction from polyclonal sera. The following procedure is used for the RGS9-1 rabbit polyclonal R4432 raised against a recombinant fragment including amino acid residues 223–484.

1. The pH of the crude serum is adjusted to 8.0 by adding 1/10 volume of 1.0 M Tris (pH 8.0), and the serum is loaded to a protein A Sepharose column (Amersham Biosciences) following the manufacturer's recommendation for capacity of the column.
2. The column is washed with ten column volumes of first 100 mM Tris (pH 8.0), then with 10 mM Tris (pH 8.0), and the antibodies eluted stepwise with 0.1 M glycine, pH 3.0, into tubes with 1/10 volume of 1.0 M Tris (pH 8.0) to neutralize the eluate. Check the protein concentrations in the elution fractions by absorbance at 280 nm or by dye-based protein assays (35), and continue eluting until the concentration of eluting protein falls below the detection limit; usually 2–3 column volumes of low pH solution is sufficient to elute the antibody. The antibody purity is checked by SDS-PAGE. Antibody samples should be treated with both DTT and 2-mercaptoethanol and boiled in SDS before application to the gel to ensure complete denaturation and reduction of disulfides.
3. Purified IgG is coupled to CNBr-activated Sepharose 4B-CL (Amersham/GE) or Affi-gel 10 (Bio-Rad) at a ratio of 5–10 mg IgG to 1 mL beads following the manufacturer's protocol. The IgG concentration is assayed before and after the reaction to determine the yield of coupled antibody. Note that this number cannot be used to determine the antigen-binding capacity of the beads, as some antigen-binding activity is usually lost in the coupling reaction. To determine the capacity, test small aliquots of the beads with varying amount of antigen (either purified or in tissue extracts), and use immunoblots to determine the amount of beads necessary to remove all of a known amount of antigen or all of the antigen in a tissue extract from solution. The amount of beads used in the following procedures described should exceed by at least twofold the minimum amount needed to bind all the antigen in the sample (the amount of antigen in the sample can be determined ahead of time by quantitative immunoblotting, protocol 9).

PROTOCOL 11: RGS9-1 IMMUNOPRECIPITATION

1. ROS membranes or retinas are solubilized in detergent. For the RGS9-1–GAP complex, a solution of 1% Nonidet P-40 in GAPN-H buffer (see protocol 3) works well. For 40 μL IgG-coupled beads, 300 μL ROS containing 10–15 μM rhodopsin are homogenized by passing through an 18 (for bovine ROS) or 23 (for mouse ROS) gauge needle in the previous buffer, or 8 to 10 retinas are sonicated in 300 μL buffer for 5 × 30 s on ice; insoluble material is removed by centrifugation at 100,000 × g for 20 min at 4°C in a TLA 100.3 rotor (Beckman).
2. The best way to promote binding of the antigen to the immobilized antibody is to pour a slurry of the antibody beads into a column of narrow diameter, and run the detergent extract slowly through this column several

times. Small samples are saved of each flowthrough fraction, as well as of the starting material to assess the course of antigen–antibody binding.

3. Alternatively, when the amount of sample and of beads is too small to make pouring a column practical, small batchwise reactions can be carried out. Solubilized ROS or retina lysate is mixed with IgG-coupled beads for 2.5 h at 4°C on a shaker. Care must be taken to make sure the shaking is sufficient to circulate the beads efficiently through the solution. The beads are separated from the supernatant by brief centrifugation.

4. The columns (or beads in microfuge tubes) are washed with ten times the column volume of the solubilization buffer (three washes using the centrifuge for the batch procedure).

5. The antigen can be eluted with peptide for antipeptide antibody (e.g., 300 µL 1 mg/mL CT215 peptide is applied to the column and incubated for >1 h followed by elution, and this step is repeated multiple times for $G_{\beta 5}$ peptide elution) or with 0.1 M glycine at pH 3.0. Eluted antigen is often of too low a concentration to be assayed by UV absorbance or by dye-binding assays, so it is usually monitored by running immunoblots on all washes and elution fractions. Usually, thrice the column volume is sufficient for efficient elution. Two to three separate washes are used for the batch procedure.

6. For analysis of coimmunoprecipitating proteins by mass spectrometry, the samples are generally loaded onto an SDS-PAGE gel and detected by Coomassie blue staining. For detection or quantification of small amounts of proteins, qualitative or quantitative (protocol 9) immunoblotting is usually used.

5.7 LOCALIZATION OF PROTEINS IN ROD OUTER SEGMENTS BY SUBCELLULAR FRACTIONATION AND IMMUNOFLUORESCENCE

Proteins must be localized to rod or cone outer segments to play a role in phototransduction. Two useful techniques for determination and confirmation of protein localization are gradient purification of rod outer segments with each fraction quantitatively assayed for the protein of interest, and immunolocalization using fluorescent secondary antibodies and confocal microscopy.

5.7.1 GRADIENT COPURIFICATION: A GENERAL WAY TO DETERMINE WHETHER A PROTEIN IS LOCALIZED TO ROS OR A CONTAMINANT FROM OTHER PARTS OF RETINA

The presence of a protein in a purified sample of rod outer segments is not sufficient to allow the conclusion to be drawn that it is a resident outer segment protein. The reason is that all purification procedures yield material that is only partially pure and always contains contamination at some level. However, it is highly unusual for a contaminating protein or organelle to display precisely the same profile across fractions of a density gradient following centrifugation as rod outer segments, especially across two different gradients, for example, one iso-osmotic gradient such

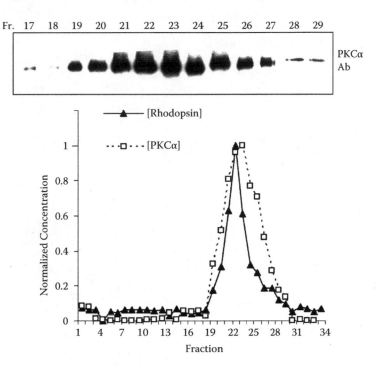

FIGURE 5.5   Fractionation of rod outer segments, and assaying for protein comigration. ROSs were purified from bovine retinas by a discontinuous sucrose gradient, and fractions were analyzed for rhodopsin content absorbance at 500 nm and for PKCα by immunoblots.

as OptiPrep, and one sucrose gradient, which induces substantial volume loss in most organelles because of high osmolarity, but much less so for rod outer segments (figure 5.5).

PROTOCOL 12: SUCROSE GRADIENT FRACTIONATION

1. ROS membranes are prepared from wild-type mouse retinas in the dark by sucrose density gradient (16,29,31) or Optiprep® (iso-osmotic) density gradient (see protocol 8).
2. The best method for fractionating the gradient is by using an automatic gradient puller such as the Auto-Densi-Flow from Labconco. This instrument uses conductance to detect the liquid surface and automatically inserts the entry hole of a collection tube just below the surface. The flow into the tube (which is usually connected to a slow peristaltic pump or a slow gravity-flow system) is horizontal, so there is virtually no mixing of vertical fractions. Fractions of 200 µL are collected for subsequent assays, and can be stored at −80°C until use.
3. The concentration of the major ROS marker protein rhodopsin in each fraction is determined by measuring the absorbance at 500 nm before and after light bleaching in 1.5% LDAO (*N,N*-dimethyldodecylamine *N*-oxide) detergent and 10 mM hydroxylamine. If very low amounts are used (e.g., from

a single mouse), it may be necessary to quantify rhodopsin by quantitative immunoblotting, as described in protocol 9.

4. Proteins in these fractions are resolved by SDS-PAGE and analyzed by immunoblotting (protocol 9) to compare the purification profile of RGS9-1, $G_{\beta 5L}$, and R9AP or other components with that of rhodopsin. The profile of a resident ROS protein should closely follow that of rhodopsin, especially with regard to the peak position, whereas the profiles of contaminants usually only partially overlap rhodopsin's but do not have coincident profiles.

5.7.2 PROTEIN LOCALIZATION BY IMMUNOFLUORESCENCE (12,14,19)

Immunolocalization experiments are essential for visualization of protein subcellular localization, but very easily produce false results owing to antibody cross-reactivity. It is useful to compare the results using different antibody preparations or using knockout mice as negative controls if possible. Useful RGS9-1 antibodies include rabbit and goat antisera raised against a C-terminal fragment of RGS9-1 (aa 226–484) (2) and a mouse monoclonal antibody that recognizes an epitope including a small part of the RGS domain and adjacent portions of the C-terminal domain (1).

PROTOCOL 13: IMMUNOFLUORESCENCE STAINING OF RETINAL SECTIONS

1. Mice are humanely euthanized (e.g., by CO_2 inhalation), and their eyes rapidly removed and placed in 4% paraformaldehyde in phosphate buffered saline (PBS, pH 7.2, GIBCO) for a 1 h fixation at 4°C. Some antigens require longer fixation times.

2. The eyes are cryoprotected by soaking in 30% sucrose in PBS at 4°C until tissue sinks, typically, for 6 h overnight.

3. The eyes are embedded in OCT (Tissue-Tek Compound) and rapidly frozen on dry ice or in liquid nitrogen, and stored at −80°C until use.

4. For cryosectioning, the embedded eyes are warmed to −20°C, cut into sections of 12–40 μm thickness using a cryomicrotome and placed on warm Superfrost Plus slides (Fisher). They are stored at −80°C until use.

5. Prior to staining, sections are thawed at −20°C for 1 h and 4°C for 1 h, then air dried for 30 min at room temperature.

6. Sections are dehydrated in methanol/acetone (1:1 v/v) at room temperature for 10 min.

7. Slides are washed in 0.1% Triton X-100 in PBST (PBS with 0.1% Tween 20). PBS alone may also be used; (results should be compared for each antigen) at room temperature for 2 × 10 min.

8. Sections are blocked with 10% sheep serum (Sigma) in PBST for 1 h at room temperature.

9. Slides are incubated with anti-RGS9 antibodies, anti-R9AP antiserum, or anti-$G_{\beta 5L}$ antibody at various dilutions (1:100–1:500) in PBST (or PBS—see note in preceding paragraph) containing 10% sheep serum for 2–3 h or overnight at a chamber humidified with PBS. Wash slides with PBST for 3 × 5 min.

10. Dye-conjugated secondary antibody is added to the slides for 1 h at the dilution recommended by the manufacturer, typically, 1:25 to 1:100 (again, it is useful when optimizing the protocol for specific antigens, antibodies, and tissues; it is best to compare different dilutions) in PBST. Then, slides are washed with PBST for 3 × 5 min.
11. The slides are mounted with a drop of Vectashield (Vector Laboratories) mounting medium and coverslipped for microscopy. Usually, the edges of the cover slip are sealed with colorless nail polish.
12. It is often useful to counterstain with a nuclear marker, such as propidium iodide, or with cell-specific markers such as peanut lectin for cone sheaths, or rhodopsin antibodies for rod outer segments.

5.8   PHOSPHORYLATION OF RGS9-1

Phosphorylation is a common mechanism for regulating protein function, including RGS proteins. RGS9-1 has been reported to be phosphorylated in vitro by either protein kinase C (PKC) (16,17,36) or protein kinase A (PKA) (37) and is phosphorylated at PKC site Ser^{475} in vivo by a kinase whose activity is inhibited by light. In vitro phosphorylation is useful for preliminary detection of potential phosphorylation reactions and identification of sites, whereas in vivo studies using phosphorylation-specific antibodies provide insight into physiological relevance of those sites identified in vitro. Candidate kinases can be readily tested in the in vitro assays by addition of inhibitors or activators of known kinases. For these experiments, it is essential to have control substrates for the kinases in question to ensure that the effective concentrations of these substances are sufficient to activate or inhibit the endogenous kinase. For example, amphipathic inhibitors may partition into membranes, reducing their effective concentrations in solution, and some activators, such as cyclic nucleotides or diacylglycerol, may be metabolized by the cell homogenates.

PROTOCOL 14: ANALYSIS IN VITRO OF PHOSPHORYLATION OF RGS9-1 BY ENDOGENOUS KINASES IN ROD OUTER SEGMENTS

1. Bovine or mouse ROSs are prepared as described earlier, typically as stocks with rhodopsin concentrations of 15–150 µm. The following procedures are all carried out in complete darkness using infrared image converters, or in dim red light.
2. Purified ROSs are homogenized as described in protocol 10 in GAPN-H buffer containing phosphatase inhibitors (and, if desired, any inhibitors or activators of specific kinases) at a dilution of 1:5, centrifuged at 8400 × g for 15 min, and then the pellet resuspended in the GAPN-H buffer to a final rhodopsin concentration of 6–60 µm.
3. NH_2OH is added to the ROS at a final concentration of 10 mM to minimize rhodopsin phosphorylation, and ATP to 2–5 mM (with $[\gamma-^{32}P]$ at a specific activity of 40–100 Ci/mol if detection will be by radioactivity). The mixture is incubated at 30°C for different times up to 20 min, and the reactions stopped by washing away the free ATP buffer (by centrifuging

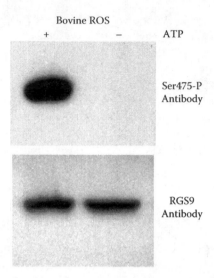

FIGURE 5.6 Detection of RGS9-1 phosphorylation by a phosphorylation-specific antibody. Purified bovine ROS was incubated with or without 2 mM ATP for 15 min as described in protocol 14. Proteins in ROS were analyzed by SDS-PAGE and immunoblotting with anti-Ser[475]-phosphate monoclonal antibody and anti-RGS9-1 antibody.

and resuspending the pellet thrice) with phosphatase-inhibitor buffer (5 mM Tris-HCl, 2 mM EDTA, 0.2 mM Na_3VO_4, 15 mM fenvalerate, 100 nM okadaic acid, 1 mM DTT). The pellets are immediately solubilized with SDS-PAGE sample buffer for detection by phospho-specific antibody, or first subjected to immunoprecipitation for detection by radioactivity.

4. ROSs are solubilized in NP-40 detergent, and RGS9-1 immunoprecipitated as described earlier and subject to SDS-PAGE, followed by autoradiography or phosphoimager analysis to detect phosphorylation. The immunoprecipitation step is critical, as RGS9-1 comigrates with tubulin in SDS-PAGE, and tubulin is a kinase substrate.

5. Alternatively, SDS-PAGE and immunoblotting with anti-Ser[475]-phosphate monoclonal antibody (at a dilution ratio of 1:500) are used to detect the specific Ser[475] phosphorylation (see section 5.5 for immunoblottting protocol) (figure 5.6).

PROTOCOL 15: ANALYSIS IN VIVO OF PHOSPHORYLATION OF RGS9-1 AND REGULATION BY LIGHT

1. Four to six wild-type mice are maintained in a dark room for a period of 16 h, followed by euthanasia and removal of retinas under dim red light or in complete darkness with the help of infrared goggles; control mice are kept for the same time in light of a specified intensity.

2. Retinas are immediately homogenized in the dark in a 1.5 mL tube using GAPN-H buffer with 1% Nonidet P-40 detergent, plus 0.2 mM Na_3VO_4, 15 μm

fenvalerate, 100 nM okadaic acid to inhibit phosphatase activities. First, a plastic pestle (Kontes) is used, and then the homogenates are sonicated on ice.

3. RGS9-1 is immunoprecipitated using rabbit polyclonal antibodies (e.g., R4432) as described earlier (protocol 10) and RGS9-1 phosphorylation is analyzed by immunoblotting with anti-Ser^{475}-phosphate-specific antibodies following SDS-PAGE. Quantitative immunoblotting is carried out as described in section 5.5. It is difficult to obtain a standard for absolute quantification of phosphorylated RGS9-1, so usually only relative amounts (e.g., light versus dark) can be determined.

REFERENCES

1. Cowan, C.W., Fariss, R.N., Sokal, I., Palczewski, K., and Wensel, T.G., *Proc Natl Acad Sci U.S.A.*, 95, 5351, 1998.
2. He, W., Cowan, C.W., and Wensel, T.G., *Neuron*, 20, 95, 1998.
3. Cowan, C.W., He, W., and Wensel, T.G., *Prog Nucl Acid Res Mol Biol*, 65, 341, 2001.
4. Pugh, E.N., Jr., *Neuron*, 51, 391, 2006.
5. Snow, B.E., Krumins, A.M., Brothers, G.M., Lee, S.F., Wall, M.A., Chung, S., Mangion, J., Arya, S., Gilman, A.G., and Siderovski, D.P., *Proc Natl Acad Sci U.S.A.*, 95, 13307, 1998.
6. Makino, E.R., Handy, J.W., Li,T., and Arshavsky, V.Y., *Proc Natl Acad Sci U.S.A.*, 96, 1947, 1999.
7. He, W., Lu, L., Zhang, X., El-Hodiri, H.M., Chen, C.K., Slep, K.C., Simon, M.I., Jamrich, M., and Wensel, T.G., *J Biol Chem*, 275, 37093, 2000.
8. Kovoor, A., Chen, C.K., He, W., Wensel, T.G., Simon, M.I., and Lester, H.A., *J Biol Chem*, 275, 3397, 2000.
9. Simonds, W.F. and Zhang, J.H., *Pharm Acta Helv*, 74, 333, 2000.
10. Hu, G. and Wensel, T.G., *Proc Natl Acad Sci U.S.A.*, 99, 9755, 2002.
11. Hu, G., Zhang, Z., and Wensel, T.G., *J Biol Chem*, 278, 14550, 2003.
12. Hu, G. and Wensel, T.G., *Methods Enzymol*, 390, 178, 2004.
13. Krispel, C.M., Chen, D., Melling, N., Chen, Y.J., Martemyanov, K.A., Quillinan, N., Arshavsky, V.Y., Wensel, T.G., Chen, C.K., and Burns, M.E., *Neuron*, 51, 409, 2006.
14. Zhang, X., Wensel, T.G., and Kraft, T.W., *J Neurosci*, 23, 1287, 2003.
15. Watson, A.J., Aragay, A.M., Slepak, V.Z., and Simon, M.I., *J. Biol. Chem.*, 271, 28154, 1996.
16. Sokal, I., Hu, G., Liang, Y., Mao, M., Wensel, T.G., and Palczewski, K., *J Biol Chem*, 278, 8316, 2003.
17. Hu, G., Jang, G.F., Cowan, C.W., Wensel, T.G., and Palczewski, K., *J Biol Chem*, 276, 22287, 2001.
18. Chen, C.K., Burns, M.E., He, W., Wensel, T.G., Baylor, D.A., and Simon, M.I., *Nature*, 403, 557, 2000.
19. Lyubarsky, A.L., Naarendorp, F., Zhang, X., Wensel, T., Simon, M.I., and Pugh, E.N., Jr., *Mol Vis*, 7, 71, 2001.
20. Keresztes, G., Martemyanov, K.A., Krispel, C.M., Mutai, H., Yoo, P.J., Maison, S.F., Burns, M.E., Arshavsky, V.Y., and Heller, S., *J Biol Chem*, 279, 1581, 2004.
21. Krispel, C.M., Chen, C.K., Simon, M.I., and Burns, M.E., *J Neurosci*, 23, 6965, 2003.
22. Nishiguchi, K.M., Sandberg, M.A., Kooijman, A.C., Martemyanov, K.A., Pott, J.W., Hagstrom, S.A., Arshavsky, V.Y., Berson, E.L., and Dryja, T.P., *Nature*, 427, 75, 2004.
23. He, W. and Wensel, T.G., *Methods Enzymol*, 344, 724, 2002.
24. Sowa, M.E., He, W., Wensel, T.G., and Lichtarge, O., *Proc Natl Acad Sci U.S.A.*, 97, 1483, 2000.

25. Skiba, N.P., Martemyanov, K.A., Elfenbein, A., Hopp, J.A., Bohm, A., Simonds, W.F., and Arshavsky, V.Y., *J Biol Chem*, 276, 37365, 2001.
26. Martemyanov, K.A., Lishko, P.V., Calero, N., Keresztes, G., Sokolov, M., Strissel, K.J., Leskov, I.B., Hopp, J.A., Kolesnikov, A.V., Chen, C.K., Lem, J., Heller, S., Burns, M.E., and Arshavsky, V.Y., *J Neurosci*, 23, 10175, 2003.
27. Baker, S.A., Martemyanov, K.A., Shavkunov, A.S., and Arshavsky, V.Y., *Biochemistry*, 45, 10690, 2006.
28. Litman, B.J., *Methods Enzymol*, 81, 150, 1982.
29. Papermaster, D.S. and Dreyer, W.J., *Biochemistry*, 13, 2438, 1974.
30. Cowan, C.W., Wensel, T.G., and Arshavsky, V.Y., *Methods Enzymol*, 315, 524, 2000.
31. Papermaster, D.S., *Methods Enzymol*, 81, 48, 1982.
32. Wensel, T.G., He, F., and Malinski, J.A., *Methods Mol Biol*, 307, 289, 2005.
33. Gill, S.C. and von Hippel, P.H., *Anal Biochem*, 182, 319, 1989.
34. Liang, Y., Fotiadis, D., Filipek, S., Saperstein, D.A., Palczewski, K., and Engel, A., *J Biol Chem*, 278, 21655, 2003.
35. Bradford, M.M., *Anal Biochem*, 72, 248, 1976.
36. Nair, K.S., Balasubramanian, N., and Slepak, V.Z., *Curr Biol*, 12, 421, 2002.
37. Balasubramanian, N., Levay, K., Keren-Raifman, T., Faurobert, E., and Slepak, V.Z., *Biochemistry*, 40, 12619, 2001.

6 Guanylate Cyclase-Based Signaling in Photoreceptors and Retina

Karl-Wilhelm Koch and Andreas Helten

CONTENTS

6.1 INTRODUCTION

The cyclic nucleotides cyclic AMP (cAMP) and cyclic GMP (cGMP) function as a second messenger in different tissues such as brain, heart, kidney, lung, eye, nose, and smooth and skeletal muscle. Cyclic nucleotides are synthesized from ATP or GTP by different isoforms of an adenylate cyclase (AC) or guanylate cyclase (GC), respectively. Nine membrane-bound and one soluble AC have been described in mammals so far, and some of these isoforms show an ubiquitous tissue distribution (1). Mammalian GCs exist also in particulate and soluble forms; they are classified in seven membrane-bound GCs (GC-A to GC-G) and six soluble GC-subunits (α_{1-3} and β_{1-3}) (2–6). Membrane-bound GCs form homodimers, whereas soluble GCs operate as heterodimers consisting of an α- and β-subunit. Membrane GCs are integrated into the membrane by one transmembrane region. By this arrangement, they are activated by extracellular signaling molecules (e.g., hormones) and transmit the primary signal to subsequent intracellular steps. Whereas three GC forms operate in this manner and function as either hormone receptors, targets for bacterial enterotoxins, or targets for intestine-derived small peptides (guanylin), other forms are named orphan receptors, because no extracellular ligands that bind and regulate these GCs have been identified so far. However, the term *orphan receptor* is misleading for at least two of these membrane-bound GCs, which are expressed in photoreceptor cells of the mammalian retina (7–9). They are regulated by small Ca^{2+}-binding proteins on their cytoplasmic domains (see following text).

Targets of cyclic nucleotides in cell signaling include cAMP- and cGMP-dependent protein kinases (PKA and PKG) (10,11), cyclic nucleotide-gated (CNG) cation channels (12,13), phosphodiesterases (PDE) (14,15), and guanine-nucleotide-exchange factors (epac, exchange protein directly activated by cAMP) (16). Synthesis of cyclic nucleotides by cyclases is counterbalanced by the activity of a class of PDEs.

6.1.1   GUANYLATE CYCLASES IN THE RETINA

The present chapter focuses on GCs in the vertebrate retina, where membrane-bound (7–9) and soluble GCs have been found and localized (17–22). Photoreceptor cells express a subset of particulate GCs named ROS-GC1 and ROS-GC2 or, alternatively, RetGC1, RetGC2, or GC-E and GC-F (7–9). The most characteristic and distinguishable feature of these GCs is that they do not respond to hormone peptides, but instead are regulated by small Ca^{2+}-binding proteins called GCAPs (guanylate cyclase-activating proteins) on their intracellular site (7–9;23–25). Although the presence of natriuretic peptides and their receptor GCs in the retina has been described for several species (20,26,27), much less is known about their specific function in this tissue. Soluble GCs are localized in several layers of the retina (17–22) and are the main target of the gaseous messenger nitric oxide (NO).

6.2   MEMBRANE-BOUND PHOTORECEPTOR GUANYLATE CYCLASES

Vertebrate photoreceptor cells respond to light with a hyperpolarization of their plasma membrane. The capture of photons by the visual pigments (rhodopsin and cone opsins) triggers a G-protein-coupled enzymatic cascade that controls the cytoplasmic level of the internal messenger cGMP in the outer segments of photoreceptor cells (28,29). CNG channels in the plasma membrane are opened by high cytoplasmic concentrations of cGMP. This so-called dark state of the cell is characterized by a constant flow of Na^+ and Ca^{2+} ions into the outer segment of the cell. Flow of Ca^{2+} into the cell is balanced by continuous extrusion via a $Na^+:Ca^{2+}$, K^+-exchanger that is also located in the plasma membrane. Light-induced hydrolysis of cGMP leads to the closure of the CNG-channels, which stops the ion flow into the cell. However, the exchanger continues to operate and expels Ca^{2+} out of the cell, leading to a net decrease of cytoplasmic Ca^{2+} concentration ($[Ca^{2+}]$). Recovery of the cell from illumination and hyperpolarization requires the shutoff of all exciting steps in the transduction cascade; in addition, it requires the refilling of the exhausted cGMP pool. GCs catalyze the synthesis of cGMP from GTP in outer segments under control of a negative Ca^{2+} feedback loop: (7–9,28,29) decreasing Ca^{2+} increases GC activity and vice versa. The action of Ca^{2+} on GC activity is not direct, but is mediated by small Ca^{2+}-binding proteins dubbed GCAPs (23–25) (figure 6.1). GCAPs belong to a group of neuronal calcium sensor (NCS) proteins (25). They are related to another Ca^{2+} sensor named recoverin, which is also part of a negative feedback loop (figure 6.1).

Regulation of ROS-GCs by GCAPs has been studied by different approaches using (a) native rod outer segment (ROS) preparations, (b) membranes prepared from ROS in combination with recombinant GCAPs, or (c) recombinant GCs reconstituted with recombinant GCAPs.

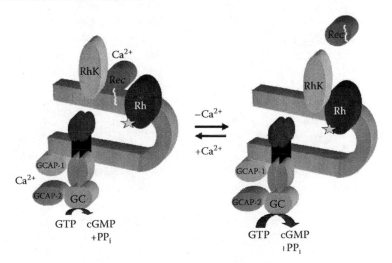

FIGURE 6.1 Ca^{2+} feedback reactions in photoreceptor cells involving NCS proteins. Disk membranes in vertebrate photoreceptor cells harbor a high density of the photopigment rhodopsin (Rh) and other membrane-bound proteins, including photoreceptor guanylate cyclases (ROS-GC1 and ROS-GC2). Membrane-bound GCs form dimers and are regulated by small Ca^{2+}-binding proteins named GCAPs. The Ca^{2+} concentration in the dark state of the cell is high and keeps the cyclase activity at a basal rate (left part of figure). Illumination leads to a decrease of cGMP and cytoplasmic $[Ca^{2+}]$, which in turn triggers several Ca^{2+}-dependent feedback reactions. For example, a rearrangement of the GCAP–GC complex leads to an increase of cyclase activity and a higher rate of cGMP synthesis. GCAPs are related to another NCS protein named recoverin (Rec). Recoverin inhibits rhodopsin kinase (RhK) at high $[Ca^{2+}]$ and thereby prevents phosphorylation of rhodopsin. Decreasing $[Ca^{2+}]$ triggers a Ca^{2+}-myristoyl switch in recoverin that facilitates the detachment of recoverin from the membrane and relieves inhibition of rhodopsin kinase.

6.2.1 GUANYLATE CYCLASE ASSAYS

Guanylate cyclase activity can be assayed by different methods; the principle of each method is described briefly in the following text:

1. Conversion of $[\alpha\text{-}^{32}P]GTP$ to $[\alpha\text{-}^{32}P]cGMP$ (30,31). The radiolabeled nucleotides are separated by thin-layer chromatography, spots containing nucleotides are identified and cut out, and radioactivity is counted in a scintillation counter. The loss of cGMP by hydrolysis due to PDE activity can be corrected by inclusion of $[^3H]cGMP$ as an internal standard.
2. A radiolabeled thio-analog of GTP, (Sp)-GTPαS, is used as substrate (32). ROS-GCs produce (Rp)-cGMPS from (Sp)-GTPαS by a cyclization reaction that involves the inversion of the configuration at the α-phosphorus atom (33). Hydrolysis of (Rp)-cGMPS by the photoreceptor PDE is very low, which makes this assay suitable for operation in the presence of activated PDE. Separation of nucleotides is performed on aluminum oxide (32).
3. The cyclization reaction performed by GCs yields, as a second reaction product, pyrophosphate (PP_i). A spectrophotometric assay measures the production of PP_i by employing several enzyme-coupled reactions that finally lead to the oxidation of β–NADH (34).
4. Classical radioimmunoassays determine the amount of cGMP produced from a competition experiment with a radiolabeled cGMP standard (35).
5. Nucleotides can be separated on a reverse-phase high-performance liquid chromatography (HPLC) column (31,33,36). Eluted peaks are detected at 254 nm, and amounts of formed cGMP are calculated from the peak area by referring to a calibration curve of precisely known cGMP standards. The assay has the advantage that it does not use radiolabeled compounds. Modern HPLC systems allow operation with high sensitivity and automatic sample injection. The detection limit can be as low as 5 pmol nucleotide. Large sample numbers can be processed in overnight runs. Working with ROS or ROS membranes, however, requires the use of specific PDE inhibitors (e.g., zaprinast) and the performance of the assay in complete darkness or under infrared illumination.

PROTOCOL 1: HPLC ASSAY OF GUANYLATE CYCLASE ACTIVITY

Samples containing ROS-GC are incubated with GC-assay buffer. The free $[Ca^{2+}]$ in the sample mix is adjusted and varied with different Ca^{2+}-EGTA buffers. Reaction products are analyzed by HPLC. The HPLC system consists of two pumps, a detector, and an autosampler and is controlled by commercial software provided by the manufacturer.

1. Prepare a GC-assay buffer stock solution (2.5 × Mg^{2+} GC buffer) containing 100 mM Hepes-KOH, pH 7.5, 140 mM KCl, 20 mM NaCl, 25 mM $MgCl_2$, 5 mM GTP, 1 mM zaprinast, 0.25 mM ATP). In cases where you plan experiments with Mn^{2+} instead of Mg^{2+}, substitute 25 mM $MgCl_2$ for 5 mM $MnCl_2$.

2. Prepare Ca^{2+}-EGTA buffer using stock solutions of $K_2CaEGTA$ and K_2H_2EGTA. A detailed description of the preparation of these stock solutions has been provided elsewhere (37). Mix the stock solutions at different ratios by always keeping the total EGTA concentration constant at 2 mM. Use a calcium buffer program (for example, WEBMAXC Standard: http://www.standford.edu/~cpatton/webmanxcS.htm) to calculate the free $[Ca^{2+}]$ in the incubation mixture. Check the free $[Ca^{2+}]$ in the mixtures with fluorescent Ca^{2+} indicators and/or a Ca^{2+} electrode as described (31,38–40).

3. Mix 20 µL of the GC assay buffer with 10 µL of bidistilled water and with 10 µL of Ca^{2+} EGTA buffer of desired free $[Ca^{2+}]$. If necessary, 10 µL of water can be substituted by other ingredients that undergo a test concerning their influence on GC activity (e.g., inhibitors, peptides, and others). GC activity is further increased by a final concentration of 100 µM ATP in the assay mixture.

4. Turn off the room light. Use only indirect and very dim red light. Keep the ROS sample in complete darkness. Start reaction by addition of 10 µL ROS, and incubate for 5 min at 30°C. Alternatively, the reaction can be started by addition of 20 µL of GC assay buffer to a premix of the other components. If this route is chosen, preincubate the premix for 5 min at room temperature before starting the reaction. Terminate the reaction by adding 50 µL of ice-cold 100 mM EDTA and heat the samples for 5 min at 95°C. All other steps can be performed under daylight conditions. Vary the incubation time according to your specific aims.

5. Centrifuge the reaction samples for 5 min at 14,000 × g in a table centrifuge. Remove carefully 50–90 µL of the supernatant. The exact volume depends on the injection syringe and injection loop of your HPLC system. Transfer the supernatant of your samples to appropriate HPLC vials.

6. Prepare 2 L of the HPLC running buffer A containing 5 mM KH_2PO_4; the pH will be around 5.0, and further adjustment of the pH is not necessary. Filtrate the buffer with a filtration device using a hydrophilic polypropylene filter (0.22 µm). Filtrate also methanol (1 L).

7. For separation of nucleotides use a reverse-phase column such as LiChro CART 250-4, LiChrospher 100 RP-18 (5 µm) provided by Merck. Set the flow rate to 1.2 mL per minute. Equilibrate the column in buffer A (5 mM KH_2PO_4). Start the run by injection of the sample. Elution of nucleotides is performed with a two-step gradient: Increase methanol from 0 to 15 % within 4 min. Then, increase methanol from 15 to 70% within 5 min. Decrease methanol from 70 to 0 % within 1 min. Reequilibrate for 4 min with buffer A before the next injection cycle can start.

8. Identify the cGMP peak in the chromatogram (figure 6.2), and determine the area of the peak with appropriate software.

6.2.2 Native Guanylate Cyclase from Purified ROS

A membrane-bound guanylate cyclase was originally purified from bovine, toad, and frog ROS by fractionation studies in combination with activity measurements (36,41). The isolated protein from bovine ROS had a molecular mass of 110–112 kDa

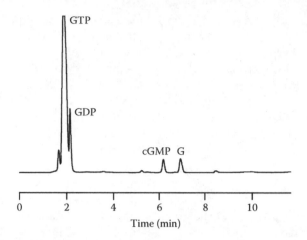

FIGURE 6.2   HPLC chromatogram of a nucleotide mixture obtained after assaying ROS-GC activity. A mixture of purified ROS was incubated with GC assay buffer at a free $[Ca^{2+}]$ of 127 nM according to protocol 1. Separation of nucleotides on a reverse-phase column yielded a cGMP peak at a retention time of 6.1 min. The peak area is linearly related to the amount of cGMP (range 5–5000 pmol) and can be calculated from a calibration curve. The GC substrate GTP is also hydrolyzed by the GTPase activity of transducin, yielding GDP. In addition, non-specific nucleotidases present in ROS produce guanosine (G) from GTP and GDP.

and was present in fractions with the highest specific activities (36), and up to five polymorphic variants of a membrane-bound GC were isolated from amphibian reti-nae (41). Subsequent partial amino acid sequencing of the isolated proteins (42,43) revealed their identity with a cDNA clone obtained from human genomic DNA (44) and a bovine retina cDNA library (45). Shortly after identification of this GC (there-after named ROS-GC1, retGC1, or GC-E) (46–48), the cloning of a second form of a photoreceptor GC was reported for several vertebrate species (named ROS-GC2, retGC2, or GC-F) and its localization in photoreceptor cells was demonstrated (23,49). However, so far this protein has not been purified from native sources, pos-sibly because it is only present in low amounts in bovine retina, the main starting material for the isolation of retinal proteins. Although the precise ratio of both GCs remains to be determined and their contribution to the overall synthesis of cGMP in rod and cone cells is still a matter of discussion (9), one can use purified bovine ROS as a source to enrich and purify the isoform ROS-GC1 (36).

PROTOCOL 2: SOLUBILIZATION AND PURIFICATION OF ROS-GC1

Starting materials for purification of ROS-GC1 are bovine ROS prepared from dark-adapted bovine retinae that were freshly obtained from a local slaugh-terhouse. Soluble cytoplasmic proteins are removed first by washing in hypo-osmotic buffer. Washed membranes are then solubilized in detergent, and proteins are selectively extracted by buffers that differ in detergent and salt content. Rhodopsin and other membrane proteins are removed first by the Tri-ton buffer. Rhodopsin is seen as the prominent protein band in the TritonX-100 extract (TE) in figure 6.3a. ROS-GC1 is solubilized in a subsequent extraction

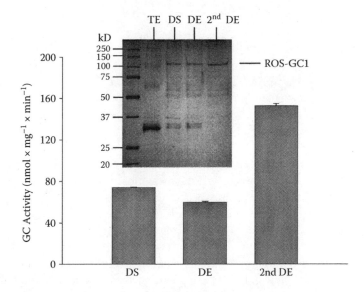

FIGURE 6.3a Solubilization of ROS-GC1 from ROS membranes. Specific guanylate cyclase activities of fractions DS, DE, and 2nd DE (see following text). Fractions were obtained by a selective detergent extraction procedure and analyzed by SDS-PAGE (inset: approx. 1 µg protein in lanes TE to 2nd DE). TE, ROS membrane proteins extracted by TritonX-100 and low salt; DS, suspension of solubilized ROS proteins after removal of proteins that are soluble in the Triton-buffer (TE); DE, n-dodecyl-β-D-maltoside extract obtained from DS after centrifugation; 2nd DE, extract obtained after a second extraction step with n-dodecyl-β-D-maltoside. A molecular weight marker is shown on the left side.

step using n-dodecyl-β-D-maltoside as detergent (DE in figure 6.3a). Repeating this extraction step leads to an already highly enriched fraction of ROS-GC1 (2nd DE in figure 6.3a). Purified ROS-GC1 is obtained after chromatography on an anion exchange column and a GTP-agarose-affinity column (36). Alternatively, ROS-GC1 can be purified on lectin columns such as Concanavalin A-Sepharose or wheat germ agglutinin (WGA)-Sepharose (43).

1. Use 4 mL of purified ROS with a rhodopsin concentration between 6 and 10 mg/mL. Scale up, if necessary. Work in dim red light.
2. Dilute ROS with low-salt buffer (10 mM Hepes-KOH, pH 7.4, 1 mM DTT, 0.1 mM PMSF, 0.1 mM EGTA) about fivefold; use, for example, four vials and centrifuge for 30 min at 50,000 × g and at 4°C. Remove the supernatant and repeat the washing step with low-salt buffer.
3. Remove the supernatant after the last washing step, and resuspend every pellet in 1 mL of Triton buffer (5% [v/v] Triton X-100, 2 mM $MgCl_2$, 10 mM Hepes-KOH, pH 7.4, 0.1 mM PMSF, 0.1 mM EGTA, 1 mM DTT). Combine all suspensions, and adjust a final concentration of 4 mg rhodopsin per milliliter by adding Triton buffer. All subsequent steps can be performed in daylight.
4. Homogenize the suspension in a homogenizer (e.g., Elvehjem homogenizer). Keep the homogenized suspension on ice for less than 30 min. Centrifuge

the suspension for 20 min at 165,000 × g at 4°C. Microultracentrifuges are most convenient at this step.

5. Remove the supernatant and resuspend the small pellet in DM-buffer (20 mM *n*-dodecyl-β-D-maltoside, 20 mM Hepes-KOH, pH 7.4, 1 M NaCl, 0.1 mM PMSF, 0.1 mM EGTA, 1 mM DTT). Adjust to 4 mg rhodopsin per milliliter (corresponding to the initial amount of rhodopsin in the starting material), and homogenize the suspension. Keep on ice for 1 h. Centrifuge for 15 min at 100,000 × g at 4°C.

6. Remove the supernatant and keep at −80°C until use. This supernatant contains an enriched fraction of photoreceptor GC. However, the small pellet also contains GC and, therefore, the extraction procedure is repeated. Use a total of 1.6 mL DM buffer, when your starting material was 4 mL ROS. Collect the supernatant after centrifugation.

7. The second DM extract prepared as described in step 6 is already highly enriched in ROS-GC1. It can be purified further by chromatography on a WGA-Sepharose column (column volume 1–2 mL). Adjust the second DM extract to the following buffer using either ultrafiltration or a spin column: 2 mM *n*-dodecyl-β-D-maltoside, 25% glycerol, 20 mM Tris-HCl, pH 7.4, 100 mM NaCl, 2 mM MnCl$_2$, 1 mM CaCl$_2$, and 1 mM DTT. Equilibrate the WGA-Sepharose with 20 mM Tris, pH 7.4, 2 mM MnCl$_2$, 1 mM CaCl$_2$ 100 mM KCl, 25% glycerol, and 1 mM DTT. Incubate the ROS-GC-containing solution with the WGA-Sepharose for 1 h. Remove the unbound material, wash the WGA-Sepharose by 3–4 column volumes of equilibrating buffer.

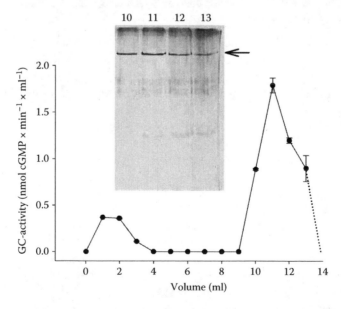

FIGURE 6.3b Purification of ROS-GC1 on WGA-Sepharose. Guanylate cyclase activities were measured in each fraction. Elution of bound ROS-GC1 by *n*-acetyl-D-glucosamine starts at an elution volume of 10 mL. Inset, SDS-Page analysis of fractions with peak of activity; see arrow that indicates ROS-GC1. Protein bands were stained with silver.

Elute-bound ROS-GC1 with the following buffer: 20 mM n-dodecyl-β-D-maltoside, 20 mM Hepes-KOH, pH 7.4, 1 M NaCl, 1 mM DTT, and 0.5 M n-acetyl-D-glucosamine.

The activity profile of fractions obtained during chromatography on WGA-Sepharose is shown in figure 6.3b. Purified ROS-GC1 was eluted from the column in fractions 10–13 mL as a protein band at 112 kDa (see arrow in figure 6.3b); its identity was verified by a ROS-GC1 specific antibody (43). The specific activities of purified ROS-GC1 preparations (36,43) were at least 240 nmol cGMP $\times$ mg^{-1} $\times$ min^{-1}. ROS-GC2 was not detectable in these fractions as tested by a ROS-GC2 specific antibody.

6.2.3 RECOMBINANT ROS-GCs

Heterologous expression of ROS-GCs in cell culture is widely used to study regulatory mechanisms by GCAPs or to investigate the biochemical consequences of site-directed mutagenesis of the ROS-GC1 or ROS-GC2 gene. COS cells, HEK293 cells, or tsA-cells (a modified version of HEK293 cells) are used for most applications (39,40,45,49).

PROTOCOL 3: HETEROLOGOUS EXPRESSION OF ROS-GCs IN TSA CELLS

1. Cultivate cells in medium M10 (minimal essential medium with 10% fetal calf serum, 2 mM L-glutamin, 1% nonessential amino acids, and 1% antibiotika/antimyotika; GIBCO BRL) at 37°C, 5% (p/p) CO_2, and about 95% humidity on petri dishes.
2. Seed 3×10^5 cells per 5 cm dish or 9×10^5 cells per 9 cm dish. Change medium (M10) after 22 h. Wait for additional 1 to 2 h before start of transfection.
3. Vector constructs pcDNA3.1-GC1 or pcDNA3.1-GC2 containing coding regions of ROS-GC1 or ROS-GC2, respectively, were used for gene transfer into cells by the calcium phosphate method, according to Chen and Okayama (50). Dissolve 30 μg of DNA in 372 μL of H_2O_{bidest}. Add 123 μL 1 M $CaCl_2$ and 495 μL 2 $\times$ BBS-buffer (50 mM BES, 280 mM NaCl, 1.5 mM Na_2HPO_4, pH 6.95). Mix and incubate for 20 min at room temperature. Pipette this solution to a 9 cm petri dish containing cells in culture medium. Incubate for 20–22 h at 35°C under 3% (p/p) CO_2 and approximately 95% humidity.
4. Remove the DNA/calcium phosphate precipitates by washing with 5 mL PBS (phosphate-buffered saline) buffer and subsequently with 3 mL PBS/0.5% (w/v) EDTA and 5 mL PBS. Refill dishes with 8 mL of prewarmed M10 medium. Incubate for 20–22 h at 37°C, 5% CO_2, and 95% humidity until harvesting.

Cell membranes of tsA cells transfected with DNA of ROS-GCs can be used for further studies on guanylate cyclase regulation. For this purpose, cell membranes need to be prepared from tsA cells.

PROTOCOL 4: PREPARATION OF TSA CELL MEMBRANES
CONTAINING HETEROLOGOUSLY EXPRESSED ROS-GC

1. Collect tsA cells in medium from petri dishes (9 cm) and pellet by centrifugation ($200 \times g$; 5 min, $4°C$). Resuspend cells in 8 mL PBS buffer and
   repeat centrifugation. Resuspend the cell pellet in 100 µL lysis buffer (10
   mM Hepes-KOH, pH 7.5, 1 mM DTT). At this stage cells can be shock-
   frozen with liquid nitrogen and stored at $-80°C$.
2. Sonify a thawed cell suspension for 4×5 s (Branson Sonifier B12, 80–100
   W). Centrifuge lysed cells ($400 \times g$, 5 min).
3. Take the supernatant and pellet the membranes by centrifugation in an
   ultracentrifuge ($125,000 \times g$, 15 min, $4°C$). Resuspend the membrane pellet
   (corresponding to cells from a 9 cm dish) in 100 µL 10 mM Hepes-KOH,
   pH 7.5, 250 mM KCl, 10 mM NaCl, 1 mM DTT. Incubate for 30–60 min.
   Homogenize by short sonification (5 s). Determine the protein concentration
   using standard methods (e.g., Amido Black method). Typically, 100–200 µg
   protein was obtained from a 9 cm petri dish resulting in a final protein concentration of tsA membranes of 1–2 mg/mL.

6.3   GUANYLATE CYCLASE-ACTIVATING PROTEINS

Outer segments of vertebrate rod and cone cells harbor a set of Ca^{2+}-binding proteins
that operate as Ca^{2+} sensors during a light response. They detect changes in intracellular $[Ca^{2+}]$ depending on the illumination conditions and regulate their targets in a
Ca^{2+}-dependent fashion. Recoverin inhibits rhodopsin kinase (GRK1) at high $[Ca^{2+}]$
in a dark-adapted cell (51). Inhibition is relieved when the $[Ca^{2+}]$ drops after illumination (figure 6.1). Calmodulin binds to the CNG channel at high $[Ca^{2+}]$ and dissociates from the binding sites when $[Ca^{2+}]$ decreases (52,53). This step increases the
affinity of cGMP for the channel, thereby facilitating the reopening of the channel.
Guanylate cyclase-activating proteins (GCAPs) activate ROS-GCs at low $[Ca^{2+}]$ and
show a modest inhibition below the basal GC activity level at high $[Ca^{2+}]$ (23–25).
The Ca^{2+}-dependent regulation of ROS-GCs by GCAPs is one of the main negative
feedback loops that adjust the light sensitivity of a photoreceptor cell during illumination. A typical example of guanylate cyclase activity as a function of the $[Ca^{2+}]$ is
shown in figure 6.4.

Mammalian photoreceptor cells express up to three, zebra fish photoreceptors
up to six, and pufferfish photoreceptors up to eight different GCAP isoforms (55).
Most functional studies in the past, however, were restricted to mammalian GCAP1
and GCAP2. These proteins were originally isolated from mammalian rod outer
segments or retinae preparations (56–58), but their easy and reliable expression in
E. coli cells allowed a large range of functional studies with recombinant proteins.
Previous work included, but was not limited to, the identification and characterization of target sites in ROS-GC1 (59–62), structure–function studies on GCAP
mutants (63–68), and the characterization of disease-related mutations in ROS-GC1

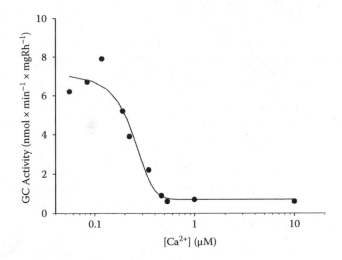

FIGURE 6.4 Ca^{2+}-dependent activation of ROS-GC in whole ROS. Typical profile of ROS-GC activities in a range of free $[Ca^{2+}]$ from 50 nM to 10 μM. Half-maximal activation (usually expressed as IC_{50}) was at 237 nM free $[Ca^{2+}]$.

and GCAP1 (23,69). Recordings from transgenic mice with altered expression levels of GCAP1 and GCAP2 confirmed the principal operational mechanism of Ca^{2+}-dependent regulation of ROS-GCs (70–73). Details of the foregoing aspects have been broadly covered in several recent reviews (8,9,23–25) and will not be discussed here. Instead, we will give a short summary of the main regulatory features of ROS-GC1 controlled by the action of GCAP1 and GCAP2. ROS-GC2 is not covered, because most previous work has focused on ROS-GC1.

6.3.1 COMPLEX OF GCAP1 AND GCAP2 WITH ROS-GC1

Results from independent experimental approaches indicated that ROS-GC1 and GCAPs form a complex at low and high $[Ca^{2+}]$ (62,63,74,75). In the Ca^{2+}-loaded form of GCAP, i.e., the resting state, the interaction between GCAP and GC leads to a conformation of GC that is only able to synthesize cGMP at a low rate. Increase in cGMP synthesis rate is triggered by a conformational change in GCAP1 and/or GCAP2, which subsequently leads to a different interaction modus operandi. The increase in GC catalytic activity is probably enabled by a better orientation of the catalytic domains in the GC dimer. Details of the activation mechanism are still missing and are a matter of current debate. For example, a "priming effect" of ATP has been suggested to be a necessary condition for ROS-GCs to achieve high levels of physiologically relevant enzymatic activities (76–78). Another line of recent research proposes that Mg^{2+}, not Ca^{2+}, is the physiologically important cation to be associated with GCAPs (39). Heterologously expressed and purified GCAPs are useful tools to test these working models.

Protocol 5: Heterologous Expression of GCAPs

1. Use *Escherichia coli* BL21-Codon Plus (DE3) cells for overexpression of GCAPs. Transform competent *E. coli* cells with plasmids containing GCAP-DNA. Use pET-11a/GCAP1 for expression of wild-type GCAP1 and pET21a/GCAP2 for expression of wild-type GCAP2. Use the plasmid pBB131 containing the gene for a yeast *N*-myristoyl-transferase (kindly provided by Dr. J. Gordon, Washington University School of Medicine, St. Louis, Missouri), when myristoylated variants of GCAPs are produced. Use a D6S mutant of GCAP1 in order to maximize degree of myristoylation. This mutation is not necessary for the heterologous expression of myristoylated GCAP2, because the amino acid sequence of GCAP2 harbors a complete consensus site for myristoylation by yeast *N*-myristoyl-transferase.

2. Thaw 100 µL of competent cells on ice, add plasmid DNA (1–10 ng) and incubate for 30 min on ice. Apply a heat pulse of 42°C for 25 s, and incubate on ice for 2 min. Add 400 µL of LB medium and grow cells for one hour at 37°C. Spread 50–500 µL on agar plates containing appropriate antibiotics (100 µg/mL ampicillin and 30 µg/mL kanamycin). Incubate overnight at 37°C.

3. Inoculate 5 mL of dYT medium supplemented with 100 µg/mL ampicillin and 25 µL/mL kanamycin with single colonies of transformed cells and grow at 37°C overnight. Add 2 mL of an overnight culture to 500 mL of dYT medium containing 100 µg/mL ampicillin and 25 µg/mL kanamycin. Grow by vigorous agitation at 37°C. Add myristic acid (100 µg/mL final concentration, dissolved in ethanol) at an OD of 0.4 (in case you plan to express a myristoylated protein). Induce expression at an OD of 0.6 by adding isopropyl-β-D-thiogalactoside (IPTG, 1 mM final concentration). Harvest cells after 4 h by centrifugation (5000 × g, 10 min, 4°C). Resuspend the pellet of 1 L bacterial culture in 40 mL of 50 mM Tris-HCl, pH 8.0.

4. Add 100 µg/mL lysozyme and 5 u/mL DNAse to the bacteria suspension. Incubate at 30°C for 30 min (water bath), and add 1 mM DTT and 0.1 mM PMSF. Centrifuge at 360,000 × g for 30 min at 4°C. GCAPs are partially present in the soluble fraction, but they are also present in high amounts in inclusion bodies. For further purification from the soluble fraction, continue at protocol 6.

5. Solubilize the pellet in 6 M guanidinium-hydrochloride, 1 mM DTT (40–60 mL per 1 L of culture). Solve by stirring at room temperature. Dissolve against 5 L of dialysis buffer (150 mM NaCl, 20 mM Tris-HCl, pH 8.0, 1 mM DTT) for 5 h. Repeat dialysis.

6. Precipitate insoluble material by centrifugation (360,000 × g, 30 min, 4°C). Take the supernatant and add 67% $(NH_4)_2SO_4$ to precipitate proteins. Pellet by centrifugation at 60,000 × g for 30 min at 4°C. The protein pellet can be stored at −20°C for later use.

Protocol 6: Purification of GCAPs

Purification of GCAP is achieved by two chromatographic steps, first, by size exclusion chromatography (SEC) and, second, by ion exchange chromatography

(IEC) (58,75,79). If GCAPs were obtained from the soluble bacterial fraction, concentrate the protein solution to the appropriate volume before applying onto the SEC column.

1. Dissolve the $(NH_4)_2SO_4$ pellet of expressed GCAP in 2–3 mL of bidistilled water. Remove undissolved material by centrifugation (100,000 × g, 15 min, 4°C). Equilibrate a HiLoad 16/60 Superdex prep grade column with gel filtration buffer (150 mM NaCl, 20 mM Tris-HCl, pH 7.5, 1 mM DTT). Add 2 mM EGTA for the purification of GCAP1 or 2 mM $CaCl_2$ for the purification of GCAP2. Adjust to a flow rate of 1 mL/min. Inject the GCAP sample. Collect the eluted fraction, and test for presence of GCAP by SDS-PAGE.

2. Combine the GCAP-containing fractions and adjust to 50 mM NaCl by dilution with 20 mM Tris-HCl, pH 7.5, 1 mM DTT. Equilibrate an IEC column (MonoQ or UnoQ) with IEC buffer A (50 mM Tris-HCl, pH 7.5, 50 mM NaCl, 1 mM DTT, EGTA, or $CaCl_2$ as indicated earlier). Set the flow rate to 0.5 mL/min, and apply the protein on the column. Wash the nonbound proteins from the column. Set the flow rate to 3 mL/min, and elute bound proteins by a gradient of 13 column volumes of 0–55% buffer B (20 mM Tris, pH 7.5, 1 M NaCl, 1 mM DTT, EGTA, or $CaCl_2$ as indicated). Collect the fractions and analyze by SDS-PAGE.

6.3.2 ANALYSIS OF GCAP PROPERTIES

GCAPs constitute a subgroup of neuronal calcium sensor (NCS) proteins. These proteins bind Ca^{2+} with moderate to high affinity and exhibit Ca^{2+}-induced conformational changes. Direct Ca^{2+} binding can be investigated by radioisotope methods using $^{45}Ca^{2+}$ or by employing fluorescent indicators to monitor free $[Ca^{2+}]$ in titration experiments (39,63,64,68). A simple and quick test for Ca^{2+}-induced conformational changes in GCAPs is a gel-shift assay (58,68,79). GCAP samples were complemented with either 1 mM $CaCl_2$ or 1 mM EGTA and were analyzed by SDS-PAGE. Protein bands of GCAP1 and GCAP2 exhibit a large change in their electrophoretic mobility. The apparent molecular weights of the Ca^{2+}-bound and Ca^{2+}-free forms are 20–21 kDa and 25–26 kDa, respectively. This simple method is a qualitative test to determine whether the Ca^{2+}-induced conformational change in a GCAP mutant is still taking place or is significantly disturbed. Another method that is widely used to study Ca^{2+}-binding proteins is to record tryptophan fluorescence as a function of $[Ca^{2+}]$ (39,63,68). Many Ca^{2+}-binding proteins exhibit a Ca^{2+}-dependent change in the maximum of the tryptophan fluorescence emission. These changes reflect changes in the conformation of the protein that are monitored by tryptophan residues. The protein sample is excited at 280–290 nm, and the emission spectrum is recorded between 300 and 450 nm. Decreasing the $[Ca^{2+}]$ from 10^{-3} to 10^{-9} M changes tryptophan fluorescence emission at 335 nm. The relative fluorescence emission at maximum is then plotted as a function of the $[Ca^{2+}]$. An example of a Ca^{2+}-dependent change in tryptophan fluorescence of GCAP1 is shown in figure 6.5. The Ca^{2+}-dependent tryptophan fluorescence emission of GCAP1 is biphasic (39,63,68).

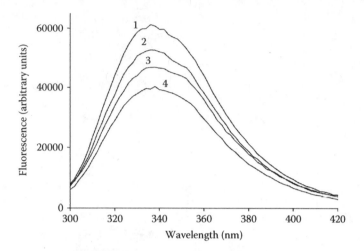

FIGURE 6.5   Tryptophan fluorescence of nonmyristoylated GCAP1. The tryptophan fluorescence emission of 1.1 µM GCAP1 was recorded in the presence of different $[Ca^{2+}]$. Trace 1 was obtained in 2 mM K_2H_2EGTA (no K_2CaEGTA added). Free Ca^{2+} concentrations in the other traces was 5 nM (trace 2), 11 nM (trace 4), and 10 µM (trace 3). Ca^{2+}-dependent changes in tryptophan fluorescence emission are biphasic (see main text for details).

PROTOCOL 7: TRYPTOPHAN FLUORESCENCE OF GCAPS

1. Dissolve a purified sample of GCAP in a buffer containing 50 mM Hepes-KOH, pH 7.4, 100 mM NaCl, and 1 mM DTT at a concentration of 2 µM.
2. Set the excitation wavelength of the fluorescence spectrometer to 280 nm.
3. Record first the emission spectrum between 300 and 450 nm of the buffer without GCAP. Then record the spectrum of the GCAP solution.
4. Correct the emission spectrum of GCAP by subtraction of the spectrum obtained with the buffer solution. This step will separate the tryptophan fluorescence signal from the Raman scattering signal of water that can interfere with the true fluorescence emission signal when low protein concentrations are used.
5. Vary the $[Ca^{2+}]$ in the medium by the use of a Ca^{2+}/EGTA buffer system as described in Protocol 1. Repeat for every $[Ca^{2+}]$ the foregoing steps. Plot the maximum fluorescence intensity as a function of the $[Ca^{2+}]$, and determine the EC_{50} value.

A Ca^{2+} titration of GCAP2 is seen in figure 6.5. The slight increase at higher $[Ca^{2+}]$ is a typical observation made with GCAP1 (39,63,68). It was recently reported that binding of Ca^{2+} to EF-hand 4 in GCAP1 causes the structural movement around Trp94, which can be detected as the increase in fluorescence emission at $[Ca^{2+}] > 10^{-6}$ M (39).

Limited proteolysis has been used to investigate the accessibility of GCAPs for trypsin. Ca^{2+}-bound and Ca^{2+}-free wild-type GCAPs have different accessibilities for trypsin (68,80). Removing Ca^{2+} induces a conformational change that opens the interior of the protein and facilitates access of the protease. The comparison of wild-type and mutant GCAP forms allows conclusions about conformational stability and accessibility of certain regions in particular GCAPs.

6.3.3 CHEMICAL MODIFICATION OF GCAPs

GCAPs contain 3–4 cysteines in their amino acid sequence. Therefore, chemical modification of cysteines opens several routes to investigating the molecular properties of these proteins. For example, introduction of spin labels and fluorescent and nonfluorescent dyes have been used to study Ca^{2+}-dependent conformational changes associated with GCAP1 and GCAP2 (65,66). Cysteines in GCAPs exhibit different accessibilities to the thiol-modifying reagent 5,5′-dithiobis(2-nitrobenzoic acid) (DTNB) as it was demonstrated by the use of cysteine mutants of GCAP1. DTNB contains a disulfide group that undergoes a disulfide exchange reaction with free cysteines in a protein. This reaction produces the chromogenic substance 5-thio-2-nitrobenzoic acid (TNB) and can be monitored by a change in absorbance at 412 nm. Modification of each thiol group in a protein produces exactly one molecule of chromogenic TNB from every DTNB and allows easy quantitation of the modification. Time-based absorbance measurements at 412 nm also enable kinetic measurements (66). For example, the four cysteines in GCAP1 react with different velocities with DTNB at different Ca^{2+} concentrations. Thus, they display different Ca^{2+} sensitivities to the molecular environment within the polypeptide chain.[66]

PROTOCOL 8: MODIFICATION OF GCAPs BY DTNB

1. Prepare a fresh DTNB solution of 12 mM DTNB in 0.1 M Tris-HCl, pH 8.0. Sonicate the solution for several seconds at 80–100 W.
2. Prepare a solution of 2–3 µM GCAP in 50 mM Hepes, pH 7.4 and 100 mM NaCl. Add either 10 µM $CaCl_2$ or 10 µM EGTA (calculate for a final volume of 2 mL in the cuvette, prepare first 1.99 mL, and degas the solution).
3. Stir the solution in the cuvette by a magnetic stirrer, and start the reaction by addition of 10 µL DTNB stock solution (60 µM final). Record the absorbance change at 412 nm for a sufficient time (5–10 min at least). Add 100 µM $CaCl_2$ to an EGTA-containing cuvette or vice versa, and record change of absorption.
4. Calculate the TNB concentration using the molar absorption coefficient for TNB of $\varepsilon_{412} = 13,600$ M^{-1} × cm^{-1} and relate to the amount of modified cysteines.

An example of thiol modification of a GCAP1 mutant is shown in figure 6.6. The GCAP1 mutant ACAA contains only one of the four cysteines present in wild-type GCAP1. This cysteine is located in the nonfunctional EF-hand 1 and is easily accessible by DTNB. A Ca^{2+} titration experiment as shown in figure 6.6 reveals that the cysteine becomes blocked at higher [Ca^{2+}] (> 100 µM), which mirrors a low affinity for Ca^{2+} binding of EF-hand 1.

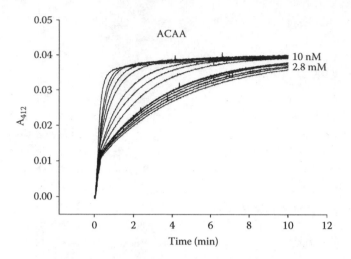

FIGURE 6.6 Thiol reactivity of the cysteine residue at position 29 in the GCAP1 mutant ACAA. Three of the four cysteines in GCAP1 were substituted by alanine; the single cysteine at position 29 is accessible to DTNB as a function of free $[Ca^{2+}]$. Recordings were started by the addition of DTNB (60 μM) to 2 μM ACAA. Free $[Ca^{2+}]$ was varied from 10 nM to 2.8 mM as indicated.

6.3.4 RECONSTITUTION OF PURIFIED GCAPs WITH MEMBRANE-BOUND ROS-GCs

The determination of guanylate cyclase activities to test modulating effects of GCAPs is performed by preparing membranes containing ROS-GCs and mixing the membranes with the corresponding GCAP solution. Sources of ROS-GC are either purified rod outer segments or cell membranes of HEK293, tsA or COS cells that contain heterologously expressed GCs. Purified samples of native ROS-GC1 in detergent solutions are not activated by GCAPs or GCAP-containing fractions (36). When ROS-GC1 samples were reconstituted in phospholipid membranes, addition of GCAPs also did not result in a Ca^{2+}-dependent activation (Koch, unpublished observation). So far it is unclear why ROS-GC preparations fail to become activated by GCAPs when they are extracted by detergents or purified to homogeneity. Loss of a photoreceptor-specific factor is unlikely, because heterologously expressed ROS-GCs in HEK293, tsA, or COS cells are specifically activated by GCAPs and exhibit high activation rates at low $[Ca^{2+}]$ (40,46,57,59,68,74). It is conceivable that detergent solutions, which are necessary to extract the membrane-bound GCs, disturb a certain conformation that is necessary to interact with GCAPs or that they simply have a shielding effect by binding to the GCAP target site. Interestingly, soluble constructs of the cytoplasmic domain of ROS-GC1 interact with GCAP2 (62) and with the Ca^{2+}-binding protein S100β (81) by similar affinities as observed for whole ROS-GC1 in membrane preparations. These constructs are also activated by GCAP2 at low $[Ca^{2+}]$ and by S100β at high $[Ca^{2+}]$. However, the basal GC activities observed with soluble ROS-GC1 constructs were approximately tenfold lower than the activities of the whole enzyme. It was recently proposed that illuminated rhodopsin initiates an ATP-dependent preincubation phase during which GCs in ROS membranes are "primed." This priming effect of ATP does not involve a phosphorylation reaction and results in a significantly enhanced stimulation of GCAP-dependent ROS-GC activation (76–78). It remains to be shown whether priming of ROS-GCs is necessary for them to become activated by GCAPs and whether disruption of the membrane structure leads to a loss of "ATP priming."

PROTOCOL 9: RECONSTITUTION OF GCAPS WITH ROS-GC

Use of ROS in subsequent reconstitution studies with GCAPs requires the removal of native endogenous GCAPs from the ROS suspension. Two washing steps under low-salt conditions are usually sufficient to remove most of the endogenous GCAPs. One or two more washing steps might be additionally necessary when the remaining guanylate cyclase activity in ROS membranes exhibits a rather high Ca^{2+} sensitivity (more than twofold activation of cyclase). However, repeated washing can result in a gradual loss of GCAP sensitivity of ROS-GCs.

1. Prepare a sample of membranes containing heterologously expressed ROS-GC from tsA cells according to protocol 4. If you use HEK293 cells, perform a similar procedure.
2. If you use ROS membranes, take an aliquot of 500 µL purified ROS (6–10 mg/mL rhodopsin). Work under dim red light. Dilute with 2 mL of a low-salt buffer (10 mM Hepes–KOH, pH 7.5, 1 mM DTT). Homogenize and centrifuge for 10 min ($\geq$ 100,000 × g). Resuspend the pellet in 2 mL of low-salt buffer, and repeat the centrifugation step. Resuspend the pellet in 250 µL resuspension buffer (50 mM Hepes-KOH, pH 7.5, 500 mM KCl, 20 mM NaCl, and 1 mM DTT).
3. Add 10 µL of a GCAP solution to 10 µL of a membrane suspension that contains one or both of ROS-GCs (e.g., tsA or ROS membranes). Vary the GCAP concentration within a range from 0 to 20 µM, depending on the properties of the specific GCAP isoform or GCAP mutant. Preincubate with 10 µL of a Ca^{2+}-EGTA buffer (see protocol 1) for at least 5 min.
4. When you plan to measure a Ca^{2+}-dependent activation profile, choose a saturating GCAP concentration and vary the free $[Ca^{2+}]$.
5. Start the incubation reaction by adding 20 µL of the GC-assay buffer (see protocol 1). Perform the incubation and analysis of data as described in protocol 1. When working with ROS membranes, perform all steps under very dim red light until you quench the reaction. When working with heterologously expressed GCs, perform all steps under room light.

GCAP1 and GCAP2 display different activation profiles (40), when the activity of ROS-GCs in ROS membranes is measured at different free $[Ca^{2+}]$. GCAP1 activates ROS-GCs at higher $[Ca^{2+}]$ than GCAP2 does (compare figures 6.7a and 6.7b). In addition to this difference in Ca^{2+} sensitivity, both GCAPs differ in the influence of their myristoyl groups on their catalytic efficiency, their monomer–dimer equilibria and their target recognition sites in ROS-GC1 (40,59–61,67,68).

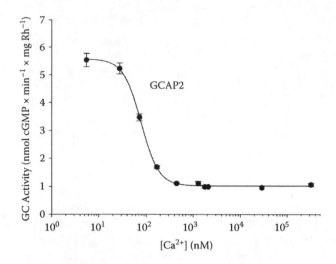

FIGURE 6.7A Reconstitution of ROS-GCs in rod outer segment membranes with 3 μM recombinant myristoylated GCAP2. GC activity was measured at the indicated free [Ca^{2+}]. Activation was half maximal at 78 nM.

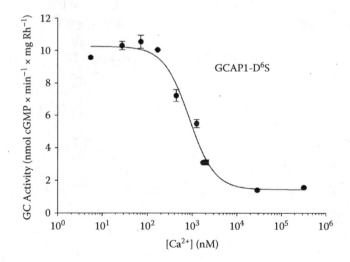

FIGURE 6.7B Reconstitution of ROS-GCs in rod outer segment membranes with 3 μM myristoylated GCAP1-D6S. GC activity was measured at the indicated free [Ca^{2+}]. Activation was half maximal at 855 nM.

ACKNOWLEDGMENTS

We thank Doris Höppner-Heitmann (IBI-1, Forschungszentrum Jülich, Germany) and Werner Säftel (University of Oldenburg) for excellent technical assistance. Research in the laboratory of Karl-Wilhelm Koch was funded by several grants of the Deutsche Forschungsgemeinschaft (DFG), an INTAS grant, and a grant from the EWE-Stiftung. We also acknowledge support from the Forschunsgszentrum Jülich.

REFERENCES

1. Sunahara, R.K. and Taussig, R., Isoforms of mammalian adenylyl cyclase: multiplicities of signaling, *Mol Interventions*, 2, 168, 2002.
2. Koesling, D., Studying the structure and regulation of soluble guanylyl cyclase, *Methods*, 19, 485, 1999.
3. Lucas, K.A., Pitari, G.M., Kazerounian, S., Ruiz-Stewart, I, Park, J., Schulz, S., Chepenik, K.P., and Waldman, S.A., Guanylyl cyclases and signaling by cyclic GMP, *Pharmacol Rev*, 52, 375, 2000.
4. Wedel, B.J. and Garbers, D.L., The guanylyl cyclase familiy at Y2K, *Annu Rev Physiol*, 63, 215, 2001.
5. Potter, L.R., Domain analysis of human transmembrane guantylyl cyclase receptors: implications for regulation, *Front Biosci*, 10, 1205, 2005.
6. Fitzpatrick, D.A., O'Halloran, D.M., and Burnell, A.M., Multiple lineage specific expansions within the guanylyl cyclase gene family, *BMC Evolut Biol*, 6, 1, 2006.
7. Pugh, E.N., Jr., Duda, T., Sitaramayya, A., and Sharma, R.K., Photoreceptor guanylate cyclases: a review, *Biosci Rep*, 17, 429, 1997.
8. Koch, K.-W., Duda, T., and Sharma, R.K., Photoreceptor specific guanylate cyclases in vertebrate phototransduction, *Mol Cell Biochem*, 230, 97, 2002.
9. Sharma, R.K., Duda, T., Venkataraman, V., and Koch, K., Calcium-modulated mammalian membrane guanylate cyclase ROS-GC transduction machinery in sensory neurons: a universal concept, *Curr Top Biochem Res*, 6, 111, 2004.
10. Francis, S.H. and Corbin, J.D., Structure and function of cyclic nucleotide-dependent protein kinases, *Annu Rev Physiol*, 56, 237, 1994.
11. Lincoln, T.M., Dey, N., and Sellak, H., Signal transduction in smooth muscle. Invited Review: cGMP-dependent protein kinase signaling mechanisms in smooth muscle: from the regulation of tone to gene expression, *J Appl Physiol*, 91, 1421, 2001.
12. Kaupp, U.B. and Seifert, R., Cyclic nucleotide-gated ion channels, *Physiol Rev*, 82, 769, 2002.
13. Kramer, R.H. and Molokanova, E., Modulation of cyclic nucleotide-gated channels and regulation of vertebrate phototransduction, *J Exp Biol*, 204, 2921, 2001.
14. Sonderling, S.H. and Beavo, J.A., Regulation of cAMP and cGMP signaling: new phosphodiesterases and new functions, *Curr Opin Cell Biol*, 12, 174, 2000.
15. Rybalkin, S.D., Yan, C., Bornfeldt, K.E., and Beavo, J.A., Cyclic GMP phosphodiesterases and regulation of smooth muscle function, *Circ Res*, 93, 280, 2003.
16. de Rooij, J., Zwartkruis, F.J.T., Verheijen, M.H.G., Cool, R.H., Nijman, S.M.B., Wittinghofer, A., and Bos, J.L., Epac is a Rap1 guanine-nucleotide-exchange factor directly by cyclic AMP, *Nature*, 396, 474, 1998.
17. Margulis, A., Sharma, R.K., and Sitaramayya, A., Nitroprusside-sensitive and intensive guanylate cyclases in retinal rod outer segments, *Biochem Biophys Res Commun*, 185, 909, 1992.
18. Ahmad, I. and Barnstable, C.J., Differential laminar expression of particulate and soluble guanylate cyclase genes in rat retina, *Exp Eye Res*, 56, 51, 1993.
19. Koch, K.-W., Lambrecht, H.-G., Haberecht, M., Redburn, D., and Schmidt, H.H.H.W., Functional coupling of a Ca^{2+}/calmodulin-dependent nitric oxide synthase and a soluble guanylyl cyclase in vertebrate photoreceptor cells, *EMBO J*, 13, 3312, 1994.
20. Haberecht, M.F., Schmidt, H.H.H.W., Mills, S.L., Massey, S.C., Nakane, M., and Redburn-Johnson, D.A., Localization of nitric oxide synthase, NADPH diaphorase and soluble guanylyl cyclase in adult rabbit retina, *Vis Neurosci*, 15, 881, 1998.
21. Gotzes, S., de Vente, J., and Müller, F., Nitric oxide modulates cGMP levels in neurons of the inner and outer retina in opposite ways, *Vis Neurosci*, 15, 945, 1998.

22. Donovan, M., Carmody, R.J., and Cotter, T.G., Light-induced photoreceptor apoptosis in vivo requires neuronal nitric-oxide synthase and guanylate cyclase activity and is caspase-3-independent, *J Biol Chem*, 276, 23000, 2001.

23. Olshevskaya, E.V., Ermilov, A.N., and Dizhoor, A.M., Factors that affect regulation of cGMP synthesis in vertebrate photoreceptors and their genetic link to human retinal degeneration, *Mol Cell Biochem*, 230, 139, 2002.

24. Palczewski, K., Sokal, I., and Baehr, W., Guanylate cyclase-activating proteins: structure, function, and diversity, *Biochem Biophys Res*, 322, 1123, 2004.

25. Koch, K.-W., GCAPs, the classical neuronal calcium sensors in the retina, *Calcium Binding Proteins*, 1, 3, 2006.

26. Rollin, R., Mediero, A., Roldán-Pallarés, M., Fernández-Cruz, A., and Fernández-Durango, R., Natriuretic peptide system in the human retina, *Mol Vision*, 10, 15, 2004.

27. Yu, Y.-Ch., Cao, L.-H., and Yang, X.-L., Modulation by brain natriuretic peptide of GABA receptors on rat retinal ON-type bipolar cells, *J Neurosci*, 26, 696, 2006.

28. Pugh, E.N., Jr. and Lamb, T.D., Phototransduction in vertebrate rods and cones: molecular mechanisms of amplification, recovery and light adaption, *Handbook of Biological Physics,* Vol. 3 (Eds. Stavenga, DeGrip, and Pugh Jr., Elsevier Science B.V.) 2000, chap. 5.

29. Burns, M.E. and Baylor, D.A., Activation, deactivation, and adaptation in vertebrate photoreceptor cells, *Annu Rev Neurosci*, 24, 779, 2001.

30. Fleischmann, D. and Denisevich, M., Guanylate cyclase of isolated bovine retinal rod axonemes, *Biochemistry*, 18, 5060, 1979.

31. Koch, K.-W. and Stryer, L., Highly cooperative feedback control of retinal rod guanylate cyclase by calcium ions, *Nature*, 334, 64, 1988.

32. Gorczyca, W.A., van Hooser, J.P., and Palczewski, K., Nucleotide inhibitors and activators of retinal guanylyl cyclase, *Biochemistry*, 33, 3217, 1994.

33. Koch, K.-W., Eckstein, F., and Stryer L., Stereochemical course of the reaction catalyzed by guanylate cyclase from bovine retinal rod outer segments, *J Biol Chem*, 265, 9659, 1990.

34. Wolbring, G. and Schnetkamp, P.P.M., Activation by PKC of the Ca^{2+}-sensitive guanylyl cyclase in bovine retinal rod outer segments measured with an optical assay, *Biochemistry*, 34, 4689, 1995.

35. Sharma, R.K., Marala, R.B., and Duda, T.M., Purification and characterization of the 180-kDa membrane guantylate cyclase containing atrial natriuretic factor receptor from rat adrenal gland and its regulation by protein kinase C, *Steroids*, 53, 437, 1989.

36. Koch, K.-W., Purification and identification of photoreceptor guanylate cyclase, *J Biol Chem*, 266, 8634, 1991.

37. Tsien, R. and Pozzan, T., Measurement of cytosolic free Ca^{2+} with Quin2, *Methods Enzymol*, 172, 230, 1989.

38. Lambrecht, H.-G. and Koch, K.-W., A 26 kd calcium binding protein from bovine rod outer segments as modulator of photoreceptor guanylate cyclase, *EMBO J*, 10, 793, 1991.

39. Peshenko, I.V. and Dizhoor, A.M., Ca^{2+}-and Mg^{2+}-binding properties of GCAP-1: evidence that Mg^{2+}-bound form is the physiological activator of photoreceptor guanylyl cyclase, *J Biol Chem*, 281, 23830, 2006.

40. Hwang, J.-Y., Lange, C., Helten, A., Höppner-Heitmann, D., Duda, T., Sharma, R.K., and Koch, K.-W., Regulatory modes of rod outer segment membrane guanylate cyclase differ in catalytic efficiency and Ca^{2+}-sensitivity, *Eur J Biochem*, 270, 3814, 2003.

41. Hayashi, F. and Yamazaki, A., Polymorphism in purified guanylate cyclase from vertebrate rod photoreceptors, *Biochemistry*, 88, 4746, 1991.

42. Margulis, A., Goraczniak, R.M., Duda, T., Sharma, R.K., and Sitaramayya, A., Structural and biochemical identity of retinal rod outer segment membrane guanylate cyclase, *Biochem Biophys Res Commun*, 194, 855, 1993.

43. Koch, K.-W., Stecher, P., and Kellner, R., Bovine retinal rod guanyl cyclase represents a new N-glycosylated subtype of membrane-bound guanyl cyclases, *Eur J Biochem*, 222, 589, 1994.
44. Shyjan, A.W., de Sauvage, F.J., Gillett, N.A., Goeddel, D.V., and Lowe, D.G., Molecular cloning of a retina-specific membrane guanylyl cyclase, *Neuron*, 9, 727, 1992.
45. Goraczniak, R.M., Duda, T., Sitaramayya, A., and Sharma, R.K., *Biochem J*, 302, 455, 1994.
46. Lowe, D.G., Dizhoor, A.M., Liu, K., Gu, Q., Spencer, M., Laura, R., Lu, L., and Hurley, J.B., *Proc Natl Acad Sci U.S.A.*, 92, 5535, 1995.
47. Yang, R.-B., Foster, D.C., Garbers, D.L., and Fülle, H.-J., Two membrane forms of guanylyl cyclase found in the eye, *Proc Natl Acad Sci U.S.A.*, 92, 602, 1995.
48. Goraczniak, R., Duda, T., and Sharma, R.K., Structural and functional characterization of a second subfamily member of the calcium-modulated bovine rod outer segment membrane guanylate cyclase, ROS-GC2, *Biochem Biophys Res Commun*, 234, 666, 1997.
49. Yang, R.-B. and Garbers, D.L., Two eye guanylyl cyclases are expressed in the same photoreceptor cells and form homomers in preference to heteromers, *J Biol Chem*, 272, 13738, 1997.
50. Chen, C. and Okayama, H., High-efficiency transformation of mammalian cells by plasmid DNA, *Mol Cell Biol*, 7, 2745, 1987.
51. Senin, I.I., Koch, K.-W., Akhtar, M., and Philippov, P.P., Ca²⁺-dependent control of rhodopsin phoshorylation: recoverin and rhodopsin kinase, *Adv Exp Med Biol*, 514, 69, 2002.
52. Hsu, Y.-T. and Molday, R.S., Modulation of the cGMP-gated channel of rod photoreceptor cells by calmodulin, *Nature*, 361, 76, 1993.
53. Weitz, D., Zoche, M., Müller, F., Beyermann, M., Körschen, H.-G., Kaupp, U.B., and Koch, K.-W., Calmodulin controls the rod photoreceptor CNG channel through an unconventional binding site in the N-terminus on the ß-subunit, *EMBO J*, 17, 2273, 1998.
54. Grunwald, M.E., Yu, W.-P., Yu, H.-H., and Yau, K.-W., Identification of a domain on the ß-subunit of the rod cGMP-gated cation channel that mediates inhibition by calcium-calmodulin, *J Biol Chem*, 273, 9148, 1998.
55. Imanishi, Y., Yang, L., Sokal, I., Filipek, S., Palczewski, K., and Baehr, W., Diversity of guanylate cyclase-activating proteins (GCAPs) in teleost fish: charcterization of three novel GCAPs (GCAP4, GCAP5, GCAP7) from zebrafish (*Danio rerio*) and prediction of eight GCAPs (GCAP1-8) in pufferfish (*Fugu rubripes*), *J Mol Evol*, 59, 204, 2004.
56. Gorczyca, W.A., Gray-Keller, M.P., Detwiler, P.B., and Palczewski, K., Purification and physiological evaluation of a guanylate cyclase activating protein from retinal rods, *Proc Natl Acad Sci U.S.A.*, 91, 4014, 1994.
57. Dizhoor, A.M., Lowe, D.G., Olshevskaya, E.V., Laura, R.P., and Hurley, J.B., The human photoreceptor membrane guanylyl cyclase, RetGC, is present in outer segments and is regulated by calcium and a soluble activator, *Neuron*, 12, 1345, 1994.
58. Frins, S., Bönigk, W., Müller, F., Kellner, R., and Koch, K.-W., Functional characterization of a guanylyl cyclase-activating protein from vertebrate rods, *J Biol Chem*, 271, 8022, 1996.
59. Lange, C., Duda, T., Beyermann, M., Sharma, R.K., and Koch, K.-W., Regions in vertebrate photoreceptor guanylyl cyclase ROS-GC1 involved in CA²⁺-dependent regulation by guanylyl cyclase-activating protein GCAP-1, *FEBS Lett*, 460, 27, 1999.
60. Sokal, I., Haeseleer, F., Arendt, A., Adman, E.T., Hargrave, P.A., and Palczewski, K., Identification of a guanylyl cyclase-activating protein-binding site within the catalytic domain of retinal guanylyl cyclase 1, *Biochemistry*, 38, 1387, 1999.
61. Krylov, D.M. and Hurley, J.B., Identification of proximate regions in a complex of retinal guanyly cyclase 1 and guanylyl cyclase-activating protein-1 by a novel mass spectrometry-based method, *J Biol Chem*, 276, 30648, 2001.

62. Duda, T., Fik-Rymarkiewicz, E., Venkataraman, V., Kishnan, R., Koch, K.-W., and Sharma, R.K., The calcium-sensor guanylate cyclase activating protein type 2 specific site in rod outer segment membrane guanylate cyclase type 1, *Biochemistry*, 44, 7336, 2005.

63. Otto-Bruc, A., Buczylko, J., Surgucheva, I., Subbaraya, I., Rudnicka-Nawrot, M., Crabb, J.W., Arendt, A., Hargrave, P.A., Baehr, W., and Palczewski, K., Functional reconstitution of photoreceptor guanylate cyclase with native and mutant forms of guanylate cyclase-activating protein 1, *Biochemistry*, 36, 4295, 1997.

64. Sokal, I., Otto-Bruc, A.E., Surgucheva, I., Verlinde, C.L.M., Wang, C.-K., Baehr, W., and Palczewski, K., Conformational changes in guanylyl cyclase-activating protein 1 (GCAP1) and its tryptophan mutants as a function of calcium concentration, *J Biol Chem*, 274, 19829, 1999.

65. Sokal, I., Li, N., Klug, C.S., Filipek, S.B., Hubbell, W.L., Baehr, W., and Palczewski, K., Calcium-sensitive regions of GCAP1 as observed by chemical modifications, fluorescence, and EPR spectroscopies, *J Biol Chem*, 276, 43361, 2001.

66. Hwang, J.-Y., Schlesinger, R., and Koch, K.-W., Calcium-dependent cysteine reactivities in the neuronal calcium sensor guanylate cyclase-activating protein 1, *FEBS Lett*, 508, 355, 2001.

67. Ermilov, A.N., Olshevskaya, E.V., and Dizhoor, A.M., Instead of binding calcium, one of the EF-hand structures in guanylyl cyclase activating protein-2 is required for targeting photoreceptor guanylyl cyclase, *J Biol Chem*, 276, 48143, 2001.

68. Hwang, J.-Y, Schlesinger, R., and Koch, K.-W., Irregular dimerization of guanylate cyclase-activating protein 1 mutants causes loss of target activation, *Eur J Biochem*, 271, 3785, 2004.

69. Duda, T. and Koch, K.-W., Retinal diseases linked with photoreceptor guanylate cyclase, *Mol Cell Biochem*, 230, 129, 2002.

70. Mendez, A., Burns, M.E., Sokal, I., Dizhoor, A.M., Baehr, W., Palczewski, K., Baylor, D.A., and Chen, J., Role of guanylate cyclase-activating proteins (GCAPs) in setting the flash sensitivity of rod photoreceptors, *Proc Natl Acad Sci U.S.A.*, 98, 9948, 2001.

71. Howes, K.A., Pennesi, M.E., Sokal, I., Church-Kopish, J., Schmidt, B., Margolis, D., Frederick, J.M., Rieke, F., Palczewski, K., Wu, S.M., Detwiler, P.B., and Baehr, W., GCAP1 rescues rod photoreceptor response in GCAP1/GCAP2 knockout mice, *EMBO J*, 21, 1545, 2002.

72. Burns, M.E., Mendez, A., Chen, J., and Baylor, D.A., Dynamics of cyclic GMP synthesis in retinal rods, *Neuron*, 36, 81, 2002.

73. Pennesi, M.E., Howes, K.A., Baehr, W., and Wu, S.M., Guanylate cyclase-activating protein (GCAP) 1 rescues cone recovery kinetics in GCAP1/GCAP2 knockout mice, *Proc Natl Acad Sci U.S.A.*, 100, 6783, 2003.

74. Duda, T., Goraczniak, R., Surgucheva, I., Rudnicka-Nawrot, M., Gorczyca, W.A., Palczewski, K., Sitaramayya, A., Baehr, W., and Sharma, R.K., Calcium modulation of bovine photoreceptor guanylate cyclase, *Biochemistry*, 35, 8478, 1996.

75. Schrem, A., Lange, C., Beyermann, M., and Koch K.-W., Identification of a domain in guanylyl cyclase-activating protein 1 that interacts with a complex of guanylyl cyclase and tubulin in photoreceptors, *J Biol Chem*, 274, 6244, 1999.

76. Yamazaki, A., Yu, H., Yamazaki, M., Honkawa, H., Matsuura, I., Usukura, J., and Yamazaki, R.K., A critical role for ATP in the stimulation of retinal guanylyl cyclase by guanylyl cyclase-activating proteins, *J Biol Chem*, 278, 33150, 2003.

77. Yamazaki, M., Usukura, J., Yamazaki, R.K., and Yamazaki, A., ATP binding is required for physiological activation of retinal guanylate cyclase, *Biochem Biophys Res Commun*, 338, 1291, 2005.

78. Yamazaki, A., Yamazaki, M., Yamazaki, R.K., and Usukura, J., Illuminated rhodopsin is required for strong activating of retinal guanylate cyclase by guanylate cyclase-activating proteins, *Biochemistry*, 45, 1899, 2006.

79. Hwang, J.-Y. and Koch, K.-W., Calcium- and myristoyl-dependent properties of guany-late cyclase-activating protein-1 and protein-2, *Biochemistry*, 41, 13021, 2002.
80. Rudnicka-Nawrot, M., Surgucheva, I., Hulmes, J.D., Haeseleer, F., Sokal, I., Crabb, J.W., Baehr, W., and Palczewski, K., Changes in biological activity and folding of guanylate cyclase-activating protein 1 as a function of calcium, *Biochemistry*, 37, 248, 1998.
81. Duda, T., Koch, K.-W., Venkataraman, V., Lange, C., Beyermann, M., and Sharma, R.K., Ca^{2+} sensor S100β-modulated sites of membrane guanylate cyclase in the photo-receptor-bipolar synapse, *EMBO J*, 21, 2547, 2002.

7 Transgenic Strategies for Analysis of Photoreceptor Function

Janis Lem and Kibibi Rwayitare

CONTENTS

7.1 INTRODUCTION

7.1.1 BACKGROUND AND AIMS

The recently completed sequencing of the human and mouse genomes has propelled research to focus on the study of gene function in normal physiology and disease processes. Transgenic mice have an extensive track record as powerful tools for elucidating gene structure and function. Early transgenic mouse studies focused on genetic dissection of gene structure, with an emphasis on identifying genetic elements of genes that regulated cell- and tissue-specific expression. Over the past two decades, with the discovery of novel genes, the focus of transgenic mouse research has increasingly shifted toward examination of gene function. Technological advances currently allow direct manipulation of the endogenous gene to examine pathological effects in vivo.

Transgenic mouse models allow an integrated approach to the study of human disease processes. Mouse models permit detailed characterization of changes in tissue morphology, biochemical pathways, and physiologic changes related to the disease state. These can be determined along a developmental time course, potentially leading to the development of diagnostic biomarkers that can define early or late stages of disease. Identifying the earliest cellular changes can help define the molecular basis of genetic disease, distinguishing secondary events that may be an effect rather than a cause of disease. Understanding the molecular mechanisms can lead to the design of rational therapies. Finally, transgenic mouse models provide a resource to test new therapies for effective treatment of human diseases.

For several reasons, the mouse is the most common transgenic animal model produced. Compared to large mammalian animal models (e.g., monkeys, sheep, and cows), mice are relatively cheap to purchase and house. Although mouse models do not always accurately model human disease, more often than not they share conserved biochemical pathways that reveal molecular mechanisms shared with other vertebrates. The most powerful reason to use mice is the well-characterized genetics of the mouse. Inbred mouse lines eliminate genetic background variations that confound interpretation of genetic inheritance in human disease and in most other mammalian models. Having transgenic mice on a defined genetic background is key to the identification of modifier genes that can influence disease susceptibility. Furthermore, a uniform genetic background is essential for definitive analysis of the contribution of epigenetic factors to disease pathology.

The term *transgene* denotes the genetic material that is introduced into an organism. The two major methods used for the introduction of genetic material into mice are by *pronuclear microinjection* or by *gene targeting* in embryonic stem (ES) cells, followed by injection of the targeted ES cells into blastocysts. The major difference between the two methods is that in pronuclear microinjections, genetic material randomly integrates into the mouse genome in an additive manner. In contrast, gene targeting uses homologous recombination to *substitute* genetic material nonrandomly into the mouse genome.

In this chapter, we present an overview of transgenic strategies for the analysis of gene function. We will discuss several commonly used transgenic strategies for analyzing gene function, including total ablation of endogenous protein expression, reduced levels of endogenous protein expression, introduction of a mutation in the endogenous gene, overexpression of normal or mutant protein, and controlled, regulated expression of transgene expression. Each of the methodologies has strengths and limitations that must be considered in the design and interpretation of a transgenic experiment. This chapter does not provide experimental details on ES cell culture methods (1,2) or detailed technical methodologies for the preparation of embryos for microinjection, as those are easily found in manuals on transgenic methodology (1,3,4). Instead, this review focuses on considerations for selecting an experimental strategy.

7.1.2 OVERVIEW OF METHODOLOGIES

7.1.2.1 Pronuclear Microinjection

In pronuclear microinjections, linearized transgene DNA is microinjected into fertilized single-cell wild-type mouse embryos (figure 7.1). Characteristic of pronuclear microinjections is the random integration of transgene DNA into the mouse genome. The integrated transgene is subject to regulation by DNA surrounding the site of insertion. Consequently, the transgene is not completely regulated in a spatial or temporal manner that models that of the endogenous gene. Transgenic mice produced in this manner can show substantial variation in spatial expression patterns and the level of transgene expression. This property can be used to advantage by correlating the level of expression with severity of phenotype. Pronuclear injection methodology lends itself to certain types of studies, such as overexpression of wild-type or dominant mutant genes (discussed in the following text). However, it is not possible to study recessive mutations on a wild-type genetic background by using this methodology.

For pronuclear microinjections, the transgene comprises three parts: an upstream regulatory promoter, the structural gene, and the polyadenylation signal. The structural gene may derive from genomic DNA, complementary DNA (cDNA), or a chimeric fusion of two genes. However, it should be noted that the presence of introns are

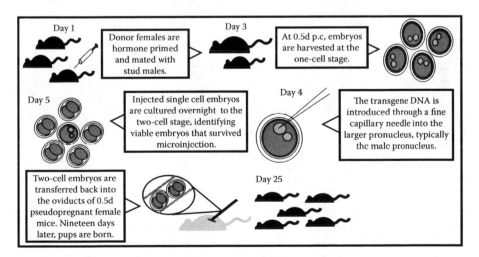

FIGURE 7.1 (See accompanying color CD.) Schematic of a pronuclear microinjection. On day one, 3- to 4-weeks-old donor females are injected with pregnant mare's serum. On day three, females receive a dose of human chorionic gonadotropin and are mated with stud males. The following day, embryos at the one-cell stage are harvested for injection. Purified linearized transgene DNA is injected into the male pronucleus of the embryo. After injection, the embryos are incubated in a 5% CO_2 incubator overnight. Embryos that develop to the two-cell stage embryos are transferred into 0.5 day postcoitus pseudopregnant females. Potential founders are born 19 days later.

reported to increase transcriptional efficiency (5). There are few restrictions on species specificity for gene expression (i.e., genes of mouse, rat, human, bovine, etc., origin may be used for microinjection into mouse embryos). The promoter defines the tissue and cell specificity as well as transgene expression level. If possible, it is desirable to use a promoter that has previously been tested in an animal model for cell specificity and promoter strength. However, regulatory elements that define temporal and spatial expression may be as far as 50 kb upstream (6,7). The unintentional deletion of an upstream regulatory element can result in transgene expression in a subset of cells that normally express the gene of interest, or show early or delayed temporal expression. The effects of these changes can also contribute to the observed phenotype.

Expression of the transgene should be tested in a cell culture system prior to microinjection whenever possible, especially when producing chimeric fusion genes, to avoid potential issues such as artificial start or splice sites or changes in translation frame that are inadvertently introduced during the cloning process. Such changes could prevent expression of a functional transgene protein.

7.1.2.2   Gene Targeting in ES Cells (Followed by Injection into Blastocysts)

Homologous recombination directly targets the endogenous gene and substitutes existing genetic material. Gene targeting may involve the introduction of a null muta-tion to produce a "knockout" transgenic mouse (8–17) (that lacks functional protein), the insertion of a dominant or recessive mutation within the endogenous gene to pro-duce a "knockin" transgenic mouse (18), or the replacement of a mutant gene with a normal wild-type gene to correct a genetic defect (19). In homologous recombinants, the transgene is regulated under its normal endogenous promoter in a manner that is temporally and spatially consistent with the wild-type endogenous gene. However, the production of homologously recombinant transgenic mice is a labor-intensive process. First, ES cells with the desired homologous recombination event must be produced (1,2,20). Most embryonic stem cells are derived from one of several 129SV mouse substrains. ES cells from other mouse strains, including C57BL/6, DBA-1, FVB/n, Balb/C, 129 × C57Bl/6, and 129 × Balb/C, are also commercially avail-able (table 7.1). ES cells are fastidious and must be meticulously cultured to retain their pluripotency. With continued passage, all ES cell lines become aneuploid and will eventually lose their ability for germ-line transmission. It is important to know for your particular ES cell line the recommended passage numbers that will retain germ-line competence. Using ES cells beyond the recommended passage number will decrease the likelihood of successful germ-line transmission.

After ES cell clones with the desired homologous recombination event are iden-tified, they are injected into the cavity of blastocyst stage embryos (figure 7.2). The injected embryos are surgically reimplanted into the uterus of foster mothers and allowed to develop to term (1). Offspring that have successfully taken up the ES cells produce chimeric founder mice. Additional rounds of breeding are required to obtain mice heterozygous or homozygous for the transgene mutation before animals are ready for phenotypic analysis.

Breeding chimeric mice to homozygosity is a time-consuming process. Recently, Valenzuela and colleagues have shown that laser-assisted injection of ES cells into

TABLE 7.1
ES Cell Lines and BAC Libraries

Source	ES Cell Type	BAC Library
Specialty Media	C57BL/6	Invitrogen, CHORI[a]
www.specialtymedia.com	DBA-1	Stratagene, CHORI
	129SvEv	
Taconic	W4 (129S6/SVEvTac)	Invitrogen
www.taconic.com		
ATCC	ES-C57BL/6	
http://stemcells.atcc.org	J1 (129S4/SvJae)	
	RW.4 (129X1/SvJ)	
	7ACS/EYFP	
	(129X1x129S1)	
	R1/E (129X1 x 129S1)	
Thrombogenics	Balb/C	Stratagene
www.thromb-x.com	129SvEv	Invitrogen
	C57BL/6	
	FVB/N	
Open Biosystems	C57BL/6	Invitrogen
www.openbiosystems.com	129	

[a] bacpac.chori.org

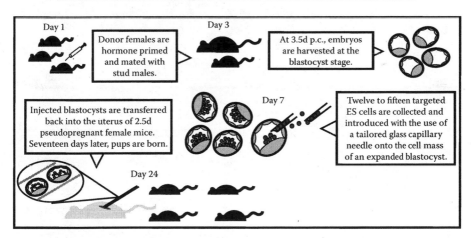

FIGURE 7.2 (See accompanying color CD.) Schematic of a blastocyst injection. On day one, 3- to 4-week-old donor females are injected with pregnant mare's serum. On day three, females receive a dose of human chorionic gonadotropin and are mated with stud males. On day seven, embryos at the blastocyst stage are harvested and injected with targeted embryonic stem cells. The injected blastocysts are transferred into the uterus of 2.5 day postcoitus pseudopregnant females. Seventeen days later, chimeric pups are born.

eight-cell embryos instead of blastocysts yielded transgenic mice that are almost wholly derived from ES cells. This permits phenotyping in the first generation of mice, because the transgene is essentially present in the heterozygous state. This represents a considerable savings in terms of time and effort (21).

7.2   EXPERIMENTAL STRATEGIES

7.2.1   KNOCKOUT MICE: ABLATION OF ENDOGENOUS PROTEIN EXPRESSION

Complete ablation of a protein is most often achieved by introduction of a null mutation within the endogenous gene. The null mutation is introduced by homologous
recombination in embryonic stem cells (1). The frequency of homologous recombination is 1 in 10^6 to 10^7 cells. Thus, it is beneficial to optimize for successful
homologous recombination in ES cells. The following conditions are recommended:
The targeting construct should be made from mouse DNA originating from the same
mouse strain as the ES cell line (22) (i.e., *isogenic* DNA). Bacterial artificial chromosome (BAC) libraries have been prepared from some 129Sv and C57Bl/6 mouse
lines, facilitating rapid cloning of genes (23,24) (table 7.1). For homologous recombination to occur, flank the mutation with isogenic DNA with total homology between
5 and 10 kb in length (25,26) (figure 7.3). For a knockout targeting construct, a positive drug-selectable marker such as neomycin, hygromycin, or puromycin is inserted
to replace exons known to be critical for gene function or to replace a region, including the translation start site. Inclusion of a positive drug-selectable marker increases
the frequency of homologous recombination to 1 in 10^2 to 10^3 cells. Insertion of a
negative selectable marker such as thymidine kinase or diphtheria toxin is reported
to increase the frequency of homologous recombinants by approximately twofold,
but also has a higher frequency of undesired rearrangements (26). Consideration
must also be given as to whether a partially functional protein product might be
produced instead of the desired null mutation.

Targeting constructs are typically introduced into ES cells by electroporation
(2,20). After 5 to 7 days of drug selection, clonal isolates are picked, and each clone

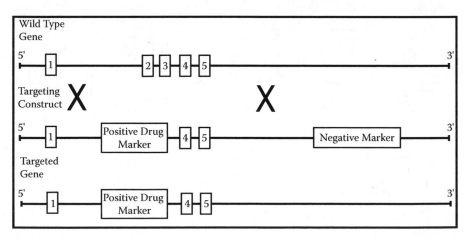

FIGURE 7.3 Schematic of a knockout strategy by homologous recombination. Using
isogenic DNA, functional domains of the wild-type gene known to be required for function are substituted with a positive drug-selectable marker such as neomycin, hygromycin, or
puromycin. A negative selectable marker such as thymidine kinase or diphtheria toxin may
be added to the 5′ or 3′ end of the targeting construct. Homology arms flanking the selectable
marker should be 5 to 10 kb in length.

tested for the desired homologous recombination event by PCR analysis and confirmed by Southern blot analysis. Clones with the desired homologous recombination event are then microinjected into blastocysts to produce chimeric mice (figure 7.2). If competent for germ-line transmission, first generation (G1) offspring derived from chimeric founders are heterozygous for the mutation and are bred to homozygosity. Several genes of the phototransduction cascade have been knocked out, including rod opsin (12,14), transducin α-subunit (8), cGMP phosphodiesterase gamma-subunit (27), rhodopsin kinase (11), and arrestin (9), to name but a few.

Complete knockouts can also reveal a developmental role for a protein, leading to embryonic or perinatal lethality or developmental abnormalities that may complicate the interpretation of function in adult tissues. To study gene function in adult tissues without the complication of embryonic lethality, it may be desirable to manipulate expression in adult tissues using conditional transgene expression methods (see following text).

7.2.2 TRANSGENE KNOCKDOWN OF ENDOGENOUS PROTEIN (HYPOMORPHS)

Heterozygous mice with a null mutation in a single allele often have a reduced level of protein expression. For example, heterozygous rhodopsin null mutant mice have approximately 50% of wild-type levels of rhodopsin (14,28). However, the reduction is not always 50% of wild-type levels, as the remaining functional allele may undergo a compensatory upregulation of gene expression. For instance, transgenic mice heterozygous for the transducin α-subunit null mutation upregulate the remaining wild-type allele to near normal levels (8).

RNA interference methods have been used successfully in transgenic mice (29,30), with germ-line transmission. In brief, expression of short hairpin RNAs (shRNA) from DNA vectors results in the production of double-stranded premicroRNAs. The endogenous Rnase III enzyme Dicer processes the double-stranded RNAs into ~22 nucleotide short interfering RNAs (siRNAs). The siRNAs produced upon cleavage by Dicer are incorporated into the *RNA-inducing silencing complex (RISC)*. The siRNAs target the RNA to be cleaved by the Argonaute protein within the RISC complex.

The shRNA DNA vectors have been successfully introduced into the germ line of mice by pronuclear injection, stable transfection into embryonic stem cells followed by injection into blastocysts, or by direct lentiviral infection of ES cells or infection of embryos (figure 7.4). These methods result in the random integration of the shRNA construct into the mouse genome. Each method has its advantages and limitations, as discussed in the following text.

The DNA vectors may be stably introduced by direct pronuclear microinjection of DNA vector shRNAs into one-cell embryos or by lentiviral infection (31) of embryos. Two methods are commonly used for lentiviral infection of embryos. Both methods are highly efficient for the production of transgenic mice (~70 to 80% of offspring are transgenic). The first method involves the removal of the *zona pellucida* by enzymatic methods, followed by lentiviral transduction. This method is relatively simple technically and does not require the use of micromanipulators. However, removal of the zona pellucida is detrimental to the survival of embryos, with significantly fewer

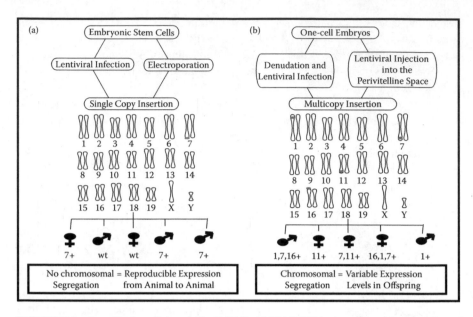

FIGURE 7.4 Introduction of short-hairpin RNA (shRNA) DNA vectors into the germ line of mice. (a) shRNA DNA vectors can be introduced by electroporation or lentiviral transduction of embryonic stem (ES) cells. The majority of clonal ES cell lines will integrate the shRNA DNA vector as a single copy. Injected ES cells will yield founders containing a single integration site. (b) ShRNA DNA vectors can also be introduced by the removal of the *zona pellucida* (denudation) by enzymatic methods followed by lentiviral transduction or by direct microinjection into the perivitelline space surrounding the embryo. These methods will yield founders with multiple integration sites. Chromosomal segregation in subsequent generations will result in offspring with multiple integration sites, each with variable levels of transgene expression.

embryos surviving. The second method involves direct injection of lentivirus into the perivitelline space surrounding the embryo. This method avoids removal of the zona pellucida, but requires the use of micromanipulators and is technically more difficult. Lentiviral transduction results in multiple integration sites of the lentiviral transgene, thus giving the high percentage of transgenic founders. However, breeding of founder mice results in the segregation of chromosomes carrying the integrated transgenes that can continue over several generations of breeding. Second- to third-generation offspring can have variable levels of transgene expression, as each has a different integration site. This complicates interpretation of the data, as there is wide animal-to-animal variability.

Transduction of embryonic stem cells with lentivirus has the advantage that a single transgene is integrated into the ES cell. Clonal ES cell lines, each with a single integration site, are subsequently injected into blastocysts to produce chimeric mice. Several different clones may be injected to obtain mouse lines with different integration sites and different levels of expression. Although more labor intensive, this method has the advantage that each founder yields offspring with the same transgene integration site and reproducibility upon phenotypic analysis.

A range of knockdowns is often produced that is dependent upon the specificity of the shRNA and its level of expression upon stable integration into the genome.

Thus, it is useful to correlate the degree of knockdown with the severity of pheno-type. Knockdown is seldom complete, although it is possible to knock down expres-sion to 5% of WT expression levels. A potential problem with this approach is that highly expressed proteins can be difficult to knock down. A phenotype will only be seen if sufficient knockdown is induced. Thus, if 10 to 20% of WT levels are suf-ficient to maintain wild-type function, a phenotype may not be observed. There is much interest in using RNA interference methods for therapeutic treatment of retinal degeneration (32). Current studies support feasibility for this approach (33–36).

7.2.3 KNOCKIN MUTATIONS

For some gene products, overexpression or underexpression of a wild-type gene results in a pathological phenotype. For instance, overexpression of wild-type rhodopsin (37) or reduced expression (19) results in retinal pathology. Thus, it is important to properly regulate the level of expression of such genes from its normal endogenous promoter to separate the effects of the mutation itself from effects arising from altered levels of transgene expression. The knockin mutation is introduced by homologous recombina-tion, followed by deletion of the drug-selectable marker. A rhodopsin palmitoylation mutation was studied using this methodology (18). These animals revealed a subtle regulatory role for palmitoylation in phototransduction that most likely would have been masked had the mutation been introduced by pronuclear microinjection. Simi-larly, farnesylation of the gamma subunit of transducin was studied, revealing regula-tory effects on light adaptation (38). Knockin mutations in several other photoreceptor genes support their role in retinal degenerative disease (39–44).

7.2.4 OVEREXPRESSION OF WT OR MUTANT PROTEIN

Information about mutant protein function can also be obtained by overexpressing either the wild-type protein or a dominant mutant of the protein. Embryo donor lines used for pronuclear microinjections should be checked for the presence of the recessive *rd* (retinal degeneration) allele (for example, the FVB/n mouse line) (45), as this could yield retinal degenerations unrelated to the transgene. The transgene is typically made using a promoter that has previously been characterized in regard to tissue-specific expression, level of expression (promoter strength), and developmen-tal time course of expression. For promoters that have not been previously character-ized, it is useful to utilize a reporter gene to define tissue specificity and timing of gene expression.

7.2.4.1 Promoter Analysis

Pronuclear microinjection techniques lend themselves to the identification of regula-tory elements within the promoter of genes. The earliest studies involved sequential truncation of portions of the 5′ upstream regulatory promoter region. Tissue or cell specificity of cloned promoters can be defined by microinjection of fusions of the promoter region with various reporter genes. Commonly used reporter genes are β-galactosidase, luciferase, and green fluorescent protein (GFP) derivatives. Alterna-tively, epitope tags such as the *myc* or *hemagglutinin (HA)* tags may be incorporated into the structural gene. Not only do the reporter genes track cell- and tissue-specific expression, they are also useful for the study of developmental gene expression. Promoters that drive cell-specific expression late in development (postnatally or in adult tissues only) are often used to define the functional role of a transgene product in adult tissue without the confounding expression of the transgene in embryonic tissues, which may produce effects arising from developmental alterations. Promoter truncation studies are also useful for identifying genetic enhancers and silencers, as well as hormone regulatory elements.

7.2.4.2 Dominant Mutants

Expression of disease-associated mutations identified from human pedigree analyses can reveal the molecular basis for disease pathology. Mutations may be expressed either as a knockin as described earlier, with appropriate spatial and temporal regulation. However, the approach is technically labor intensive. Alternatively, dominant mutants can be studied by expressing the mutation from a tissue-specific promoter by injection into embryos derived from wild-type mice. This approach yields results more rapidly than by homologous recombination and does not require that isogenic DNA be isolated. This approach often produces a disease phenotype that appropriately models features seen in the human disease. Despite differences in gene expression level, the same phenotypic trend is observed with differences in severity of phenotype. Because the mutant transgene is usually expressed on a wild-type genetic background, the endogenous protein product is also present. For transgenic mutations that produce partially functional gene products, the normal endogenous protein may compensate for the defect. Genetic mutations in numerous genes, including rhodopsin (37,46–49), rds/peripherin (50), apoptotic genes (51,52) and others (53,54) have been studied using this methodology.

Alternatively, dominant mutations may be introduced by pronuclear microinjection into embryos derived from null mutant mice, if such mice exist for the gene of interest. This can be somewhat riskier than creating the mice on a wild-type background and backcrossing onto the null mutant genetic background, as different strains of mice vary in their suitability to generate embryos for microinjection. As a rule of thumb, null mutant mice on a hybrid genetic background typically are more resilient to microinjection than are mice on a pure-bred genetic background.

7.2.4.3 Wild-Type Gene Expression

Expression of the WT gene in a null mutant mouse line or a spontaneously occurring mutant mouse strain can also be used to verify the identity of a candidate gene, resulting in rescue of the pathological phenotype. For instance, a mutation in the phosphodiesterase β-subunit gene was confirmed as the cause of degeneration in the *rd* mouse by expressing the wild-type gene from the rod opsin promoter in the *rd* mouse line (19). Degeneration was also rescued in the *rds* (50) (retinal degeneration slow) mouse line by expression of the *rds/peripherin* gene.

In the case where isoforms arise from alternative splicing, it can be useful to study isoform specific functions by the expression of specific splice isoforms in tissues normally expressing the isoform. For instance, the difference in the long and short forms of retinal arrestin were examined by microinjection of one of the forms onto a null arrestin genetic background (55). Alternatively, the effect of promiscuous stimulation of signaling or regulatory pathways may be studied by introducing a transgene that activates a biochemical pathway. For instance, persistent activation of the rod G-protein transducin causes degeneration (56).

7.2.4.4   Loss-of-Function Mutations

Introduction of a null mutation is the most straightforward type of loss-of-function mutation, already described earlier. However, the production of a null mutation by homologous recombination in embryonic stem cells is labor intensive. If the gene to be investigated belongs to a characterized gene family or contains functional elements that have been defined in other genes, it is often possible to design dominant negative mutations that will impair gene function and interfere with the function of the normal endogenous protein. For instance, mutations can be made in domains that are known to be required for enzymatic activity, but retain a binding domain that competes with the normal endogenous protein. When little is known about functional domains of the gene, a comparison of the gene across different species can often identify highly conserved amino acids or domains that are likely to play an important role in gene function. Making such mutations can be useful for defining a specific functional role. An excellent example of such a study involved mutation of the phosphorylation sites of the rhodopsin gene (57,58). Such phosphorylation sites are highly conserved in the large family of G-protein-coupled receptors and has provided much information about the mechanism by which rhodopsin activity is terminated by phosphorylation. Again, pronuclear injection of mutants will produce founders with a range of expression levels. It is useful to correlate the level of transgene expression with the severity of phenotype.

7.2.5   CONDITIONAL OR REGULATED TRANSGENE EXPRESSION

Quite often, a gene product is expressed in more than one tissue type. In such cases, a null mutation or knockin mutation results in lethality, preventing study of gene function. Transgenic methods have been developed that allow "conditional" expression of the transgene in selected tissues.

The simplest conditional knockouts are produced using a Cre/LoxP (59,60) (figure 7.5) or Flp/Frt system (61). For the null mutant gene to be produced, LoxP sites are introduced by homologous recombination into introns flanking a critical exon, creating what is referred to as a "floxed" gene. Embryonic stem cell clones containing the floxed gene are microinjected into blastocysts to produce transgenic mice. These mice are phenotypically normal until excision of DNA between the LoxP sites is induced by the expression of Cre recombinase. The Cre recombinase is expressed in a tissue-specific manner by producing a second mouse that expresses Cre recombinase from a tissue-specific promoter. This mouse is typically produced by pronuclear microinjection. Cross-breeding transgenic mice carrying the floxed gene with transgenic mice with tissue-specific expression of Cre recombinase produces transgenic mice in which a null mutation is present only in Cre-expressing tissues. This approach circumvents the confounding contribution of multisystemic

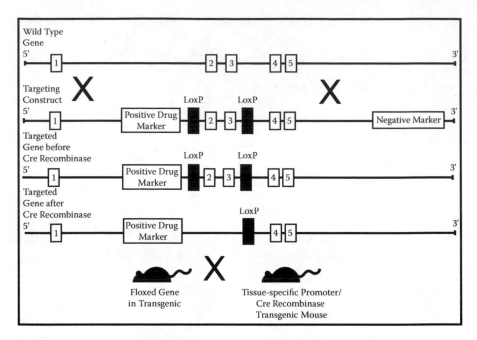

FIGURE 7.5 Gene-targeting strategy for a conditional knockout using a Cre-LoxP system. LoxP sites and a positive drug-selectable marker such as neomycin, hygromycin, or puromycin are introduced by homologous recombination. A negative selectable marker such as thymidine kinase or diphtheria toxin may be attached to the 5′ or 3′ end of the targeting construct. A conditional knockout is achieved by crossing the mouse carrying the floxed gene with a second transgenic mouse expressing Cre recombinase driven from a tissue-specific promoter. This yields offspring that have the excised gene only in the targeted tissue type.

effects on other tissues. Transgenic mice targeting Cre recombinase to either rod (62–64) or cone (65,66) photoreceptor or other retinal cell types (67–70) have been developed. This methodology has been applied to the study of kinesin-2 function in photoreceptor cells (64).

Transgene expression may also be regulated by introducing response elements regulated by tetracycline (71,72), rapamycin (73), progesterone (74) antagonist, or the insect hormone ecdysone (75). For example, these regulatable response elements can be incorporated into the Cre recombinase transgene construct and allow regulation of the timing and level of expression in a dose-dependent manner (76–78).

7.3 CONCLUSIONS

We have provided a brief overview of the most commonly used strategies for the production of transgenic mouse models. The itemization is far from complete and combinations of the different features of each strategy are constantly evolving. In selecting a strategy to use, it is most important to carefully assess how stringently the animal model that you produce will answer the experimental questions proposed. The type of strategy taken for genetic mutations associated with dominantly inherited human diseases will be quite different from a strategy to study a highly conserved element (binding site, modification site, etc.) that has not been linked to human disease, but nevertheless is likely to play an important role in gene function. For the former, regulation of the transgene from the endogenous gene is less important than it is for the latter, where the physiological effects may be subtler. The availability of time, and correspondingly money, are also considerations. Although it may be desirable to knockin a mutation to study its function, the production of a targeting vector and homologous recombinant ES cell clones to the first generation of phenotypable offspring can take as long as 9 to 12 months, depending upon the complexity of the targeting design. For pronuclear injections, construction of the transgene to the initial analysis of mouse phenotype may take from 5 to 7 months. Thus, producing a knockout or knockin mouse takes nearly twice as long as producing a mouse by standard pronuclear microinjection. Regardless, with careful planning, the benefits of having a transgenic mouse model are a powerful tool that can provide unique information about in vivo gene function.

ACKNOWLEDGMENTS

JL acknowledges financial support from the National Eye Institute, Research to Prevent Blindness, Massachusetts Lions Eye Research Fund, and the Foundation Fighting Blindness. Thanks to Katherine Malanson for helpful comments on the manuscript.

REFERENCES

1. Nagy, A., Gertsenstein, M., Vintersten, K., and Behringer, R. *Manipulating the mouse embryo: A laboratory manual* (Cold Spring Harbor Laboratory Press, Cold Spring Harbor, 2003).
2. Ramirez-Solis, R., Davis, A.C., and Bradley, A. Gene targeting in embryonic stem cells. *Methods Enzymol* 225, 855–891 (1993).
3. Jackson, I.J. and Abbott, C.M. *Mouse genetics and transgenics: a practical approach* (Oxford University Press, Oxford, New York, 2000).
4. Hofker, M.H. and van Deursen, J. *Transgenic mouse methods protocols* (Humana Press, Totowa, NJ, 2003).
5. Brinster, R.L., Allen, J.M., Behringer, R.R., Gelinas, R.E., and Palmiter, R.D. Introns increase transcriptional efficiency in transgenic mice. *Proc Natl Acad Sci U.S.A.* 85, 836–840 (1988).
6. Qin, Y., Kong, L.K., Poirier, C., Truong, C., Overbeek, P.A., and Bishop, C.E. Long-range activation of Sox9 in Odd Sex (Ods) mice. *Hum Mol Genet* 13, 1213–1218 (2004).
7. Carter, D., Chakalova, L., Osborne, C.S., Dai, Y.F., and Fraser, P. Long-range chromatin regulatory interactions in vivo. *Nat Genet* 32, 623–626 (2002).
8. Calvert, P.D., Krasnoperova, N.V., Lyubarsky, A.L., Isayama, T., Nicolo, M., Kosaras, B., Wong, G., Gannon, K.S., Margolskee, R.F., Sidman, R.L., Pugh, E.N., Jr., Makino, C.L., and Lem, J. Phototransduction in transgenic mice after targeted deletion of the rod transducin alpha-subunit. *Proc Natl Acad Sci U.S.A.* 97, 13913–13918 (2000).
9. Xu, J., Dodd, R.L., Makino, C.L., Simon, M.I., Baylor, D.A., and Chen, J. Prolonged photoresponses in transgenic mouse rods lacking arrestin. *Nature* 389, 505–509 (1997).
10. Chen, C.K., Burns, M.E., He, W., Wensel, T.G., Baylor, D.A., and Simon, M.I. Slowed recovery of rod photoresponse in mice lacking the GTPase accelerating protein RGS9-1. *Nature* 403, 557–560 (2000).

158 Signal Transduction in the Retina

11. Chen, C.K., Burns, M.E., Spencer, M., Niemi, G.A., Chen, J., Hurley, J.B., Baylor, D.A., and Simon, M.I. Abnormal photoresponses and light-induced apoptosis in rods lacking rhodopsin kinase. *Proc Natl Acad Sci U.S.A.* 96, 3718–3722 (1999).
12. Humphries, M.M., Rancourt, D., Farrar, G.J., Kenna, P., Hazel, M., Bush, R.A., Sieving, P.A., Sheils, D.M., McNally, N., Creighton, P., Erven, A., Boros, A., Gulya, K., Capecchi, M.R., and Humphries, P. Retinopathy induced in mice by targeted disruption of the rhodopsin gene. *Nat Genet* 15, 216–219 (1997).
13. Liou, G.I., Fei, Y., Peachey, N.S., Matragoon, S., Wei, S., Blaner, W.S., Wang, Y., Liu, C., Gottesman, M.E., and Ripps, H. Early onset photoreceptor abnormalities induced by targeted disruption of the interphotoreceptor retinoid-binding protein gene. *J Neurosci* 18, 4511–4520 (1998).
14. Lem, J., Krasnoperova, N.V., Calvert, P.D., Kosaras, B., Cameron, D.A., Nicolo, M., Makino, C.L., and Sidman, R.L. Morphological, physiological, and biochemical changes in rhodopsin knockout mice. *Proc Natl Acad Sci U.S.A.* 96, 736–741 (1999).
15. Mendez, A., Burns, M.E., Sokal, I., Dizhoor, A.M., Baehr, W., Palczewski, K., Baylor, D.A., and Chen, J. Role of guanylate cyclase-activating proteins (GCAPs) in setting the flash sensitivity of rod photoreceptors. *Proc Natl Acad Sci U.S.A.* 98, 9948–9953 (2001).
16. Ramamurthy, V., Niemi, G.A., Reh, T.A., and Hurley, J.B. Leber congenital amaurosis linked to AIPL1: a mouse model reveals destabilization of cGMP phosphodiesterase. *Proc Natl Acad Sci U.S.A.* 101, 13897–13902 (2004).
17. Maeda, A., Maeda, T., Imanishi, Y., Kuksa, V., Alekseev, A., Bronson, J.D., Zhang, H., Zhu, L., Sun, W., Saperstein, D.A., Rieke, F., Baehr, W., and Palczewski, K. Role of photoreceptor-specific retinol dehydrogenase in the retinoid cycle in vivo. *J Biol Chem* 280, 18822–18832 (2005).
18. Wang, Z., Wen, X.H., Ablonczy, Z., Crouch, R.K., Makino, C.L., and Lem, J. Enhanced shutoff of phototransduction in transgenic mice expressing palmitoylation-deficient rhodopsin. *J Biol Chem* 280, 24293–24300 (2005).
19. Lem, J., Flannery, J.G., Li, T., Applebury, M.L., Farber, D.B., and Simon, M.I. Retinal degeneration is rescued in transgenic rd mice by expression of the cGMP phosphodiesterase beta subunit. *Proc Natl Acad Sci U.S.A.* 89, 4422–4426 (1992).
20. Turksen, K. Embryonic stem cells: methods and protocols. In *Methods in Molecular Biology*, Vol. 185, 499. ed. Turksen, K. Humana Press, Totowa, 2002.
21. Poueymirou, W.T., Auerbach, W., Frendewey, D., Hickey, J.F., Escaravage, J.M., Esau, L., Dore, A.T., Stevens, S., Adams, N.C., Dominguez, M.G., Gale, N.W., Yancopoulos, G.D., DeChiara, T.M., and Valenzuela, D.M. F0 generation mice fully derived from gene-targeted embryonic stem cells allowing immediate phenotypic analyses. *Nat Biotechnol* 25, 91–99 (2007).
22. te Riele, H., Maandag, E.R., and Berns, A. Highly efficient gene targeting in embryonic stem cells through homologous recombination with isogenic DNA constructs. *Proc Natl Acad Sci U.S.A.* 89, 5128–5132 (1992).
23. Ohtsuka, M., Ishii, K., Kikuti, Y.Y., Warita, T., Suzuki, D., Sato, M., Kimura, M., and Inoko, H. Construction of mouse 129/Ola BAC library for targeting experiments using E14 embryonic stem cells. *Genes Genet Syst* 81, 143–146 (2006).
24. Adams, D.J., Quail, M.A., Cox, T., van der Weyden, L., Gorick, B.D., Su, Q., Chan, W.I., Davies, R., Bonfield, J.K., Law, F., Humphray, S., Plumb, B., Liu, P., Rogers, J., and Bradley, A. A genome-wide, end-sequenced 129Sv BAC library resource for targeting vector construction. *Genomics* 86, 753–758 (2005).
25. Hasty, P., Rivera-Perez, J., and Bradley, A. The length of homology required for gene targeting in embryonic stem cells. *Mol Cell Biol* 11, 5586–5591 (1991).
26. Hasty, P., Rivera-Perez, J., Chang, C., and Bradley, A. Target frequency and integration pattern for insertion and replacement vectors in embryonic stem cells. *Mol Cell Biol* 11, 4509–4517 (1991).

27. Tsang, S.H., Gouras, P., Yamashita, C.K., Kjeldbye, H., Fisher, J., Farber, D.B., and Goff, S.P. Retinal degeneration in mice lacking the gamma subunit of the rod cGMP phosphodiesterase. *Science* 272, 1026–1029 (1996).

28. Liang, Y., Fotiadis, D., Maeda, T., Maeda, A., Modzelewska, A., Filipek, S., Saperstein, D.A., Engel, A., and Palczewski, K. Rhodopsin signaling and organization in heterozygote rhodopsin knockout mice. *J Biol Chem* 279, 48189–48196 (2004).

29. Snove, O., Jr. and Rossi, J.J. Expressing short hairpin RNAs in vivo. *Nat Methods* 3, 689–695 (2006).

30. Snove, O., Jr. and Rossi, J.J. Toxicity in mice expressing short hairpin RNAs gives new insight into RNAi. *Genome Biol* 7, 231 (2006).

31. Singer, O., Tiscornia, G., Ikawa, M., and Verma, I.M. Rapid generation of knockdown transgenic mice by silencing lentiviral vectors. *Nat Protocols* 1, 286–292 (2006).

32. Campochiaro, P.A. Potential applications for RNAi to probe pathogenesis and develop new treatments for ocular disorders. *Gene Ther* 13, 559–562 (2006).

33. Cashman, S.M., Binkley, E.A., and Kumar-Singh, R. Towards mutation-independent silencing of genes involved in retinal degeneration by RNA interference. *Gene Ther* 12, 1223–1228 (2005).

34. Tessitore, A., Parisi, F., Denti, M.A., Allocca, M., Di Vicino, U., Domenici, L., Bozzoni, I., and Auricchio, A. Preferential silencing of a common dominant rhodopsin mutation does not inhibit retinal degeneration in a transgenic model. *Mol Ther* 14, 692–699 (2006).

35. Kiang, A.S., Palfi, A., Ader, M., Kenna, P.F., Millington-Ward, S., Clark, G., Kennan, A., O'Reilly, M., Tam, L.C., Aherne, A., McNally, N., Humphries, P., and Farrar, G.J. Toward a gene therapy for dominant disease: validation of an RNA interference-based mutation-independent approach. *Mol Ther* 12, 555–561 (2005).

36. Palfi, A., Ader, M., Kiang, A.S., Millington-Ward, S., Clark, G., O'Reilly, M., McMahon, H.P., Kenna, P.F., Humphries, P., and Farrar, G.J. RNAi-based suppression and replacement of rds-peripherin in retinal organotypic culture. *Hum Mutat* 27, 260–268 (2006).

37. Olsson, J.E., Gordon, J.W., Pawlyk, B.S., Roof, D., Hayes, A., Molday, R.S., Mukai, S., Cowley, G.S., Berson, E.L., and Dryja, T.P. Transgenic mice with a rhodopsin mutation (Pro23His): a mouse model of autosomal dominant retinitis pigmentosa. *Neuron* 9, 815–830 (1992).

38. Kassai, H., Aiba, A., Nakao, K., Nakamura, K., Katsuki, M., Xiong, W.H., Yau, K.W., Imai, H., Shichida, Y., Satomi, Y., Takao, T., Okano, T., and Fukada, Y. Farnesylation of retinal transducin underlies its translocation during light adaptation. *Neuron* 47, 529–539 (2005).

39. Vasireddy, V., Uchida, Y., Salem, N., Jr., Kim, S.Y., Mandal, M.N., Reddy, G.B., Bodepudi, R., Alderson, N.L., Brown, J.C., Hama, H., Dlugosz, A., Elias, P.M., Holleran, W.M., and Ayyagari, R. Loss of functional ELOVL4 depletes very long-chain fatty acids (>=C28) and the unique {omega}-O-acylceramides in skin leading to neonatal death. *Hum Mol Genet* 16, 471–482 (2007).

40. Lentz, J., Pan, F., Ng, S.S., Deininger, P., and Keats, B. Ush1c216A knock-in mouse survives Katrina. *Mutat Res* 616, 139–144 (2007).

41. Vasireddy, V., Jablonski, M.M., Mandal, M.N., Raz-Prag, D., Wang, X.F., Nizol, L., Iannaccone, A., Musch, D.C., Bush, R.A., Salem, N., Jr., Sieving, P.A., and Ayyagari, R. Elovl4 5-bp-deletion knock-in mice develop progressive photoreceptor degeneration. *Invest Ophthalmol Vis Sci* 47, 4558–4568 (2006).

42. Gross, A.K., Decker, G., Chan, F., Sandoval, I.M., Wilson, J.H., and Wensel, T.G. Defective development of photoreceptor membranes in a mouse model of recessive retinal degeneration. *Vision Res* 46, 4510–4518 (2006).

43. Chan, F., Bradley, A., Wensel, T.G., and Wilson, J.H. Knock-in human rhodopsin-GFP fusions as mouse models for human disease and targets for gene therapy. *Proc Natl Acad Sci U.S.A.* 101, 9109–9114 (2004).

44. Weber, B.H., Lin, B., White, K., Kohler, K., Soboleva, G., Herterich, S., Seeliger, M.W., Jaissle, G.B., Grimm, C., Reme, C., Wenzel, A., Asan, E., and Schrewe, H. A mouse model for Sorsby fundus dystrophy. *Invest Ophthalmol Vis Sci* 43, 2732–2740 (2002).

45. Errijgers, V., Van Dam, D., Gantois, I., Van Ginneken, C.J., Grossman, A.W., D'Hooge, R., De Deyn, P.P., and Kooy, R.F. FVB.129P2-Pde6b(+) Tyr(c-ch)/Ant, a sighted variant of the FVB/N mouse strain suitable for behavioral analysis. *Genes Brain Behav* (2006).

46. Naash, M.I., Hollyfield, J.G., al-Ubaidi, M.R., and Baehr, W. Simulation of human autosomal dominant retinitis pigmentosa in transgenic mice expressing a mutated murine opsin gene. *Proc Natl Acad Sci U.S.A.* 90, 5499–5503 (1993).

47. Li, T., Franson, W.K., Gordon, J.W., Berson, E.L., and Dryja, T.P. Constitutive activation of phototransduction by K296E opsin is not a cause of photoreceptor degeneration. *Proc Natl Acad Sci U.S.A.* 92, 3551–3555 (1995).

48. Li, T., Snyder, W.K., Olsson, J.E., and Dryja, T.P. Transgenic mice carrying the dominant rhodopsin mutation P347S: evidence for defective vectorial transport of rhodopsin to the outer segments. *Proc Natl Acad Sci U.S.A.* 93, 14176–14181 (1996).

49. Li, T., Sandberg, M.A., Pawlyk, B.S., Rosner, B., Hayes, K.C., Dryja, T.P., and Berson, E.L. Effect of vitamin A supplementation on rhodopsin mutants threonine-17 --> methionine and proline-347 --> serine in transgenic mice and in cell cultures. *Proc Natl Acad Sci U.S.A.* 95, 11933–11938 (1998).

50. Travis, G.H., Groshan, K.R., Lloyd, M., and Bok, D. Complete rescue of photoreceptor dysplasia and degeneration in transgenic retinal degeneration slow (rds) mice. *Neuron* 9, 113–119 (1992).

51. Joseph, R.M. and Li, T. Overexpression of Bcl-2 or Bcl-XL transgenes and photoreceptor degeneration. *Invest Ophthalmol Vis Sci* 37, 2434–2446 (1996).

52. Nir, I., Kedzierski, W., Chen, J., and Travis, G.H. Expression of Bcl-2 protects against photoreceptor degeneration in retinal degeneration slow (rds) mice. *J Neurosci* 20, 2150–2154 (2000).

53. Salchow, D.J., Gouras, P., Doi, K., Goff, S.P., Schwinger, E., and Tsang, S.H. A point mutation (W70A) in the rod PDE-gamma gene desensitizing and delaying murine rod photoreceptors. *Invest Ophthalmol Vis Sci* 40, 3262–3267 (1999).

54. Olshevskaya, E.V., Calvert, P.D., Woodruff, M.L., Peshenko, I.V., Savchenko, A.B., Makino, C.L., Ho, Y.S., Fain, G.L., and Dizhoor, A.M. The Y99C mutation in guanylyl cyclase-activating protein 1 increases intracellular Ca^{2+} and causes photoreceptor degeneration in transgenic mice. *J Neurosci* 24, 6078–6085 (2004).

55. Burns, M.E., Mendez, A., Chen, C.K., Almuete, A., Quillinan, N., Simon, M.I., Baylor, D.A., and Chen, J. Deactivation of phosphorylated and nonphosphorylated rhodopsin by arrestin splice variants. *J Neurosci* 26, 1036–1044 (2006).

56. Raport, C.J., Lem, J., Makino, C., Chen, C.K., Fitch, C.L., Hobson, A., Baylor, D., Simon, M.I., and Hurley, J.B. Downregulation of cGMP phosphodiesterase induced by expression of GTPase-deficient cone transducin in mouse rod photoreceptors. *Invest Ophthalmol Vis Sci* 35, 2932–2947 (1994).

57. Mendez, A., Burns, M.E., Roca, A., Lem, J., Wu, L.W., Simon, M.I., Baylor, D.A., and Chen, J. Rapid and reproducible deactivation of rhodopsin requires multiple phosphorylation sites. *Neuron* 28, 153–164 (2000).

58. Doan, T., Mendez, A., Detwiler, P.B., Chen, J., and Rieke, F. Multiple phosphorylation sites confer reproducibility of the rod's single-photon responses. *Science* 313, 530–533 (2006).

59. Yu, Y. and Bradley, A. Engineering chromosomal rearrangements in mice. *Nat Rev Genet* 2, 780–790 (2001).

60. Kos, C.H. Cre/loxP system for generating tissue-specific knockout mouse models. *Nutr Rev* 62, 243–246 (2004).

61. Marszalek, J.R., Liu, X., Roberts, E.A., Chui, D., Marth, J.D., Williams, D.S., and Gold-stein, L.S. Genetic evidence for selective transport of opsin and arrestin by kinesin-II in mammalian photoreceptors. *Cell* 102, 175–187 (2000).

62. Le, Y.Z., Zheng, L., Zheng, W., Ash, J.D., Agbaga, M.P., Zhu, M., and Anderson, R.E. Mouse opsin promoter-directed Cre recombinase expression in transgenic mice. *Mol Vis* 12, 389–398 (2006).

63. Li, S., Chen, D., Sauve, Y., McCandless, J., Chen, Y.J., and Chen, C.K. Rhodopsin-iCre transgenic mouse line for Cre-mediated rod-specific gene targeting. *Genesis* 41, 73–80 (2005).

64. Jimeno, D., Feiner, L., Lillo, C., Teofilo, K., Goldstein, L.S., Pierce, E.A., and Williams, D.S. Analysis of kinesin-2 function in photoreceptor cells using synchronous Cre-loxP knockout of Kif3a with RHO-Cre. *Invest Ophthalmol Vis Sci* 47, 5039–5046 (2006).

65. Le, Y.Z., Ash, J.D., Al-Ubaidi, M.R., Chen, Y., Ma, J.X., and Anderson, R.E. Targeted expression of Cre recombinase to cone photoreceptors in transgenic mice. *Mol Vis* 10, 1011–1018 (2004).

66. Akimoto, M., Filippova, E., Gage, P.J., Zhu, X., Craft, C.M., and Swaroop, A. Trans-genic mice expressing Cre-recombinase specifically in M- or S-cone photoreceptors. *Invest Ophthalmol Vis Sci* 45, 42–47 (2004).

67. Matsuda, T. and Cepko, C.L. Controlled expression of transgenes introduced by in vivo electroporation. *Proc Natl Acad Sci U.S.A.* 104, 1027–1032 (2007).

68. Zhang, X.M., Chen, B.Y., Ng, A.H., Tanner, J.A., Tay, D., So, K.F., Rachel, R.A., Copeland, N.G., Jenkins, N.A., and Huang, J.D. Transgenic mice expressing Cre-recombinase spe-cifically in retinal rod bipolar neurons. *Invest Ophthalmol Vis Sci* 46, 3515–3520 (2005).

69. Gelman, D.M., Noain, D., Avale, M.E., Otero, V., Low, M.J., and Rubinstein, M. Transgenic mice engineered to target Cre/loxP-mediated DNA recombination into cat-echolaminergic neurons. *Genesis* 36, 196–202 (2003).

70. Campsall, K.D., Mazerolle, C.J., De Repentigny, Y., Kothary, R., and Wallace, V.A. Characterization of transgene expression and Cre recombinase activity in a panel of Thy-1 promoter-Cre transgenic mice. *Dev Dyn* 224, 135–143 (2002).

71. Morgan, W.W., Richardson, A., Sharp, Z.D., and Walter, C.A. Application of exog-enously regulatable promoter systems to transgenic models for the study of aging. *J Gerontol A Biol Sci Med Sci* 54, B30–40; discussion B41–42 (1999).

72. Albanese, C., Hulit, J., Sakamaki, T., and Pestell, R.G. Recent advances in inducible expression in transgenic mice. *Semin Cell Dev Biol* 13, 129–141 (2002).

73. Harvey, D.M. and Caskey, C.T. Inducible control of gene expression: prospects for gene therapy. *Curr Opin Chem Biol* 2, 512–518 (1998).

74. Tsai, S.Y., O'Malley, B.W., DeMayo, F.J., Wang, Y., and Chua, S.S. A novel RU486 inducible system for the activation and repression of genes. *Adv Drug Deliv Rev* 30, 23–31 (1998).

75. No, D., Yao, T.P., and Evans, R.M. Ecdysone-inducible gene expression in mammalian cells and transgenic mice. *Proc Natl Acad Sci U.S.A.* 93, 3346–3351 (1996).

76. Hasan, M.T., Friedrich, R.W., Euler, T., Larkum, M.E., Giese, G., Both, M., Duebel, J., Waters, J., Bujard, H., Griesbeck, O., Tsien, R.Y., Nagai, T., Miyawaki, A., and Denk, W. Functional fluorescent Ca^{2+} indicator proteins in transgenic mice under TET con-trol. *PLoS Biol* 2, e163 (2004).

77. Okoye, G., Zimmer, J., Sung, J., Gehlbach, P., Deering, T., Nambu, H., Hackett, S., Melia, M., Esumi, N., Zack, D.J., and Campochiaro, P.A. Increased expression of brain-derived neurotrophic factor preserves retinal function and slows cell death from rhodopsin mutation or oxidative damage. *J Neurosci* 23, 4164–4172 (2003).

78. Chang, M.A., Horner, J.W., Conklin, B.R., DePinho, R.A., Bok, D., and Zack, D.J. Tetracycline-inducible system for photoreceptor-specific gene expression. *Invest Oph-thalmol Vis Sci* 41, 4281–4287 (2000).

2

*Vertebrate Nonvisual
Phototransduction*

8 Melanopsin Signaling and Nonvisual Ocular Photoreception

*Sowmya V. Yelamanchili, Victoria Piamonte,
Surendra Kumar Nayak, Nobushige
Tanaka, Quansheng Zhu, Kacee Jones,
Hiep Le, and Satchidananda Panda*

CONTENTS

8.1 INTRODUCTION

Organisms, ranging from cyanobacteria to mammals, use an intrinsic circadian clock and photosensors to regulate behavior and physiology in resonance with geophysical time. In mammals, an endogenous master clock functions in the *suprachiasmatic nuclei* (SCN) of the hypothalamus, and is entrained to the day–night cycle by direct light input from the retina. Appropriate photoentrainment is essential for general health, as shift work and similar chronic circadian desynchronization have been shown to be major risk factors in several sleep disorders, metabolic syndromes, and in cancer (reviewed in (1)). Understanding the molecular processes underlying photoentrainment will therefore help identify new strategies and targets for therapeutic intervention in these disorders.

In addition to circadian photoentrainment, light received through mammalian eyes also mediates several other photoadaptive behaviors and physiologies. These include dynamic pupillary light responses (PLRs), and light modulation of the neuroendocrine system such as light-induced suppression of pineal melatonin synthesis and release. Finally, ocular photoresponse also accounts for temporal niche-dependent modulation of activity-rest state, i.e., general suppression of activity in nocturnal species, and improvement in alertness in diurnal mammals. Persistence of these photic responses in many blind human patients and in blind animal models deficient in rod/cone signaling components or with outer retinal degeneration has led to a collective description of these responses as nonvisual, non-image-forming, or adaptive photic responses [reviewed in Van Gelder (2)]. Complete loss of these responses in experimental bilateral enucleation animal studies established that a novel nonrod, noncone inner retina photopigment may play a significant role in adaptive light responses. The spectral nature of the novel photopigment was predicted from the action spectra of circadian photoentrainment and PLR assays in rodents (3,4).

A novel combination of retrograde labeling to mark retinal ganglion cells (RGCs) that directly project to the circadian brain center, the SCN, and electrophysiological characterization of light responsiveness of these neurons led to the discovery of intrinsically photosensitive retinal ganglion cells (ipRGCs) (5,6). The ipRGCs demonstrated some unique properties distinct from those of rod/cone photoreceptors: light-evoked membrane depolarization, high threshold sensitivity, long response latency, slow deactivation, and resistance to bleaching (5). Homology cloning of melanopsin (7), colocalization of the protein product with ipRGCs (6), and phenotypic analysis of melanopsin-deficient mice has established that melanopsin functions as a true photopigment in the ipRGCs and significantly contributes to adaptive photoresponses (6,8–11).

The discovery of melanopsin's role in adaptive photoresponses has opened an entry point to understand the molecular bases of nonvisual photoresponses. Genetic analysis has established that both rod/cone and melanopsin photopigments participate in these processes, albeit with varying magnitudes specific for the individual response (9,12). Currently, the relative contribution of rod or cone photoreceptors to various non-image-forming responses is unclear. The neuronal circuitry and the underlying molecular signaling events conveying light information from rod/cone and ipRGCs to the brain centers regulating different adaptive responses have yet to be clearly mapped out. In parallel, molecular and biochemical characterization of melanopsin photopigment will unravel which properties of the ipRGCs are encoded in the photopigment and how photoexcited melanopsin transmits light information along the downstream signaling cascade. Several methods and techniques at whole-animal, cellular, and biochemical levels combined with genetic and pharmacological perturbations will prove invaluable in addressing these questions. We will describe some techniques widely used by several labs and a few recently developed in our lab. In each section we will also highlight how use of the specific approach or technique has made a seminal contribution to our understanding of adaptive photoresponses in mammals. The first half of the chapter will describe methods using whole-animal or animal tissue, and in the second part we will focus on heterologously expressed melanopsin.

8.2   ANIMAL BEHAVIORAL STUDIES AND EXPRESSION PATTERNS

Photoadaptive behavior assays in small rodents have been extremely useful in defining the nature of the photopigments driving these responses, and genetically dissecting their relative contributions. The widely used assays include: (a) PLR, (b) light-induced suppression of activity in nocturnal rodents (or negative masking), (c) light-induced suppression of pineal melatonin synthesis and release, and (d) circadian photoentrainment. These four behaviors are regulated by different brain centers and differ in underlying neuronal signaling network, threshold sensitivity, dynamic range of sensitivity, and response under prolonged illumination. With careful planning, all four assays can be performed on the same group of animals. Pupillary responsiveness to light is a relatively sensitive and quick photoadaptive response in which the pupil constricts within milliseconds of sudden increase in irradiance and the constriction is maintained under prolonged illumination. The assay has an excellent dynamic range of sensitivity over several orders of irradiance levels, and the results are relatively independent of the time of the day the assay is performed. Similar to PLR, negative masking is an assay with significant dynamic range of sensitivity. Within a few seconds of bright illumination during the subjective night, mice usually reduce their overall activity and under prolonged illumination over hours the activity partially recovers. Light-induced suppression of pineal melatonin synthesis and release is yet another assay conducted during the subjective night, when the melatonin synthesis usually rises. Within several minutes of illumination, the pineal melatonin levels, and subsequently, circulating melatonin levels, drop. Finally, the circadian photoentrainment assay in rodents is a resource-intensive procedure that is often used to measure several aspects of the effect of light on circadian rhythms, including general entrainment to an imposed light–dark cycle, the phase-shifting effect of light on circadian behavior under constant darkness, and circadian behavior under constant light.

The light sensitivity of these four behavioral responses appropriately reflects the relative contributions of the rod/cone pathway and melanopsin. A rapid response such as PLR at all light intensities is dominantly regulated by the rod/cone pathway, so that the loss of outer retina leads to a log unit reduction in photosensitivity of PLR, whereas the loss of melanopsin causes a modest change in PLR only at high light intensity. The melanopsin pathway, however, plays a dominant role in negative masking and circadian photoentrainment that exhibit high threshold sensitivity, where the loss of melanopsin causes a significant change in photic sensitivity of these responses, whereas complete outer retina degeneration does not cause any detectable change in sensitivity. Although outer retina degeneration causes no significant reduction in the sensitivity of photic suppression of pineal melatonin synthesis, the effect of the loss of melanopsin has not been extensively studied. Such differential sensitivities will allow fine molecular dissection of adaptive photoresponses using whole-animal behavior assays.

8.2.1 PUPILLARY LIGHT RESPONSE (PLR)

PLR is a simple yet powerful assay that has been widely used in characterizing photopigments and signaling components driving nonvisual photoresponses in animals. Persistence of PLR in mice with retina degeneration (rd) was the first observation suggesting the existence of nonvisual photoreceptors in mammals (13).

Although the PLR in some birds (14) can be driven by iris-resident photopigment ex vivo, PLR in small rodents is usually performed in live anesthetized animals. The mouse is handheld by the scruff along an illuminated light path and a video image of the pupil is acquired for up to 3 min. For detailed descriptions of measuring mouse PLR, please refer to Van Gelder (15).

PROTOCOL 1: MEASUREMENT OF PLR

Instrument setup

1. Camera: A consumer-grade video camera with infrared recording ability is commonly used. The camera should be fitted with a macro and close-up lens. If the video image is to be processed by an image processing software such as Eye Tracker (Arrington Research, Arizona), outfitting the camera with an additional infrared filter (89C or 89B filter) usually improves the contrast.
2. Infrared light source: The built-in infrared LEDs in the video camera usually provide enough illumination. However, in some camera models, mounting additional filters may partially block the infrared LED. An extra infrared LED source, easily available from several Internet vendors, usually improves image quality.
3. Visible light source: A halogen or xenon light source fitted with a fiber-optic or liquid light guide. Either the light source or the light guide should be fitted with a filter holder to hold narrow band-pass interference and/or neutral density filters to regulate spectral quality. A calibrated photometer is used to measure irradiance level from various filter combinations. The light source is preferably fitted with a foot-pedal-activated shutter.
4. Ancillary equipment: We typically use several flexible arm magnetic bases, mounting screws to position and secure the video camera, infrared LED, light guide, and the irradiance detector on a piece of steel plate. The display monitor of the video camera is covered with a red acetate filter (Lee filter #029 or #106). The researcher uses a headlamp covered with the same red acetate filter to handle mice during the assay. All visible light leaks from the light source and other instruments in the room are blocked with opaque material following the institute's safety guideline.
5. Ergonomics: All instruments and the camera lenses should be appropriately positioned and adjusted to obtain the best-quality infrared video of

the mouse eye. We use some additional magnetic bases and flexible arms for proper positioning of the mouse without causing too much stress on the animal or the researcher.

Pupil measurement

6. Familiarize the mice to human handling for up to 1 week prior to the experiment to reduce stress.
7. Dark-adapt the mice for at least 2 h before the commencement of the experiment so as to completely dilate the pupil.
8. Hold the mouse by its scruff and position its eye to a point where a sharp infrared video image is captured by the camera. Capture a dark baseline video for up to 1 min.
9. Illuminate the mouse eye and collect the video image for another 1 min. In wild-type mice under bright illumination, pupil constriction commences within a few milliseconds, and maximum constriction is achieved within 10 s.
10. After termination of the light pulse, record an additional 30 s to 1 min infrared video to measure pupil relaxation.
11. Measure repeatedly pupil constrictions on the same animal after ~2 h of dark adaptation between measurements. Change the interference filter and/ or neutral density filters after one batch of animals is examined at a given light quality. Examine the mice for their PLR to a different light quality after the dark adaptation.
12. Measure the pupil diameter of the enlarged video images on the display screen. Calculate the ratio of cornea diameter to pupil diameter. Alternatively, use automated image analysis software, such as the pupil measurement feature of Eye Tracker (Arrington Research) to measure pupil diameter at a defined interval. In our experience the image processing software is typically optimized for larger pupils and adapting them for the mouse eye needs a significant amount of optimization of the light, video quality, contrast, and image size.

Several factors including stress level can significantly interfere with PLR, thereby slowing the pupil constriction rate and preventing complete pupil constriction in wild-type mice even at high irradiance level. Therefore, noise level and human activity in the room should be kept to a minimum. Occasionally, mice show anterior chamber dysgenesis leading to attachment of the iris to the cornea, which causes incomplete constriction. In *rd;opn4*$^{-/-}$ and comparable genotypes with no rod/cone and melanopsin function, the pupil does not constrict even in response to bright light (9,12). To ensure the lack of pupil constriction is due to a deficiency in the light signaling pathway and not a result of abnormal development of the neuromusculature responsible for pupil constriction, at the end of PLR a drop of dilute cholinergic agonist such as 1% pilocarpine may be applied to each eye. The drug usually completely constricts the pupil within a few minutes of application.

8.2.1 Circadian Photoentrainment Assay

Circadian activity measurement in rodents has been extensively used to characterize photoreceptor properties (4,16) and for genetic dissection of light input pathway
(9,12) that synchronizes the circadian clock to the ambient lighting. Action spectra
of circadian photoentrainment in hamsters predicted that a vitamin-A-based photopigment with peak response of around 480–500 nm (4) may be involved. However,
intact photoentrainment in *rd* mice (16) and in various other outer retina degeneration mouse models (17,18) established that rod/cone photoreceptors are dispensable.
Normal entrainment of melanopsin-deficient mice to an imposed light–dark cycle
and the attenuated phase shift of the oscillator in response to discrete light pulse
demonstrated the complex interplay between the rod/cone and melanopsin pathways
in photoentrainment, where the loss of melanopsin can be partially compensated by
rod/cone photoreceptors whereas the loss of rod/cone function can be fully compensated by melanopsin. Complete loss of entrainment in mice deficient in rod/cone and
melanopsin function proved the necessity and sufficiency of these photopigments in
circadian photoentrainment (9,12). As genetic perturbation models targeting specific
molecules or cell types become available, the photoentrainment assay will continue
to be a powerful tool for examining this process.

The assay also serves to monitor photic suppression of activity in nocturnal
rodents. Negative masking behavior in mice deficient in rod/cone and/or melanopsin function closely parallels circadian behavior in these mice. The loss of rods/
cones have little to no effect on negative masking in response to bright light, whereas
$opn4^{-/-}$ mice become more active than wild-type littermates under prolonged illumination (19). Mice with neither functional rod/cone nor melanopsin fail to show any
negative masking (9,12).

Although small-animal activity can be monitored by various means such as
infrared beam breaks, temporal monitoring of drinking or eating, and body temperature telemetry, wheel running activity measurement often produces better quality
data and is widely used in various circadian labs. For a thorough description of the
room setup, apparatus, data collection and analysis software, we strongly encourage
readers to refer to a recent article (20). Mice are individually housed in specially
designed cages equipped with running wheels. Up to 24 wheel cages are placed
inside a light-tight box with independent light control, ventilation, and sensors for
humidity, temperature, and light. A constant level of humidity and temperature is
maintained inside the room. The ventilation fans provide constant white noise that
masks all other temporal auditory cues, such as routine human activities in the room.
Individual wheel cages are connected via switches to the Clocklab™ acquisition
software (Actimetrics, Chicago, Illinois) that continuously records wheel rotation
over days and weeks. Wheel rotations are recorded in 1–10 min bins.

Protocol 2: Circadian Wheel Running Assay

1. House mice individually in wheel running cages under 12 h light and 12 h
   dark (LD) during the first week of the experiment to entrain their activity

rhythm to the imposed lighting regime. Mice of C57Bl/6 background usually consolidate their wheel running activity to the night, so that the onset
of wheel running activity coincides with the dark onset. They run on the
wheel almost continuously for the first 6–8 h of the night, after which the
wheel running is intermittent for the rest of the night. Very little wheel running activity is observed during the day (figure 8.1).

2. For measurement of negative masking, switch on lights in the wheel running chamber for 1–3 h, starting from 2 h after the beginning of the dark
phase to assess the effect of light on activity (if any) by comparing the
activity in 1–10 min bins during the light pulse with activity during the
comparable time for the previous 2–3 nights.

3. Because the light pulse used to assess photic suppression of activity can
also reset the circadian oscillator, to ensure proper entrainment, mice are
again returned to light–dark cycles for one additional week.

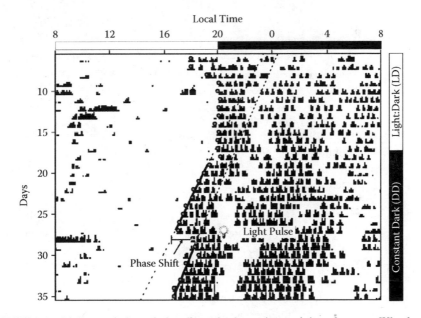

FIGURE 8.1 Light regulation of circadian wheel running activity of mouse. Wheel running activity record of a mouse showing entrainment to a light–dark cycle and phase shift in
response to a brief pulse of light. Vertical bars along each horizontal line show wheel running
activity over 24 h. Local time and period of light and dark during entrainment are represented
by white and dark boxes on the top. The mouse was placed in a light–dark cycle for almost
2 weeks during which the circadian clock is entrained to the imposed LD cycle and the time
of activity onset maintains a constant phase relationship with the time of dark onset. After 2
weeks of entrainment, the mouse was released into constant darkness. Under constant darkness, the activity rhythm is under the control of the endogenous circadian oscillator, which in
this mouse runs with a period length of <24 h. A brief 15 min light pulse administered at ~4 h
after activity onset phase delays the circadian clock, so that the new activity onset is delayed
relative to that prior to the light pulse.

4. Turn off the lights, and let the animals free-run under constant darkness (DD). Under such conditions, wheel running activity is entirely under the control of the endogenous clock, so that the onset of wheel running activity each day is a good measure of the phase of the oscillator. C57Bl/6 mice usually free run with a period length of ~23.5 h, which reflects the pace of the endogenous oscillator. The average wheel running period length is normalized to 24 h and the phase of the oscillator is denoted by circadian time (CT), where CT12 usually marks the onset of wheel running activity.

5. Administer a pulse of light at a predetermined CT time after 1–2 weeks of free running under DD to assess the effect of a brief light pulse on the phase · of the oscillator. All organisms exhibit a time-of-the-day-specific effect of light on phase resetting, where light administered during the subjective day (inactivity period in nocturnal rodents) usually results in little to no phase change, whereas a light pulse during the early subjective night causes significant phase delays or the same pulse in late subjective night advances the phase, depending on the time of the light pulse. As a result, a light pulse administered at a fixed CT during the subjective night is commonly used to assess light entrainment of the clock. A small variation in the pace of the endogenous clock among individual mice often translates into relatively significant differences in time of activity onset (CT12) after 1–2 weeks in constant darkness, the local time at which light pulse to be administered is calculated for individual mice.

6. Transfer the mice or the wheel cages to a separate chamber, where light pulses of defined wavelength, intensity, and duration are administered, and subsequently return them to their original light tight chamber for further data collection. Use infrared night vision goggles or low-intensity red headlamps to handle mice.

7. Complete resetting of the oscillator usually occurs within 2–3 days. However, monitor the wheel running activity of mice under DD for 1–2 weeks after the light pulse to obtain an accurate estimate of the phase shift. Determine linear regressions of activity onset for ~1 week each prior to light pulse and after light pulse. Extrapolate the two lines, and the phase difference between these two regression lines on the day after light pulse is considered the magnitude of phase shift. A delay in activity onset after the light pulse is considered a phase delay and designated with a negative sign, whereas an advance in activity onset is considered a phase advance.

8. After ~2 weeks of the first light pulse, the same animals in wheel cages running under DD can be used to assess circadian phase shift in response to another pulse of light of different intensity, duration, or wavelength. In multiple light pulse experiments, the phase of the light pulse, intensity, or duration should be randomized.

9. Circadian wheel running period length assessed under constant light (LL) is also useful in assessing factors affecting light input to the circadian clock. At the end of the light-pulse-induced phase shift assay, re-entrain the mice to LD cycle for 2 weeks and then transfer them to LL. LL has

a period-lengthening effect on the rodent circadian oscillator, such that wild-type mice free-run with a period length longer than 24 h. Mice with an altered photoentrainment pathway may become arrhythmic or free-run with a different period length. For example, $opn4^{-/-}$ mice free-run with a period length slightly shorter than that of the wild-type littermates (10,11), whereas the $rd;opn4^{-/-}$ or $cng3^{-/-};gnat^{-/-};opn4^{-/-}$ mice free-run with a period length indistinguishable from that under DD (9,12).

8.2.2 PHOTIC SUPPRESSION OF PINEAL MELATONIN SYNTHESIS AND RELEASE

Photic suppression of pineal melatonin synthesis and release is a nonvisual ocular photoresponse widely preserved throughout evolution in both diurnal and nocturnal species. In humans and other large animals serum melatonin level is often measured to evaluate the effect of nocturnal illumination on pineal melatonin release (21). However, in small rodents the limited serum quantity and naturally occurring mutations in melatonin synthesis pathway (22) often make it impractical to measure serum melatonin level; therefore, a surrogate measure is often used. Light also causes transcriptional downregulation of a gene encoding for a key enzyme aralkylamine N-acetyltransferase (AANAT) in the melatonin biosynthesis pathway in the pineal. The change in AANAT mRNA level can be measured by quantitative RT-PCR from the total RNA extract of the pineal. We often use this terminal assay after completion of PLR and other wheel-cage-based assays for nonvisual photoresponses (9). Mice are entrained to LD cycle for 2 weeks and on the day of the assay, the light phase is extended for 9 h, while the control group of animals transition into the regular dark phase. Both groups of mice are sacrificed and enucleated under dim red or infrared light, and the brain is dissected under normal white light. The pineal gland in mouse is a small (~1 mm diameter) transparent gland, which is located just beneath the junction between the sagittal and transverse sutures of the skull. Therefore, care should be taken to make the incision in the scalp along its periphery. The ventral surface of the scalp is viewed under a dissecting microscope to locate and dissect the pineal gland. Total RNA from individual pineal glands is extracted using RNeasy micro (Qiagen, California) or any equivalent kit and quantitative RT-PCR for AANAT message levels are performed as described in Panda et al. (9).

Several modifications of this approach are found in the literature. Discreet light pulses of defined spectral quality during the night can be substituted for the 9 h extension of the light phase.

8.3 IDENTIFICATION AND CHARACTERIZATION OF MELANOPSIN-EXPRESSING CELLS

Methods to specifically mark or stain a functional group of fixed or unfixed neurons have been a critical aspect toward understanding their ontogeny, cellular architecture, dendritic fields, and axonal projections. Several methods have been employed to study melanopsin-expressing RGCs, and we will allude to a few of them with specific emphasis on the strengths and potential use of these approaches. The most critical technique used in identifying the ipRGCs is retrograde labeling of RGCs projecting to the SCN (5,6). In this approach, the animal is deeply anesthetized and, using stereotaxic injection instruments, a small amount of fluorescent dye is injected to the ventral region of the SCN. A few days later the retina is harvested to test whether the label has retrogradely traveled to a small subpopulation of retinal ganglion cells. This allows for subsequent histological or electrophysiological recordings from the SCN projecting RGCs. The technique requires a good knowledge of stereotaxic coordinates of the species (and strain) and extensive practice to achieve specific staining of the SCN projecting RGCs. Despite these challenges, the retrograde labeling approach has been used in rats, mice, and monkeys (5,6,23) to identify ipRGCs. This approach holds promise to characterize the cellular properties of melanopsin-expressing RGCs in diverse mammalian species that are not amenable to genetic encoding of melanopsin-expressing RGCs as has been achieved in mouse.

Comprehensive descriptions of the ontogeny and axonal projections of melanopsin-expressing RGCs in mice have come from transgenic knockin mice expressing tau:lacZ fusion protein from the native melanopsin locus (6,24). The fusion protein reliably reports transcriptional regulation of melanopsin and via binding of the *tau* domain of the fusion protein to the microtubules reports axonal projection of melanopsin-expressing RGCs. A similar approach to generate a GFP knockin mouse in our, and in many other, labs that would allow live cell imaging and electrophysiological recording from ipRGC has not been successful because of low or undetectable levels of GFP protein. Better GFP variants and/or a transgenic approach may ultimately prove successful. However, adeno-associated virus-expressing GFP from CMV promoter has been successfully used to transduce melanopsin-expressing RGCs in rodents (25). As the tropism of a virus may vary in a species- or strain-specific manner, extension of this approach to other species may need extensive trials with various serotypes. Immunostaining for a coexpressed marker, such as PACAP, has also been used to characterize the melanopsin-expressing RGC. Results from all of these strategies to identify mRGCs have shown (a) the distribution and number of ipRGCs in mouse retina (6,24), (b) that a majority of mRGCs project to the SCN (6,25–28), (c) melanopsin expression and function in mice precede that of rod/cone photoreceptors (29–31), (d) loss of melanopsin does not alter the overall development and projections of these cells (8), and finally, (e) a comprehensive map of brain regions receiving direct axonal projections of the ipRGCs (24).

One simple method to identify and characterize melanopsin RGCs has been immunostaining with antimelanopsin antibodies. A rabbit polyclonal antibody raised against an N-terminal extracellular epitope of mouse melanopsin and subsequently several other polyclonal antibodies raised against melanopsin from other mammals have been extensively used to describe the cellular architecture of ipRGCs as well as the subcellular distribution of melanopsin in these cells (reviewed in Berson (34)). Unlike the rod/cone opsins, melanopsin immunostaining is observed along the dendrites, soma, and axons of the ipRGCs; most of the protein is diffusely distributed throughout the plasma membrane of the ipRGCs. Immunostaining of the retina sections shows most melanopsin-positive soma reside in the RGC layer, whereas the dendrites heavily arborize in the off-sublayer of the inner plexiform layer. Immunostaining in other mammalian species has revealed an overall similarity in cellular architecture; however, the distribution of ipRGCs across the retina shows some variability. Although in rodents, which lack a defined macula, melanopsin-expressing RGCs are almost evenly distributed throughout the retina, in primates with a distinct macula, including humans, some discrepancies have been described. Melanopsin immunostaining in human, as in rodents, is almost uniformly distributed throughout the retina (24), whereas in monkey (23) no staining was found in the macula region.

The melanopsin immunostaining approach has major applications in testing whether candidate signaling components, channels, and neurotransmitters are expressed in ipRGCs or conversely test whether mutation in any candidate gene affects the expression and cellular architecture of the ipRGCs. For example, costaining for melanopsin and PACAP established that the major PACAP-expressing cells that were known to constitute the RHT also express melanopsin (34). We will briefly describe the melanopsin immunostaining protocol using a rabbit polyclonal antibody raised against the first 17 amino acids of mouse melanopsin (NM_013887). We have not tested this antibody for cross-reactivity with rat, hamster, or human melanopsin. Because melanopsin expression is limited to the RGC sublayer of the retina, both retina flat mounts and retina cross sections can be used as starting material.

Two basic approaches are used to visualize cells of interest within the retina. First, a cross section of the retina can be made that allows observation of melanopsin-expressing cells in the context of the various cell layers of the retina. This is done by cutting approximately 8–10 μm thick sections on a cryostat. Sliced sections mounted onto glass slides are then subject to immunostaining. The second approach, which will be described in the following text, is a flat mount of the retina allowing visualization of the distribution of cells across the retina. The same staining steps can be followed for staining retina sections.

PROTOCOL 3: IMMUNOSTAINING ON FLAT MOUNT RETINA

1. Following an approved protocol to euthanize mice, enucleate the eyes using a pair of curved tip scissors. Fix the tissue by placing the eye in 4% PFA at 4°C for 4 h to overnight.
2. Transfer the eye to a 35 mm Petri dish filled with phosphate-buffered saline (PBS), and isolate the retina under a dissecting microscope:
   2a. Using Vannas scissors, remove the cornea and lens by making an incision along the equator of the eye (just below the border where the sclera meets the cornea).
   2b. Holding the outer edge of the eyecup with forceps, gently separate the retina from the adjacent choroids around the entire eyecup until only the optic nerve holds the two together. Clip the optic nerve to release the retina.

3. Using a camel hair number 0 paintbrush, gently transfer the retina to a 2 mL round-bottomed plastic tube containing 4% PFA for further fixation for 12–24 h at 4°C. Better staining is obtained by fixing the eye cup in 4% PFA after step 2a. Perform the steps (from step 4–13) using the paintbrush to transfer between wells of a 24-well cell culture dish containing 2 mL volume for each step (with the exception of antibody incubations, which can be done in volumes as small as 200 µL).

4. Wash thrice for 10 min at room temperature (RT) in PBS.

5. Block for 1 h at RT with gentle agitation in blocking buffer (10% normal donkey serum, 0.3% Triton X-100 in PBS).

6. Wash thrice for 10 min at RT in PBS.

7. Incubate 12 to 24 h at 4°C with gentle agitation in primary antibody (rabbit antimelanopsin, 1:1000) in blocking buffer.

8. Wash thrice for 10 min at RT in PBS. For steps 9 and 10, cover the dish with aluminum foil to protect the fluorescent secondary antibody from photobleaching.

9. Incubate for 1 h at RT in incubating buffer with secondary antibody (Cy3 conjugated donkey antirabbit, 1:500).

10. Wash thrice for 10 min at RT in PBS.

11. Transfer the retina to a glass slide; make 3–4 cuts, equidistant apart, from the periphery of the retina toward the center.

12. Open the retina so that it lies flat with the vitreous side up (this puts the ganglion cells up), and blot excess PBS.

13. Coverslip with an antifade mounting medium, and then store at 4°C in the dark before viewing with fluorescent microscopy.

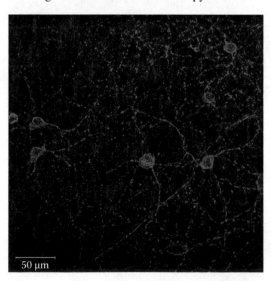

FIGURE 8.2 (See accompanying color CD.) Melanopsin immunostaining in mouse retina. A flatmount of retina from a 10-week-old mouse stained with a polyclonal antimelanopsin antibody. Melanopsin staining is detected in the soma, dendrites, and in the axons of a small subset of retinal ganglion cells.

As the soma, dendrites, and the axons of the melanopsin-expressing RGCs often lie in different planes, we typically examine the slides under a confocal microscope and collect Z-stack images. A typical flatmount image is shown in figure 8.2.

8.4 HETEROLOGOUS EXPRESSION AND FUNCTIONAL ASSAYS

Electrophysiological recordings from intact ipRGCs have established some unique properties of these photoreceptor cells that are distinct from those of rod/cone photoreceptors. The ipRGCs exhibit the following characteristics: long latency for activation, depolarization upon photostimulation, prolonged deactivation after cessation of light pulse, resistance to bleaching even after repeated photostimulation, and properties of both dark and light adaptation (reviewed in Berson (32)). Distinguishing which of these properties are encoded in the photopigment versus the cell and finding the molecular components and signaling events that underlie these properties is crucial to understanding the melanopsin/ipRGC phototransduction mechanism. However, exclusive expression of melanopsin in a few ipRGCs of the inner retina poses a major hurdle for extensive biochemical and cell biological characterization of its function in native ipRGCs. Therefore, heterologous expression of melanopsin in nonnative cells has been a critical tool.

The opsin class of photopigments has been functionally expressed in both mammalian and *Xenopus* oocyte expression systems (35,36). Rod/cone opsins from vertebrates lacking an intrinsic photoisomerase activity to regenerate their chromophore are typically expressed in cultured mammalian cells, immunoaffinity purified, reconstituted with retinoid photopigments, and used in biochemical studies. Opsins with an intrinsic photoisomerase activity such as chlamydominal rhodopsins upon expression offer nondestructive cell-based assays to assess their function (37). Because melanopsin immunostaining in the retina suggested that unlike rod/cone opsins, melanopsin was not localized to any specialized membrane structure (38) and the photoelectrical properties of the ipRGCs implicated an intrinsic photoisomerase activity (5,6), it was presumed that heterologous expression of melanopsin might provide a viable tool.

Successful biochemical reconstitution of spectrally active melanopsin expressed in COS cells (39) has been followed by several reports of its functional expression in other cell types where nondestructive cellular assays have been established. In native ipRGCs, photoactivation of melanopsin leads to increase in intracellular calcium (40), and membrane depolarization (5,6). Therefore, the heterologous cellular assays for melanopsin are based on photoactivation of intracellular calcium increase (41), changing the properties of native (42), or coexpressed candidate channels (43,44). Additionally, isolation and culturing of frog melanophores expressing native melanopsin also offers a simple cell culture system and a convenient densitometric assay to interrogate the molecular bases of melanopsin function. Methods to isolate, culture, and assay light responses from *Xenopus* dermal melanophores have been extensively described in Rollag et al. (45). We will allude to expression of melanopsin in mammalian cells and in *Xenopus* oocytes. In both systems, light-dependent function of heterologously expressed melanopsin can be monitored by electrical recording.

Human or mouse melanopsin can be expressed either as native protein or as fusion protein with a C-terminal epitope tag without significantly affecting the expression level and function. A C-terminus His (42) or 1D4 tag (39) facilitates (a) monitoring expression by immunohistology or by Western blot analysis and (b) immunoaffinity purification or immunoprecipitation using specific antibodies. The choice of using a mammalian expression system or *Xenopus* oocytes is primarily driven by the relative advantage of the expression system and the researcher's goals.

8.4.1 MAMMALIAN EXPRESSION SYSTEM

Several immortalized mammalian cell lines can be relatively easily transfected using calcium phosphate or lipid-based transfection reagents. Coexpression of melanopsin along with a GFP-based live cell expression marker has been used to generate melanopsin-expressing cells for subsequent electrophysiological studies (42,44). Standard patch clamp techniques for cultured cells along with controlled lighting are employed in electrophysiological recording to monitor light-evoked changes in membrane potential. This approach has established several key spectral properties of the ipRGCs: slow latency to activation, resistance to bleaching, and peak spectral sensitivity, which are not only unique to the ipRGCs, but can also be reconstituted in any cell type expressing melanopsin. This has raised the possibility that these properties are encoded in the melanopsin photopigment itself or are results of its functional interaction with other signaling components. To address these questions, the high-level expression and purification of melanopsin from mammalian cells are essential.

The first report of melanopsin expression in COS cells by transient transfection and affinity purification resulted in a functionally active, but low opsin yield (39). To achieve stable and high expression levels, we have established a lentiviral expression system. Oncoretroviral vectors have the potential for stable gene expression by integrating a transgene into the host genome. Lentivirus-based viral vectors have the distinguishing property of being able to escape gene silencing, a common problem with the first-generation oncoretroviral vectors, thus achieving stable transgene expression in both dividing and nondividing cells (46,47). In addition, the new generation of self-inactivating lentiviral vectors, characterized by transcriptional inactivation of the integrated virus because of a 3'LTR deletion mutation (48), carry a central polypurine tract (cPPT) from the *pol* gene of human immunodeficiency virus-1 (HIV-1) that increases nuclear translocation of viral genomic DNA (49,50) and a woodchuck hepatitis virus posttranscriptional regulatory element (wPRE) that enhances transgene expression (51). The full-length melanopsin cDNA is fused with a synthetic peptide coding for a 9–15 amino acid tag corresponding to the C-terminal end of bovine rhodopsin that is recognized by a monoclonal antibody 1D4 (52). This c-terminal tag facilitates subsequent immunoaffinity purification. A lentiviral gene delivery system also allows for expression of melanopsin in primary cells, mouse fibroblasts, and in cell types that are resistant to transient transfection by traditional methods.

PROTOCOL 4: GENERATION OF CELL LINES STABLY EXPRESSING MELANOPSIN USING LENTIVIRAL VECTORS

Generation of lentiviral vectors for melanopsin expression allows high-level expression of the protein in both primary and transformed cultured cells. The protocol describes virus production and generation of HEK293T cells stably expressing melanopsin.

1. Production of vesicular stomatitis virus glycoprotein (VSVG) envelope protein-pseudotyped lentiviral vector is well described elsewhere (Dull et al., 1998). In brief, HEK 293T cells are transfected with the expression vector, packing plasmids, and a plasmid coding for the VSVG envelope protein by the calcium-phosphate method. Virus is harvested over the following 2 days and concentrated by ultracentrifugation (68,000 × g). The titer of lentiviral particles is determined by measuring the amount of HIV p24 gag antigen by ELISA following the vendor's protocol (NENTM Life Science Products, Boston, Massachusetts). The multiplicity of infection (MOI) of lentiviral vector is estimated by the following equation: 1 ng of p24 = 105 infectious units (IUs).
2. Seed ~10^4 HEK 293T cells/well in a 12-well plate with Dulbecco's Modified Eagle Media (D-MEM) (Invitrogen) supplemented with fetal calf serum (FCS) (10%, v/v), streptomycin (100 g/mL), and penicillin (100 U/mL). Culture the cells overnight at 37°C in a humidified incubator with 95% air/10% to allow the cells to adhere.
3. Transduce HEK 293T cells by lentiviral vector (LV-CMV-mOpn4) at an MOI of 10 and incubate for 4 days to get maximum transgene expression. We have observed the optimum MOI is cell type specific. Therefore, for expression in other cell types, the optimum MOI should be empirically determined by transducing cells with a series of dilution of the viral stock.
4. Sparsely subculture the transduced HEK 293T cells at ~10^3 in a culture dish (15 mm in diameter) and incubate until each single cell forms a colony.
5. Randomly pick ~10 colonies into a 96-well plate, and let them grow.
6. Subculture each clone into two 12-well plates.
7. Confirm expression levels of transgene in each clone by both Western blot and immunohistochemical analysis (figure 8.3a).
8. Confirm expression of functional melanopsin either by patch clamp recording or FLIPR assay.

8.4.2 MELANOPSIN FLIPR ASSAY

We tested several high-throughput assays for G-protein-coupled receptors and found the use of calcium-sensitive cell-permeable Fura-based calcium dye produces the best signal-to-noise ratio. Other approaches involving calcium-sensitive fluorescent proteins and membrane localization of GFP-tagged arrestin can also be adapted for assessing melanopsin function.

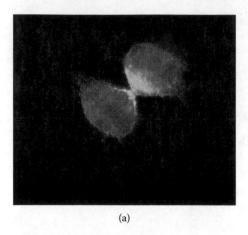

(a)

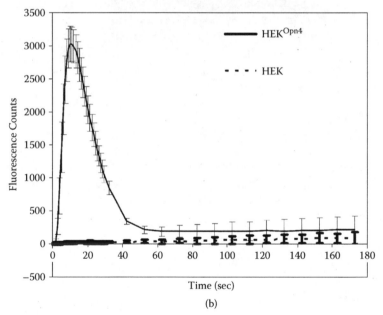

Time (sec)

(b)

FIGURE 8.3 (See accompanying color CD.) Functional expression of melanopsin in cultured mammalian cells. (a) HEK293T cells stably expressing melanopsin show largely membrane-localized immunostaining for melanopsin. (b) Photoactivation of HEK293T cells expressing melanopsin exhibit transient increase in intracellular calcium level. The cells were plated in FLIPR compatible 384-well plates, preincubated with a Fura-2-based dye for an hour prior to simultaneous light activation and fluorescence measurement in a FLIPR-3 machine. Time-lapsed fluorescence measurements (average + SD, n = 15 wells) show transient increase in fluorescence in cells expressing melanopsin and no significant change in fluorescence in normal HEK293T cells.

We employ Fluorimetric Light Imaging Plate Reader (FLIPR) and a Fura-2 acetoxymethyl ester-based calcium-sensitive dye (such as FLIPR Calcium 3 kit from Molecular Devices, California) for our assays. The 488 nm excitation laser or LED light source of the FLIPR machine that activates the dye also acts as a light source for melanopsin. This limits the use of the assay to a single saturating intensity of 488 nm as the light source for activating melanopsin. However, the assay can be used in high-throughput format in 384- or 1536-well format for (a) quick evaluation of functional expression of melanopsin in a given cell type, (b) pharmacological intervention of the signaling pathway leading to increase in intracellular calcium release, (c) evaluation of a series of targeted or randomly mutagenized melanopsin clones for structure–function studies, (d) identifying novel components required for functional expression, signaling, and regeneration of melanopsin by a functional genomics screen, and (e) screening for small molecule antagonists of melanopsin. Some of these applications require highly consistent transient transfection, whereas the others depend on cell lines stably expressing melanopsin.

To transiently express melanopsin in mammalian cells, a mouse melanopsin cDNA was cloned into a pcDNA3.1 vector under the transcriptional control of a CMV promoter. An IRES-EGFP sequence was inserted downstream of melanopsin for coexpression of EGFP, which can be used to assess transfection efficiency. The expression vector can also be used to generate cell lines stably expressing melanopsin, although the expression level is slightly less than when using lentiviral vectors. For generation of the mouse melanopsin stable cell line, the same construct was transfected using Lipofectamine2000 (Invitrogen, California) and selected for neomycin (Invitrogen, California) resistance over the course of 2 weeks. The cells were then FACS-sorted (fluorescence-activated cell sorting) using GFP as a fluorescent marker. Several single GFP-positive cell clones were grown and selected. For transient transfection assays, this construct was transiently transfected into HEK cells using Fugene 6 (Roche, California) following the vendor's protocol.

PROTOCOL 5: FLIPR ASSAY FOR MELANOPSIN FUNCTION IN CELLS STABLY OR TRANSIENTLY EXPRESSING MELANOPSIN

1. For transient transfection, normalize concentration of sample DNAs to 40 ng/µL.
2. Grow HEK293T cells in T-75 cm² flasks at ~90% confluent on the day of the transfection.
3. Resuspend calcium dye (FLIPR Calcium 3 kit, Molecular Devices) in 1× HBSS buffer (Hank's Balance Salt Solution diluted with 20 mM HEPES).
4. Day 1. Make Fugene-OptiMEM (Roche, Switzerland; Invitrogen, California) solution at 3:1 (vol/vol) ratio. Incubate for 5 min. Spot-plasmid DNA (1 µL/well) to 384-well black-walled clear-bottomed FLIPR compatible plate (Greiner Bio-One, North Carolina).
5. Add 10 µL of Fugene-OptiMEM solution (60 µL Fugene in 5 mL OptiMEM), incubate 30 min at RT (Caution: Bring everything to RT first, and only then add Fugene to OptiMEM; not the other way around).
6. While incubating Fugene-OptiMEM solution, trypsinize HEK 293T cells.
7. Adjust the cell density to 6000 cells/30 µL in DMEM medium with 10% FBS, 1 mM sodium pyruvate, and 1× antibiotic/antimycotic (invitrogen, California).

8. After incubation of DNA-Fugene mix, plate 30 µL of the cell suspension into each well using a matrix Impact2 Multichannel Pipettor (Matrix Technologies, New Hampshire). For subsequent steps, protect the cells from light by placing two opaque bottom plates, one on the top and on the bottom of the FLIPR plate. For assessing melanopsin function in cell lines stably expressing the protein, plate 10,000–20,000 cells/well in 384-well FLIPR compatible plate.

9. Incubate the cells at 37°C in a standard tissue culture incubator for 2 days (48 h).

10. Day 3. Under dim red light (Kodak 1A filter) remove media by inverting the plate and padding on paper towels, and carefully add 30 µL of calcium dye (Molecular Devices, California). For some cell types, including CHO cells, addition of 5 mM probenecid in the dye improves the signal.

11. Incubate at RT for 1 h, then read on FLIPR (Molecular Devices, California). As mentioned before, the laser that activates the dye also photoactivates melanopsin. So, unlike a typical FLIPR assay in which baseline data of a few seconds is collected before addition of the GPCR agonist, the melanopsin FLIPR assay produces a fluorescence signal as soon as the cells are exposed to the excitation light.

In a typical experiment the fluorescence signal reaches its peak within 5 s and slowly declines afterward (figure 8.3b). The assay is extremely sensitive and can detect melanopsin-mediated calcium response.

8.4.3 FUNCTIONAL EXPRESSION OF MELANOPSIN IN *XENOPUS* OOCYTE EXPRESSION SYSTEM

Xenopus oocyte is a common system to study electrophysiological and signaling properties of membrane receptors, channels, and transporters. In a typical experiment, stage 4 oocytes are injected with capped mRNA and after 2–4 days of protein synthesis, folding, and targeting, the oocytes are used in electrophysiological studies. The mRNA injection step allows for coinjection of several subunits or components of a signaling cascade where the relative amount of individual components can be titrated. Additionally, as in other electrophysiological studies, pharmacological perturbation agents can be used to characterize signaling events. Importantly, cell-impermeable agents can be injected into the oocytes. For our specific application, the relatively large size of the oocytes makes it easy to stage and impale the cell under dim red light conditions.

Both vertebrate and invertebrate opsins have been successfully expressed in *Xenopus* oocytes, where the light-activated opsin typically triggers a cascade of signaling events leading to the opening of native membrane channels and membrane depolarization (36,54). Voltage clamp whole-cell recording is used to monitor the change in membrane potential. We have developed two different assays for melanopsin phototransduction in *Xenopus* oocytes. In one assay, only capped

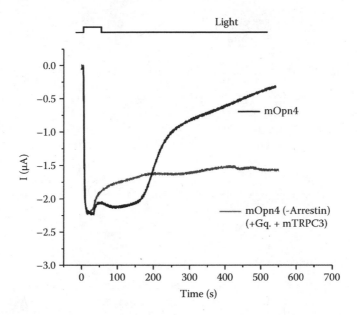

FIGURE 8.4   Functional expression of melanopsin in *Xenopus* oocytes. Light-evoked membrane depolarization in *Xenopus* oocytes expressing melanopsin. Stage 4 *Xenopus* oocytes were injected with capped poly(A)+ mRNA coding for *Opn4*, *TrpC3*, *Gnaq*, *Arr1b*, and incubated at 18°C for 2 days. On the day of recoding the oocyte was impaled with recording electrode, perfused with *11-cis* retinal, and exposed to a brief light pulse. Light evokes slow depolarization and slow turnoff of a photocurrent. In the absence of arrestin, the photocurrent persists for the entire duration of the recording session.

mRNA encoding mouse or human melanopsin is injected into the oocytes. On the day of recording, healthy oocytes are selected and incubated with *11-cis* retinal in the recording chamber. Upon illumination, melanopsin activates the $G_{\alpha q}/G_{\alpha 11}$-PLC pathway, leading to release of intracellular Ca^{2+}. Light-evoked increase in intracellular Ca^{2+} triggers the opening of calcium-activated native chloride channels, leading to membrane depolarization, which peaks within 3–5 s of commencement of the light pulse and returns to baseline immediately afterward. This assay is highly responsive to both light quality and quantity and has been used to generate action spectra of melanopsin-mediated photocurrent. The array has also been used to assess the chromophore use of melanopsin, and potential interaction with arrestin during the photoisomerization step. The second assay involves coexpression of melanopsin with arrestin, $G_{\alpha q}$, and the TrpC3 ion channel. With the right $G_{\alpha q}$ expression level, light triggers the opening of both the calcium-activated native chloride channels and the ectopically expressed TrpC3 channels. The TrpC3 channel remains open during the light pulse and returns to baseline several seconds after cessation of light pulse. This assay has established the role of arrestin in desensitizing the photoactivated melanopsin after light pulse and can potentially be used to

characterize signaling events and molecular interactions involving the photoactivation of melanopsin to the opening of TrpC channel [figure 8.4, (43)].

PROTOCOL 6: PREPARATION OF CAPPED MRNA FOR OOCYTE EXPRESSION

For optimum poly(A)$^+$ mRNA synthesis by in vitro transcription (IVT), we
cloned our genes of interest into a pOX expression vector (55) that has a T3/T7
promoter and a C-terminal polyadenylation site. The vector has a few unique
restriction sites, including NotI, in its backbone for linearizing the plasmid prior
to IVT. Any other vector with these features will also be suitable. Precautions
should be taken to minimize RNase contamination at each step. Dedicated
pipettes, barrier tips, water baths, or incubators should be used. The genes used
for mOpn4 studies are mOpn4 (NM_013887.1), Arrb1a (NM_177231), Arrb2
(NM_145429), Gαq (NM_008139), and mTrpC3 (NM_019510).

Synthesis of capped mRNA

We typically use the AmpliCap-Max T3 high-yield message maker kit (EPI-
CENTRE, Cat. ACM04033) to synthesize capped mRNA from linearized
plasmids. Any standard IVT kit with a capping reaction can also be used.

1. Linearize the constructs cloned in *Xenopus* oocyte expression vector (pOX)
   by NotI digestion. Make sure the DNA is completely digested by running a
   small aliquot on an agarose DNA gel.
2. Purify the linearized DNA using Qiagen PCR DNA purification kit. Adjust
   the DNA final concentration to 1.0 μg/μL. Avoid purifying cut bands from
   agarose gel, as a small amount of residual agarose can significantly inhibit
   IVT reaction and lower the RNA yield.
3. At room temperature, set up an IVT reaction following the vendor supplied
   protocol:

   5 μL RNAse-free water
   1 μL (1 μg/μL) linearized template DNA
   2 μL 10× transcription buffer
   8 μL Cap/NTP Premix
   2 μL 100 mM DTT
   2 μL AmpliCap-Max T7 or T3 enzyme solution
   _____
   20 μL Total reaction volume

   Incubate at 37°C for 30 min when making transcripts >500 bases. Incubate
   at 37°C for 2 h when making transcripts <500 bases.
4. After the IVT reaction, bring up the reaction volume to 100 μL by adding
   RNase-free water into the reaction mix. Follow the vendor's instruction to
   column-purify RNA using the RNAeasy Mini Kit (Qiagen, California).
5. Elute the capped mRNA with 40 μL of RNase-free water. To ensure maximum elution of RNA, reload the first elution onto the column and spin for 1
   min at top speed in a bench top centrifuge.

6. Measure the mRNA concentration with a standard spectrophotometer. Typical yield of capped RNA is around 40–50 µg/reaction.
7. (Optional). Run a small aliquot of mRNA on a RNA or DNA gel along with appropriate ladder. Successful full-length mRNA synthesis produces a single distinct band of appropriate size on the gel, whereas incomplete IVT or RNA degradation will produce a smear. One batch of capped mRNA can be used for up to a year without repeated freezing and thawing.
8. For long-term storage, add 3 µL of RNAse inhibitor (Superase-In from Ambion, Texas) to the mRNA sample and mix by gentle pipetting. Aliquot 1–3 µL of mRNA into PCR tubes. Transfer mRNAs to −80°C freezer.

Preparation of Xenopus oocytes

Follow your institutional guidelines for maintenance of frogs and for harvesting stage 4 oocytes. Alternatively, you can order stage 4 oocytes from commercial vendors, who ship dissected ovaries. Wipe the working area with 70% alcohol, and flame-sterilize all dissecting tools to avoid any bacterial contamination of the oocytes. Filter-sterilize all buffers and media.

1. Place the *Xenopus* ovary in a 10 cm dish half filled with ND96 (-Ca^{2+}) (96 mM NaCl, 2 mM KCl, 1 mM $MgCl_2$, HEPES, pH 7.5, no calcium). Under a dissecting microscope, carefully slice apart the sacs containing oocytes using two curved forceps. Transfer the small sliced pieces from half of the ovary to a 50 mL Falcon tube containing 25 mL ND96 (-Ca^{2+}). You can store the other half of the ovary at 18°C in ND96 (+Ca^{2+}) for up to a week.
2. Wash sliced pieces twice with ND96 (-Ca^{2+}), drain the excessive solution, and pour the sliced pieces into a new 50 mL Falcon tube with 25 mL collagenase (1 mg/mL freshly dissolved in ND96 (-Ca^{2+}), from Invitrogen, California).
3. For better digestion, place the tube on a horizontal orbital shaker. Set at slow motion for 90 min at room temperature. Usually, the collagenase digests the outer lobes and releases individual oocytes. If big undigested lobes are still found after 90 min, continue the digestion for another 30 min. The length of digestion often depends on the vendor and the quality of collagenase.
4. Rinse the oocytes with ND96 (-Ca^{2+}) several times, and then pour the oocytes into a 10 cm dish. Carefully select the oocytes that look healthy and have well-defined animal and vegetal poles with a presterilized glass Pasteur transfer pipette, and transfer to a new Falcon tube with ND96 (-Ca^{2+}). If most of the oocytes are attached with follicles, set up second digestion with 25 mL of collagenase solution (1 mg/mL) for 10 min at room temperature.
5. Pour off the collagenase solution, and wash oocytes several times with ND96 (-Ca^{2+}) to remove the follicles that come off the oocytes. Add 25 mL of ND96 (+Ca^{2+}) (96 mM NaCl, 2 mM KCl, 1 mM $MgCl_2$, 1.8 mM $CaCl_2$, 5 mM HEPES, supplemented with 100 U/mL penicillin, 100 µg/mL streptomycin, and 2.5 mM sodium pyruvate) into the tube. Gently pour the oocytes into a 10 cm tissue culture dish. Store the prepared oocytes at 18°C. Check the oocytes every day to remove sick or dead oocytes. If there are too many

dead oocytes, transfer the live ones to a new dish with fresh ND96 (+Ca^{2+}). Oocytes can stay alive for up to a week at 18°C.

PROTOCOL 7: MICROINJECTION OF mRNA INTO *XENOPUS* OOCYTES

mRNA sample preparation for oocyte injection

1. On ice, thaw the mRNA samples for injection.
2. Mix 1 μL (1 μg/μL) of mRNA from each sample: mouse melanopsin, G$_{\alpha q}$, and Arrβ1a. In parallel, you can prepare additional mRNA mixes lacking arrestin, TrpC3, or G$_{\alpha q}$ to evaluate their contribution to the melanopsin-mediated photocurrent.
3. Add 1 μL of RNase inhibitor (Ambion, California). Top up the mixture to 10 μL with RNase-free water.
4. Mix the mRNAs with a pipette, and place the samples on ice.

Microinjection and protein expression

1. Use a micromanipulator and a Drummond microinjector for mRNA injection. Back-fill the glass micropipette (3.5 in. Drummond #3-000-2-3-G/X from Drummond Scientific Company, Florida; prepared with P-97 from Sutter Instrument) with mineral oil using a microfill syringe, and place the micropipette in position of Drummond microinjector. Set the injection volume of the microinjector to ~50 nL.
2. Centrifuge the RNA mix in a tabletop centrifuge at maximum speed for 1–2 min to pellet any precipitate or insoluble material that can potentially clog the injection needle. Carefully pipette out 4–5 μL of the content, and pipette onto a clean, disposable 35 mm dish.
3. Carefully break the tip of the micropipette with a forceps under microscope. Empty the mineral oil from inside of the micropipette. Using backpressure, fill it with 2–3 μL of prepared mRNA sample from the dish.
4. Line up the healthy oocytes in a dish containing ND96 (+Ca^{2+}) with a mesh on the bottom to hold the oocytes in place, and inject each oocyte with 50 nL of mRNA sample. You can inject up to 50 oocytes before running out of RNA in the microinjector. For each additional RNA mix, use a separate glass micropipette.
5. Collect the injected oocytes with a sterilized glass Pasteur pipette, and transfer them to a new dish with fresh ND96 (+Ca^{2+}). Place the dish in an 18°C incubator, and allow protein expression for 48 h. Saturating amounts of melanopsin, G$_{\alpha q}$, Arrβ1 are usually synthesized within 24 h of injection. TrpC proteins can take up to 3 days to reach saturating level. During the incubation period check the dishes every day under a dissecting microscope to remove dead or sick oocytes.

PROTOCOL 8: FUNCTIONAL ASSAYS ON *XENOPUS* OOCYTE

Upon light pulse, melanopsin will activate its coupled G$_{\alpha q}$ protein, and subsequently activates PLCβ to produce IP3 and DAG, which further activates the native chloride channel and/or the ectopically expressed TrpC3 channel. The resulting inward current can be recorded in voltage clamp mode with holding potentials at around −70 mV. Positioning and impalement of oocytes for recording should be made under dim red light, and the subsequent recordings should be performed under dark conditions.

Electrical recording

1. Fill one of the perfusion lines with Barth's saline (87.5 mM NaCl, 2.4 mM NaHCO₃, 2 mM KCl, 1 mM CaCl₂, 1 mM MgCl₂, 5 mM Tris-HCl, pH 7.2), and the other with 40 μM *all-trans* retinal (dissolved in DMSO) in Barth's saline.
2. Flush the perfusion lines with Barth's saline.
3. Place two electrodes with recording micropipettes in position, and dip both of the micropipettes under the Barth's saline. Zero the baseline, and perform a "Z" test on both recording needles.
4. Carefully pick a good oocyte under dim red light, and transfer it to the recording chamber prefilled with Barth's saline.
5. Impale the oocyte with the two recording needles under dim red light, and place a white light source on top of the oocyte.
6. Clamp the membrane potential at −70 mV, and perfuse the oocytes with *all-trans* retinal for 4 min followed with 1 min Barth's saline.
7. Pulse the oocyte with saturating white light for 60 s. Typically, an inward 500 nA current will be detected. TrpC3 is a slow-opening channel and takes about 35 to 50 s to reach the maximum current.
8. After cessation of light pulse, continue monitoring the current until it returns to baseline.

8.5 PURIFICATION OF MELANOPSIN FOR SPECTRAL AND BIOCHEMICAL STUDIES

Monoclonal 1D4 antibody has been used to purify melanopsin with satisfactory results (39). Therefore, we also used this 1D4 antibody to purify mouse melanopsin (mOpn4). Lentiviral vectors containing the full-length mOpn4 coding sequence fused to a nonapeptide antigenic tag from bovine rhodopsin (52) under the transcriptional regulation of CMV promoter was constructed, and then stable HEK293T cell lines expressing mOpn4 were established as described earlier.

PROTOCOL 9: mOPN4₁D₄ AFFINITY PURIFICATION

1. Seed ~10^8 HEK 293T cells expressing mOpn4₁D₄ in a 500 cm² dish at 37°C, and wait for 12 h until the cells adhere. After 12 h, cover the dish with aluminum foil to prevent the cells from getting exposed to light for another 24 h till they become confluent. Note: All steps should be carried out under dark or dim red light (Kodak 1A).
2. Harvest the cells by flushing PBS containing 5 mM EDTA over the loosely adherent cells and pipetting up and down. Spin down the cells at 700 × g for 10 min.

3. Homogenize the cell pellet in 9 volumes of ice-cold HEPES buffer (20 mM HEPES (N-(2-hydroxyethyl) piperazine-N'-(2-ethanesulfonic acid), pH 7.4, 1 mM EDTA, and protease inhibitor cocktail (ROCHE)) with a dounce homogenizer (Heidolph, Germany). Centrifuge the homogenate at 2000 × g for 15 min at 4°C to collect the supernatant (S1). The cell pellet (P1) is discarded.

4. Spin S1 at 25,000 × g for 1 h at 4°C. Use the pellet (P2 or the plasma membrane fraction) for the extraction procedure, and discard the supernatant (S2).

5. Resuspend P2 thoroughly with the help of a 23 gauge needle in 2 mL of phosphate buffer (10 mM sodium phosphate buffer, pH 6.4/150 mM NaCl, and protease inhibitors).

6. Extract P2 with 1% DM (n-Dodecyl-β-D-maltoside) (made in PBS) for 1 h at 4°C under end over rotation. Centrifuge the extracted membrane at 14000 × g for 1 h at 4°C to remove the unextracted material.

7. Prepare NHS-Sepharose beads coupled with Rho 1D4 purified monoclonal mouse antibody (prepared according to manufacture's protocol, Amersham Biosciences). Typically, 5 mg of antibody is coupled with 1 mL to obtain 95% binding efficiency. Incubate ~1.5 mL of extract with 150 µL of coupled NHS-Sepharose beads overnight at 4°C on end-to-end rotation.

8. Spin down the beads in a benchtop centrifuge (2000 × g/1 min at 4°C) and wash (2000 × g/1 min) consecutively in 5 times volume of high-salt buffer (150 mM NaCl, 0.1% DM in 10 mM sodium phosphate buffer, pH 6.4) followed by 5 times volume of low-salt buffer (0.1% DM in 10 mM sodium phosphate buffer, pH 6.4) at RT.

9. Incubate the beads with 200 µL of the low-salt buffer containing 100 µM of epitopic peptide (VSKTETSQVAPA) of 1D4 antibody for 30 min at room temperature. The peptide is usually prepared fresh in double-distilled water. Spin down and collect the supernatant. Repeat this step twice to elute the rest of the protein from the beads. Elute contains melanopsin in its nonreconstituted (without the chromophore) form.

10. For reconstitution or spectral studies, the pellet P2 (from step 4) is reconstituted with 25 µM of *11-cis* retinal for 2 h at 4°C under end over rotation. The rest of the purification procedure is carried out in the same way as described in this section.

11. The final yield of the pure protein obtained from ~6*10⁸ cells is 5–6 µg. For quantitation, linear range of BSA (50–500 ng) is run along with the pure protein on SDS PAGE and silver-stained. The amount of pure protein is calculated by comparing BSA bands to the protein band, 10 µL of pure protein (containing approximately 100 ng) is usually loaded for Western blotting or for silver staining.

12. The foregoing purification procedure is fairly good for obtaining protein in its pure form to conduct spectral studies. The amount of protein purified by this procedure is also in a good range for mass spectroscopic analysis.

REFERENCES

1. Hastings, M.H., Reddy, A.B., and Maywood, E.S., (2003) A clockwork web: circadian timing in brain and periphery, in health and disease. *Nat Rev Neurosci* 4: 649–61.
2. Van Gelder, R.N., (2001) Non-visual ocular photoreception. *Ophthalmic Genet* 22: 195–205.
3. Lucas, R.J., Douglas, R.H., and Foster, R.G., (2001) Characterization of an ocular photopigment capable of driving pupillary constriction in mice. *Nat Neurosci* 4: 621–6.
4. Takahashi, J.S., et al., (1984) Spectral sensitivity of a novel photoreceptive system mediating entrainment of mammalian circadian rhythms. *Nature* 308: 186–8.
5. Berson, D.M., Dunn, F.A., and Takao, M., (2002) Phototransduction by retinal ganglion cells that set the circadian clock. *Science* 295: 1070–3.
6. Hattar, S., et al., (2002) Melanopsin-containing retinal ganglion cells: architecture, projections, and intrinsic photosensitivity. *Science* 295: 1065–70.
7. Provencio, I., et al., (2000) A novel human opsin in the inner retina. *J Neurosci* 20: 600–5.
8. Lucas, R.J., et al., (2003) Diminished pupillary light reflex at high irradiances in melanopsin-knockout mice. *Science* 299: 245–7.
9. Panda, S., et al., (2003) Melanopsin is required for non-image-forming photic responses in blind mice. *Science* 301: 525–7.
10. Panda, S., et al., (2002) Melanopsin (Opn4) requirement for normal light-induced circadian phase shifting. *Science* 298: 2213–6.
11. Ruby, N.F., et al., (2002) Role of melanopsin in circadian responses to light. *Science* 298: 2211–3.
12. Hattar, S., et al., (2003) Melanopsin and rod-cone photoreceptive systems account for all major accessory visual functions in mice. *Nature* 424: 76–81.
13. Keeler, C.E., (1927) Iris movements in blind mice. *American Journal of Physiology* 81: 107–112.
14. Tu, D.C., et al., (2004) Nonvisual photoreception in the chick iris. *Science* 306: 129–31.
15. Van Gelder, R.N., (2005) Nonvisual ocular photoreception in the mammal. *Methods Enzymol* 393: 746–55.
16. Ebihara, S. and Tsuji, K., (1980) Entrainment of the circadian activity rhythm to the light cycle: effective light intensity for a Zeitgeber in the retinal degenerate C3H mouse and the normal C57BL mouse. *Physiol Behav* 24: 523–7.
17. Foster, R.G., et al., (1991) Circadian photoreception in the retinally degenerate mouse (rd/rd). *J Comp Physiol [A]* 169: 39–50.
18. Lucas, R.J., et al., (2001) Identifying the photoreceptive inputs to the mammalian circadian system using transgenic and retinally degenerate mice. *Behav Brain Res* 125: 97–102.
19. Mrosovsky, N. and Hattar, S., (2003) Impaired masking responses to light in melanopsin-knockout mice. *Chronobiol Int* 20: 989–99.
20. Siepka, S.M. and Takahashi, J.S., (2005) Methods to record circadian rhythm wheel running activity in mice. *Methods Enzymol* 393: 230–9.
21. Rollag, M.D. and Niswender, G.D., (1976) Radioimmunoassay of serum concentrations of melatonin in sheep exposed to different lighting regimens. *Endocrinology* 98: 482–9.

22. Roseboom, P.H., et al., (1998) Natural melatonin 'knockdown' in C57BL/6J mice: rare mechanism truncates serotonin N-acetyltransferase. *Brain Res Mol Brain Res* 63: 189–97.
23. Dacey, D.M., et al., (2005) Melanopsin-expressing ganglion cells in primate retina signal colour and irradiance and project to the LGN. *Nature* 433: 749–54.
24. Hattar, S., et al., (2006) Central projections of melanopsin-expressing retinal ganglion cells in the mouse. *J Comp Neurol* 497: 326–49.
25. Gooley, J.J., et al., (2003) A broad role for melanopsin in nonvisual photoreception. *J Neurosci* 23: 7093–106.
26. Hannibal, J. and Fahrenkrug, J., (2004) Target areas innervated by PACAP-immunoreactive retinal ganglion cells. *Cell Tissue Res* 316: 99–113.
27. Morin, L.P., Blanchard, J.H., and Provencio, I., (2003) Retinal ganglion cell projections to the hamster suprachiasmatic nucleus, intergeniculate leaflet, and visual midbrain: bifurcation and melanopsin immunoreactivity. *J Comp Neurol* 465: 401–16.
28. Sollars, P.J., et al., (2003) Melanopsin and non-melanopsin expressing retinal ganglion cells innervate the hypothalamic suprachiasmatic nucleus. *Vis Neurosci* 20: 601–10.
29. Fahrenkrug, J., Nielsen, H.S., and Hannibal, J., (2004) Expression of melanopsin during development of the rat retina. *Neuroreport* 15: 781–4.
30. Hannibal, J. and Fahrenkrug, J., (2004) Melanopsin containing retinal ganglion cells are light responsive from birth. *Neuroreport* 15: 2317–20.
31. Sekaran, S., et al., (2005) Melanopsin-dependent photoreception provides earliest light detection in the mammalian retina. *Curr Biol* 15: 1099–107.
32. Berson, D.M., (2003) Strange vision: ganglion cells as circadian photoreceptors. *Trends Neurosci* 26: 314–20.
33. Hannibal, J., et al., (2004) Melanopsin is expressed in PACAP-containing retinal ganglion cells of the human retinohypothalamic tract. *Invest Ophthalmol Vis Sci* 45: 4202–9.
34. Hannibal, J., et al., (2002) The photopigment melanopsin is exclusively present in pituitary adenylate cyclase-activating polypeptide-containing retinal ganglion cells of the retinohypothalamic tract. *J Neurosci* 22: RC191.
35. Oprian, D.D., et al., (1987) Expression of a synthetic bovine rhodopsin gene in monkey kidney cells. *Proc Natl Acad Sci U S A* 84: 8874–8.
36. Khorana, H.G., et al., (1988) Expression of a bovine rhodopsin gene in Xenopus oocytes: demonstration of light-dependent ionic currents. *Proc Natl Acad Sci U S A* 85: 7917–21.
37. Nagel, G., et al., (2003) Channelrhodopsin-2, a directly light-gated cation-selective membrane channel. *Proc Natl Acad Sci U S A* 100: 13940–5.
38. Belenky, M.A., et al., (2003) Melanopsin retinal ganglion cells receive bipolar and amacrine cell synapses. *J Comp Neurol* 460: 380–93.
39. Newman, L.A., et al., (2003) Melanopsin forms a functional short-wavelength photopigment. *Biochemistry* 42: 12734–8.
40. Sekaran, S., et al., (2003) Calcium imaging reveals a network of intrinsically light-sensitive inner-retinal neurons. *Curr Biol* 13: 1290–8.
41. Isoldi, M.C., et al., (2005) Rhabdomeric phototransduction initiated by the vertebrate photopigment melanopsin. *Proc Natl Acad Sci U S A* 102: 1217–21.
42. Melyan, Z., et al., (2005) Addition of human melanopsin renders mammalian cells photoresponsive. *Nature* 433: 741–5.
43. Panda, S., et al., (2005) Illumination of the melanopsin signaling pathway. *Science* 307: 600–4.
44. Qiu, X., et al., (2005) Induction of photosensitivity by heterologous expression of melanopsin. *Nature* 433: 745–9.

45. Rollag, M.D., et al., (2000) Cultured amphibian melanophores: a model system to study melanopsin photobiology. *Methods Enzymol* 316: 291–309.
46. Naldini, L., et al., (1996) In vivo gene delivery and stable transduction of nondividing cells by a lentiviral vector. *Science* 272: 263–7.
47. Somia, N.V., et al., (2000) Retroviral vector targeting to human immunodeficiency virus type 1-infected cells by receptor pseudotyping. *J Virol* 74: 4420–4.
48. Miyoshi, H., et al., (1998) Development of a self-inactivating lentivirus vector. *J Virol* 72: 8150–7.
49. Follenzi, A., et al., (2000) Gene transfer by lentiviral vectors is limited by nuclear translocation and rescued by HIV-1 pol sequences. *Nat Genet* 25: 217–22.
50. Zennou, V., et al., (2000) HIV-1 genome nuclear import is mediated by a central DNA flap. *Cell* 101: 173–85.
51. Ramezani, A., Hawley, T.S., and Hawley, R.G., (2000) Lentiviral vectors for enhanced gene expression in human hematopoietic cells. *Mol Ther* 2: 458–69.
52. MacKenzie, D., et al., (1984) Localization of binding sites for carboxyl terminal specific anti-rhodopsin monoclonal antibodies using synthetic peptides. *Biochemistry* 23: 6544–9.
53. Dull, T., et al., (1998) A third-generation lentivirus vector with a conditional packaging system. *J Virol* 72: 8463–71.
54. Zemelman, B.V., et al., (2002) Selective photostimulation of genetically chARGed neurons. *Neuron* 33: 15–22.
55. Jegla, T. and Salkoff, L., (1997) A novel subunit for shal K+ channels radically alters activation and inactivation. *J Neurosci* 17: 32–44.

3

Invertebrate Visual Phototransduction

9 Phototransduction in *Drosophila*
Use of Microarrays in Cloning Genes Identified by Chemically Induced Mutations Causing ERG Defects

Hung-Tat Leung, Lingling An, Julie Tseng-Crank, Eunju Kim, Eric L. Harness, Ying Zhou, Junko Kitamoto, Guohua Li, Rebecca W. Doerge, and William L. Pak

CONTENTS

ABSTRACT

One of the main reasons for the power of *Drosophila* in addressing problems of biological importance is that it allows forward genetic screens of chemically induced mutations to be carried out relatively easily. The problem of phototransduction was addressed by forward genetic screens of *Drosophila* mutants with defective electroretinogram (ERG). However, identifying and isolating the gene that carries the lesion responsible for the mutant phenotype are difficult and time consuming. Recently, a number of strategies have been proposed to improve the accuracy and resolution of mutation mapping so that the isolation of the lesion-bearing genes may be facilitated. However, these strategies generally involve fine mapping steps that are also labor intensive and time consuming, particularly for mapping mutants whose phenotypes can only be detected by electroretinogram (ERG) recording. Here, we describe a strategy for gene cloning that is based on DNA microarrays and that does not require fine mapping of mutations. It is based on the observation that almost all ERG defect-causing mutations also cause alterations in the mRNA levels. DNA microarrays are used to detect all annotated genes that are altered in mRNA levels in a target mutant compared to a wild-type control. Using information based solely on deficiency mapping, we then look for genes within the mapped interval that show the largest and most statistically significant alterations in mRNA levels. We describe applications of this strategy to five genes, two previously identified ones and three new ones. In all cases, it was possible to identify one or two candidate genes within the mapped interval reliably and rapidly. It generally took 4 months or less from the time of fly stock expansion for RNA isolation to the time of candidate gene identification. If there were more than one candidate gene in the mapped interval, the correct gene was identified by sequencing.

9.1 BACKGROUND AND INTRODUCTION

Almost everything we know about *Drosophila* phototransduction has come from studies based on genetic and molecular genetic approaches. A forward genetic approach to *Drosophila* phototransduction was initiated in the winter months of 1966/1967 by chemically mutagenizing fruit flies, *Drosophila melanogaster*, to generate mutants that are defective in the electroretinogram (ERG), extracellularly recorded, light-elicited, mass response of the eye. The objective was to isolate phototransduction-defective mutants, but the mutagenesis screen targeted ERG-defective mutants in the hope that phototransduction-defective mutants would appear as a subset of such mutants (reviews in references 1,2,3,4). In the early years of mutagenesis, we screened the flies for nonphototaxis first, in an attempt to enrich the sample for impaired phototactic behavior before testing them for their ERGs (5). However, phototactic behavioral screens turned out to be very inefficient at identifying ERG-defective mutants. Ultimately, we decided to screen for ERG defects directly, and most of our mutants were isolated this way. Over a period of about 10 years, we isolated over 200 mutants falling into about 50 complementation groups. We never carried out our mutagenesis to saturation. When we felt that we had reached a point of diminishing returns, we decided to stop. Several other groups were also engaged in mutagenesis of *Drosophila* for the isolation of mutants of neurobiological interests at the time and isolated some ERG-defective mutants. The Benzer group's main focus was behavior, although it also isolated some X-chromosome ERG-defective mutants (6,7). The Heisenberg group's interest was on visual behavior, and it too isolated some X-chromosome ERG-defective mutants (8,9). The John Meriam group at UCLA also contributed to the isolation of autosomal ERG-defective mutants (10).

With the development of gene cloning techniques for *Drosophila* in subsequent years, it became possible to identify the protein products that are altered in the mutants, paving the way for the determination of key players in phototransduction. However, cloning genes that are identified only by chemically induced mutations has always been a notoriously time-consuming and laborious process. Largely for this reason, about a half of the genes corresponding to the 50 complementation groups into which our mutants fall have not yet been characterized. In the absence of any other information, these genes are cloned on the basis of their chromosomal location. Gene cloning thus begins with mapping of the target mutants. One then has to identify the gene corresponding to the mutant among the many genes residing within the mapped interval. Obviously, the more accurate the mapping, the easier it is to identify the target gene. Consequently, most of the recent efforts to facilitate cloning have been aimed at improving the accuracy and resolution of mapping.

In the remainder of this chapter, we will first discuss some of the recently developed methods of mapping *Drosophila* mutants. We will then describe the microarray-based gene cloning strategy that we have been employing recently with considerable success. It makes much of the time-consuming fine mapping unnecessary. Moreover, it identifies a very small number of candidate genes within the mapped interval, allowing one to concentrate on these genes.

9.2 MAPPING

9.2.1 TRADITIONAL METHODS

Two most commonly used methods of mapping are recombination mapping and deficiency mapping. Meiotic recombination mapping consists of determining frequencies of recombination between known visible markers and the target mutation, which causes a detectable phenotype. Its accuracy depends on the availability of the markers and the number of recombinants scored. It is generally not a method of choice because, even under the best of conditions, the scarcity of suitable markers makes it difficult to achieve a resolution of less than several hundred kilobases (11). Moreover, in the case of ERG mutants, the phenotypes of recombinants must be scored by ERG recording, making it impractical to score a large number of recombinants. However, we have often used this method to make a rough initial determination of the chromosomal region to which a mutation maps.

In deficiency mapping, the chromosomal position of the target mutation is determined from its failure to complement previously characterized deficiencies (deletions), when placed in trans with the deficiencies. The accuracy of this method depends on the extent to which the genome is covered by deficiencies and the availability of molecular information on the breakpoints of the deficiencies. It is highly desirable that any given region of the genome be covered by multiple deficiencies. Deficiencies are usually large, and failure of a mutation to complement a single deficiency would map it within the large region encompassed by the deficiency. If, on the other hand, a mutation fails to complement several deficiencies with different breakpoints, the mutation would map within the region of overlap among the deficiencies, providing mapping of much higher resolution. There have been on-going efforts to obtain complete coverage of the *Drosophila* genome with chromosomal deficiencies whose breakpoints are determined at the nucleotide level. Elexis, Inc., a private company, and DrosDel, a publicly funded European consortium, have independently generated a total of about 1000 molecularly defined deficiencies, which average 140 (Elexis) to 400 kb (DrosDel) in size (12,13). These are estimated to cover ~80% of the genome (14). A current effort at the Bloomington *Drosophila* Stock Center is aimed at ~1500 more deletions of ~200 kb average size to obtain >95% genome coverage (cited in Reference 14). In addition, a large number of deficiencies whose breakpoints have not been determined are also available. However, the coverage is still quite uneven. Some regions are densely covered with deficiencies, whereas for some others, coverage is relatively poor. Moreover, for most of the deficiencies available, the breakpoints have not been determined at the molecular level. Thus, even with the improved coverage and improved molecular characterization of deficiency breakpoints, our experience has been that, with some exceptions, for most mutations of interest it is difficult to map reliably with a resolution of less than ~50–100 kb. This is still too large a region to identify the target gene readily. The recent improvements in mapping to be discussed in the next section are intended to improve the resolution further.

9.2.2 Recent Improvements in Mapping Strategy

Several different mapping strategies that would considerably improve the resolution have been proposed recently. They are P-element-mediated male recombination mapping (15,16), SNP (single-nucleotide polymorphism)-based mapping (17), and molecularly defined P-element insertion-based mapping (11). P-element-mediated male recombination mapping relies on the fact that male recombination, which is normally rare in *Drosophila*, can occur at a frequency of up to 1% in crosses involving fly strains carrying P elements. Because recombination occurs primarily at the site of P element insertion (15), from the examination of recombinants identified with the help of flanking markers, one can determine if any given P element chosen for mapping lies to the right or left of the mutation. Thus, using a series of P elements in the region of interest, any given target mutation can be mapped to an accuracy that depends only on the density of P elements in the region (16). As in the case of deficiencies, there have been ongoing efforts to increase the coverage of the genome by P element insertions. The Bloomington *Drosophila* Stock Center lists nearly 10,000 P insertion lines that are mapped at the nucleotide level (Flybase) (http://www.flybase. org). Perhaps, the largest collection of molecularly defined P element insertion lines currently in existence is that generated by a group headed by Jaeseob Kim of Korea Advanced Institute of Science and Technology, Daejon, Korea, and marketed by GenExel-Sein, Inc. This group generated nearly 100,000 P insertion lines, ~25,000 of which are said to be unique. If one includes those that are inserted immediately upstream of a gene on the sense strand and those immediately downstream of a gene on the antisense strand,[1] as well as those within the gene, the genome coverage is estimated to be about 52–70% (Eunkyung Bae, GenExel, private communication). Genome coverage by P element insertions is not expected to reach 100% because a substantial fraction of genes in the *Drosophila* genome is thought to be refractory to P insertions. Nevertheless, P element insertions are sufficiently abundant to make this an effective tool. However, it has a number of drawbacks (see also Reference 11) and does not eliminate many of the time-consuming steps in mapping. First of all, the method is not applicable to mutations on the X chromosome because it requires that the chromosome carrying the mutation and the homologous chromosome carrying the P element be in trans to each other in male flies. This is not possible, because male flies have only one X chromosome. Second, the method requires that visible markers be recombined into the chromosomes carrying the mutations being mapped. Third, to obtain sufficient accuracy, several rounds of mapping using chromosomes with different P element insertions need to be performed. Fourth, at each round of mapping, this method maps the mutation either to the left or right of the P insertion and does not provide any further clues to the approximate map position to guide the choice of the next P insertion. Finally, the method does not necessarily identify the gene carrying the mutation. It ultimately maps the mutation between two adjacent P insertions in the chromosome. In a region densely covered with P insertions, there may only be a small number of genes within the mapped interval. In a sparsely covered region, however, there could be tens of, or even over a hundred, genes in the interval.

The second method is one based on SNPs (17). It is basically meiotic recombination mapping that utilizes SNPs as molecular markers for recombination events. SNPs are single-nucleotide differences between homologous chromosomes that have been shown to exist in sufficient numbers in the *Drosophila* genome for this method to be practicable. Crosses are set up to allow meiotic recombinations to take place between the chromosome bearing the mutation to be mapped and a homologous chromosome that bears a visible marker. The visible marker is often a P insertion into which a visible mutation has been engineered. Recombinations occur in females heterozygous for the aforementioned two chromosomes, and recombinants are detected in the following generation with the help of the visible marker. This method of mapping requires that one first construct an SNP map in the region of the chromosome to be tested. SNPs used for the map must be polymorphic between the marker chromosome and the mutant chromosome such that their chromosomal origin can be identified either from restriction fragment length polymorphisms or sequence differences. The recombinant offspring that are identified with the help of the P insertion marker are then tested for both their mutant phenotype (by ERG recording in our case) and the genotypes of the SNP markers. The genotypes of SNPs in any particular recombinant establish the site of the crossover event in relation to the SNP map, and the phenotype (mutant or wild type) of the recombinant establishes the site of the mutation in relation to the site of the crossover (to the left or right of the crossover event). If a sufficient number of such recombinants is analyzed, one can map the mutation between two neighboring SNPs in the SNP map. The recombinant events most informative regarding the mapping would be the rarest ones that occur between the mutation and the SNP nearest to it. Among the major drawbacks of this method is the necessity for constructing an SNP map each time a mutation in a different region of the genome is to be mapped. If preexisting information on SNPs is used in the construction of the map, it will still have to be confirmed. Moreover, the greater the density of SNP markers and the greater the number of recombinants generated and tested, the greater will be the accuracy of mapping, and each recombinant will have to be tested for the mutant phenotype by ERG recording and for the genotype of SNP markers by either restriction digestion or sequencing. All these are potential time-consuming steps. In addition, the final outcome, in general, will not necessarily be the identification of a single target gene but a number of candidate genes.

Still another recently proposed method of mapping is that based on the use of molecularly defined P element insertions (11). This method is essentially classical meiotic recombination mapping that uses, instead of visible markers, P element insertions as markers to determine recombination rates. Potentially, this can be a powerful method because a large number of P insertions, for which the sites of insertions have been determined at the molecular level, are already available, and more are being generated (see previous section). A major drawback of this method of mapping, however, is that because of the large number of recombinants that need to be scored, its use is restricted to mapping mutations that produce easily scorable phenotypes. Zhai et al. (11), for example, used this method to map lethal mutations. They carried out their mapping in two steps, rough mapping followed by fine mapping. Both utilized the same recombination mapping strategy, but in fine mapping they used closely spaced P insertions that span the interval defined by rough mapping. They recommend scoring >1000 recombinant progeny for each P insertion cross to attain an accuracy of <1 cM in rough mapping and scoring 10,000 progeny to achieve an accuracy of 0.1 cM for each P insertion in fine mapping. Because they were mapping lethal mutations, mutations could be scored readily in large numbers. For mutants with phenotypes that are more difficult to score, scoring large number of progeny will be a problem. For ERG mutants, the method is too labor intensive to be practical. Moreover, the 0.1 cM accuracy attained by fine mapping still corresponds, on the average, to about 50 kb, and the gene corresponding to the mutant will still have to be identified among all the genes in that interval.

9.3 MICROARRAY-BASED GENE CLONING

To circumvent many of these problems, we have been employing a gene cloning strategy based on DNA microarrays with considerable success. This strategy is based on the observation that most mutations that cause defective ERGs also cause alterations in steady-state cellular mRNA levels, which are referred to as "gene expression" levels hereafter. The mutant of interest is first mapped by deficiency mapping. Microarrays of the mutant are carried out along with those of wild type to detect all genes that are altered in RNA levels in the mutant in comparison to wild type. One then focuses on those genes that localize within the mapped interval. Candidate genes show up as those with the largest and most statistically significant alterations in gene expression within the mapped interval. This approach, in our hands, has enabled us to bypass the time-consuming and labor-intensive intermediate steps of fine mapping and focus immediately on a very small number (usually one or two) of candidate genes.

In basic terms, DNA microarrays are massive, parallel Northern blotting. Microarrays were first used to assess the simultaneous expression of a large number of genes over 10 years ago (18,19). However, their application to the entire genome awaited the completion of genome sequencing. It was then possible to examine the expression of all genes in the sequenced genome simultaneously. For *Drosophila*, nearly complete sequencing of the euchromatic portion of the genome was announced only 7 years ago (20). In microarrays, oligonucleotide probes representing all genes in the annotated genome are immobilized on the surface of a chip. RNA isolated from the experimental tissue is converted to cRNA and cRNA is hybridized to these probes. Oligonucleotide arrays (chips) covering the entire sequenced *Drosophila* genome are commercially available from Affymetrix. The Affymetrix GeneChip uses 25-mer oligonucleotides as probes, and, for *Drosophila*, 14 oligonucleotide probes, known as a probe set, are used to measure the transcriptional abundance of each gene sequence. The latest version of Affymetrix *Drosophila* genome array (version 2.0) is based on *Drosophila* Genome Annotation release 3.1 (released in 2003, Flybase) and consists of 18,952 probe sets. A probe set usually corresponds to a gene, but in some cases more than one probe set is used to probe the transcriptional feature of a gene.

For the microarray strategy to work, the following two conditions must be met: (a) that most chemically induced ERG-defect-causing mutations alter the transcriptional levels of the genes harboring the mutations, and (b) that very few other genes within the mapped limits of the target mutation show transcriptional alterations of similar magnitude and statistical significance as the target gene. Evidence that these suppositions indeed are correct ultimately came from the actual tests of the technique, to be described later. However, some data bearing on the first of these suppositions were available from Northern results on ERG-defective mutants. Whenever Northern blot analyses were reported on ERG-defective mutants, most showed transcriptional alterations [e.g., *norpA*: (21); *trp*: (22); *ninaA*: (23); *inaF*: (24)]. However, in virtually all these studies, Northerns were carried out only on a subset of alleles available. To avoid any selection bias, we carried out, in four separate studies (on four different genes), Northern blots on all available allelic mutants with the exception of those whose poor viability precluded harvesting enough flies for RNA extraction. Figure 9.1 is a semiquantitative representation of the results on 15 ERG-defective mutants from these four studies (25–28). A striking finding is that only 2 of the 15 mutants showed no change in the RNA level. All these mutants showed transcriptional downregulation, but we have also seen mutants with transcriptional upregulation, though they seem to be relatively rare. These results would suggest that even if one had only one mutant allele available in a target gene, the likelihood of seeing a transcriptional alteration is rather high $[1 - (2/15) = 0.87]$. If multiple mutant alleles are available in a given target gene, the likelihood rapidly approaches 1.0, e.g., $[1 - (2/15)(2/15) = 0.98]$ for two allelic mutants. In reality, mutations are not all equal in inducing transcriptional alterations. Those that cause strong mutant phenotypes also tend to induce strong transcriptional alterations. Thus, for strong

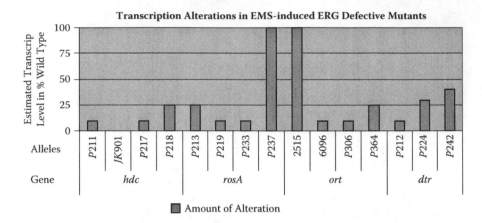

FIGURE 9.1 Transcriptional alterations in chemically induced ERG-defective mutants. The amount of alterations was estimated from the published Northern results in the case of the first three genes, *hdc* (25), *rosA* (26), and *ort* (27). The *dtr* data were obtained by quantitative real-time RT-PCR by Hosuk Lee [Personal communications (28)].

mutants, the likelihood of detecting transcriptional alterations is much higher than the preceding simple calculation would suggest.

The second requirement, that there be very few other genes that show significant transcriptional alterations through pleiotropic effects within the mapped limits of the target gene, is much more difficult to assess without actually carrying out microarrays on ERG-defective mutants. One can argue through a simple numerical argument that that is likely to be the case. Because the sequenced *Drosophila* genome consists of 120 Mbp (29) and one cytogenetic band, on the average, represents about 20 kb of DNA, one band corresponds to about 1.7×10^{-4} of the sequenced genome. Suppose one assumes that a mutation causes about 100 genes to be differentially expressed, through pleiotropic causes, with magnitude and statistical significance above a preset criterion. If one assumes that these genes are distributed randomly in the genome, the chance of seeing these genes within an average-sized chromosomal band is 1.7×10^{-2}. If a target mutation has been mapped within five chromosomal bands, the likelihood of one of these genes falling within the mapped limits of the target gene is ~0.09, and for a mutant mapped within 10 chromosomal bands, the corresponding number is 0.2. These numbers seem small and manageable. However, we assumed that only about 100 genes are pleiotropically altered in expression with sufficient magnitude and significance. Actually, any given mutation causes a thousand or more genes to be altered in expression pleiotropically. We reasoned that most of these alterations are likely to be small and low in statistical significance and that ones that are large and significant are much smaller in number. We then tested the validity of this argument by actual experiments. We first carried out microarray studies on two known genes to see if these could be reisolated using this strategy and then applied the strategy to three new genes. Results of these studies suggest that the basic assumptions appear to be correct. There are not that many genes, within the mapped limits, that are altered in expression pleiotropically with sufficient magnitude and statistical significance to confuse the identification of the correct gene. Results from these studies are described in the following text after the descriptions of the experimental procedures and statistical analysis.

9.4   EXPERIMENTAL PROCEDURES

The first step in microarray experiments is the preparation of good-quality total RNA. For each mutant complementation group targeted for study, we prepare three independent replicate RNA samples of about 5 µg total RNA from each of at least two different allelic mutants, except when multiple alleles are not available, plus a wild-type control. The samples are taken to the Purdue Genomics Core Facility (PGCF), and the personnel there carry out all the experimental steps associated with the arrays, including target preparation and hybridization, and washing, staining, and scanning of the arrays. The Doerge group then carries out statistical analysis of the data, and the Pak group carries out the identification of candidate genes from the statistically analyzed data and validation of identification.

RNA is extracted from heads of adult flies within a few hours of eclosion. Young fly heads are used as a precaution against the possibility that the mutations being examined cause age-dependent degeneration. *Drosophila* heads can be harvested in large numbers relatively easily following a standard protocol. Flies are lightly anesthetized with CO_2, collected into a 50 mL Falcon tube, and quickly frozen by dipping the tube into liquid nitrogen. Vortexing the tube several times separates body parts, allowing heads to be isolated from other body parts by sieving (Fisher Scientific no. 25 and no. 40). All these steps are carried out in a $-20°C$ room to ensure that the fly parts remain frozen throughout. Total RNA is extracted using the RNeasy mini Kit (Qiagen). Although we only need about 5 µg total RNA per sample, we usually prepare as much as 10 times more than that to compensate for losses during manipulations and to have enough available for use in quality control and Northern blots or RT-PCR of selected mutants. It is also possible to carry out microarray runs with as little as 10 ng total RNA by adding an extra cycle of cRNA synthesis/amplification in target preparation (see following text). This procedure can be used, for example, for mutants that cannot be harvested in large numbers for RNA extraction because of poor viability. Moreover, one can implement single-fly-head microarrays using this protocol because one can extract ~100 µg total RNA from 1000 heads, or ~100 ng per head.

Quality control consists of 1% agarose-formaldehyde gel separation and a spectral scan in the 210 to 320 nm region. The gel scan will confirm the absence of DNA contamination and degradation products and test the integrity of RNA from the presence of the 28S and 18S ribosomal RNA bands. The spectral scan will determine the sample concentration and purity. The test for RNA integrity from the presence of the

28S and 18S bands is not too useful for *Drosophila* because the 28S band is often missing or indistinct even in good-quality *Drosophila* total RNA. The *Drosophila* 28S molecule has a heat-sensitive break near its midpoint that causes the molecule to be processed into two fragments, and these two fragments migrate in a manner similar to 18S rRNA (30).

The Purdue Genomics Core Facility processes the RNA samples by following the Affymetrix protocol (http://www.affymetrix.com/Auth/support/downloads/manuals/expression_s2_manual.pdf). Briefly, the facility staff checks the quality of the total RNA once again using a Nanodrop spectrophotometer and Agilent 2100 Bioanalyzer (Agilent Technologies, Palo Alto, California). The Agilent Bioanalyzer allows electrophoretic and chromatographic separations of 1 μL quantities of RNA samples to assess RNA quality. From total RNA, mRNA is reverse-transcribed to cDNA using reverse transcriptase and a T7-oligo(dT) primer. The resulting cDNA is biotinylated and amplified using T7 RNA polymerase in the presence of biotin-labeled uridine analogs to yield amplified biotin-labeled cRNA. To achieve optimum sensitivity, the cRNA target is then fragmented by heating to 94°C for 35 min. Fragmented cRNA is mixed with the hybridization cocktail and hybridized to the chip for 16 h. The chip is washed to wash away the nonhybridized material and stained with streptavidin/phycoerytherin. After a poststain wash, the chip is stained with an antibody solution containing goat IgG and biotinylated antistreptavidin antibody to amplify the signal on the chip. The probe array is stained once again with streptavidin/phycoerytherin, and washed. The chips are then scanned, and the data obtained are uploaded onto a designated Web site and made available for analysis.

In the two-cycle protocol, RNA is amplified further by adding an extra cycle of cRNA synthesis/amplification. This protocol allows the use of as little as 10 ng total RNA for a microarray run. After the cDNA synthesis from total RNA, the resulting cDNA is amplified to cRNA as before, but without biotin labeling. The amplified unlabeled cRNA is then used to synthesize cDNA once again, and this cDNA is used to go through one more cycle of cRNA synthesis/amplification, but this time in the presence of biotin-labeled nucleotides. The remaining steps are the same as before.

9.5 STATISTICAL ANALYSIS

An analysis of variance (ANOVA) is employed to statistically test for differential expression between mutants and the control (as well as between mutants) using normalized data (31). Because the number of probes may be different for each (biologically replicated) gene, an ANOVA model is fitted for each gene:

$$\log(y_{ijk}) = \mu + S_i + P_j + (SP)_{ij} + e_{ijk}$$

where y_{ijk} is the perfect match for the k-th replicate for probe P_j under strain/mutant S_i; μ is the average of logarithm of the perfect match over all probes, strains, and replicates; S and P are the strain/mutant and probe main effects, respectively; SP is the interaction between strain/mutant and probe; and e_{ijk} is the error term, which is normally distributed with mean 0 and variance σ^2.

For each gene we are interested in testing for a strain/mutant (S) effect (the difference between mutant and its wild type), and whether it is different from zero. If this null hypothesis is rejected, we conclude the specific gene being tested exhibits statistically significant differential expression for the comparison at hand. Because we have about 19,000 genes, it is necessary to adjust the type I error rate to accommodate multiple testing issues. Both false discovery rate (FDR) (32) approach and Holm's sequential Bonferroni correction procedure (33) are used. The significance level is chosen as 0.05.

9.6 TESTS OF MICROARRAY STRATEGY

Because, to our knowledge, the microarray strategy for gene cloning had not been utilized previously, we first carried out feasibility studies on two known ERG-defective mutants to see if the corresponding genes could be reisolated using the microarray strategy. We then tested the strategy on three other genes that had not been previously cloned. Although the identification of these three genes is at different stages of validation, results clearly indicate that the strategy efficiently and speedily identifies the genes corresponding to ERG-defective mutants.

9.6.1 Previously Identified Genes

For reisolation of previously identified genes, we used the following two mutants: trp^{P343} and $inaF^{P106}$. They are null mutants in the known phototransduction genes, *trp* (22,34–37, reviewed in references 38,39,40) and *inaF* (24).

The microarray data are organized in terms of probe sets. As explained elsewhere (9.3 microarray-based gene cloning), a probe set usually corresponds to a gene, although occasionally more than one probe set is used to probe a gene. The statistical analysis identifies all those probe sets that detect differences in mRNA levels between a mutant and a corresponding wild-type control with a significance level of 0.05 (see section 9.5). We did not set a threshold for the amount of difference and included all probe sets satisfying the preceding criterion regardless of the amount of difference detected. Thus, the spreadsheet of final results contains the probe set ID number, log-fold change (plus or minus) in RNA level, P value, and standard deviation for each probe set satisfying the foregoing criterion (see, for example, table 9.1).

We first examined the trp^{P343} versus wild-type data for differential expression of the *trp* gene in the mutant compared to wild type. The identity of the probe set corresponding to the *trp* gene was obtained from the Affymetrix *Drosophila* 2.0 annotation file (http://www.affymetrix.com/support/technical/byproduct.affx?product=fly-20), by first sorting the data according to chromosomal locations. The data showed that the *trp* gene (probe set ID: 1622920_at; map position: 99C6-7) was altered in expression in the trp^{P343} mutant by −1.49 log-fold compared to wild type (table 9.1). Moreover, the log-fold change for this gene had a P value of 0, indicating 0% probability of being wrong. To mimic relatively crude mapping of this gene, we expanded the cytogenetic region of search to 99B5-99D7. This is approximately a 530 kb region flanking the *trp* gene (99C6-7) containing over 100 genes. Table 9.1 displays only those genes show-

TABLE 9.1

Genes Showing Expression Changes in *trp* (99B5-D7)

Probe Set ID	Log-Fold Change	P Value	Alignments	Gene Symbol	Cytol Position
1630575_at	0.11612	6.00E-06	3R:25434121-25435846	*CG1907*	99B5
1639887_s_at	−0.10821	0.010553	3R:25447054-25449150		99B6
1633427_at	−0.098574	0.012257	3R:25495348-25497129	*CG7582*	99B7
1623332_at	0.47929	1.33E-15	3R:25529626-25530040	*Obp99d*	99B8
1623675_at	0.45882	4.59E-06	3R:25531338-25531898	*Obp99b*	99B8
1629822_a_at	0.34093	6.66E-16	3R:25533766-25537477	*CG15506*	99B9
1630167_at	0.12574	7.27E-06	3R:25538248-25539314	*Pcd*	99B9
1640024_at	0.070686	0.0010157	3R:25552057-25553158	*CG7598*	99B9
1637778_a_at	0.10236	0.0010479	3R:25554931-25556997	*CG1969*	99B9
1636543_a_at	0.13988	2.48E-06	3R:25555180-25555836	*CG1969*	99B9
1632838_at	0.10319	0.0038958	3R:25562789-25563930	*ATPsyn gamma*	99B10
1627018_s_at	0.10654	0.00018303	3R:25562941-25564224	*ATPsyn gamma*	99B10
1634668_at	−0.1126	0.0016858	3R:25614740-25616673	*Ice*	99C1
1630912_at	0.115	0.00017594	3R:25618325-25619437	*CG7789*	99C1
1633455_at	0.10183	0.0021423	3R:25630150-25631538	*CG18112*	99C2
1632209_at	0.16193	0.0042651	3R:25632003-25632852	*CG7829*	99C2
1641539_a_at	0.21964	1.71E-06	3R:25632208-25632852	*CG7829*	99C2
1630399_at	0.24161	2.05E-12	3R:25660691-25669791	*Cad99C*	99C4
1638836_at	0.10095	0.0048446	3R:25684457-25686258	*CG31035*	99C5
1625178_at	−0.06986	0.0010353	3R:25692170-25718319	*CG31038*	99C5-6
1629136_at	−0.09551	0.0022417	3R:25728072-25729416	*capa*	99C6
1622920_at	−1.4916	0	3R:25729607-25735377	*trp*	99C6-7
1639317_at	−0.1712	0.00055827	3R:25798539-25805880	*CG7896*	99D1
1631208_at	0.11056	0.00055662	3R:25812047-25812662	*CG15525*	99D1
1635589_at	0.063056	0.010539	3R:25825554-25830452	*CG7920*	99D1
1628339_a_at	0.14609	1.73E-05	3R:25829629-25830452	*CG7920*	99D1
1639413_at	0.48241	0.00032653	3R:25852082-25852818	*ocn*	99D3
1629270_at	0.11243	0.0002206	3R:25858433-25859979	*sry α*	99D3
1638101_at	0.20822	2.43E-10	3R:25863896-25864619	*CG15528*	99D3
1630992_at	0.063252	0.01189	3R:25867601-25869786	*CG7946*	99D3
1640738_a_at	0.15503	0.00031284	3R:25869778-25871133	*spn-A*	99D3
1641206_at	0.1798	0.00014812	3R:25873401-25936479	*sima*	99D3-7
1623800_at	0.13497	0.0070582	3R:25937141-25939823	*CG18682*	99D7

Note: There are over 100 annotated genes in this cytological interval. Only those genes whose expression changes are detected at a significance level of 0.05 are listed. (9.5 statistical analysis). Alignments refer to molecular coordinates of probe sets in bp's. These coordinates used by Affymetrix differ slightly from those used in Flybase in absolute value. However, the relative values are the same.

ing expression changes that are significant at a 0.05 level. Within this region, only four other probe sets detected log-fold changes greater than 0.3 in absolute value, ranging between 0.341 and 0.482, and none of them had a P value of 0 (table 9.1). A very similar story unfolded for *inaF*[FP106]. Expression of this mutant was downregulated by −2.36 log-fold compared to wild type, with a P value of 0. Searching a >200 kb region centered around the location of *inaF*, the next highest log-fold change detected by any probe set was −0.765, and the gene corresponding to this probe set was located about 100 kb away from *inaF* (data not shown). We thus concluded that, with null mutants, the microarray strategy will pinpoint the corresponding genes immediately and unequivocally. Mapping need not even be highly refined.

9.6.2 Previously Unidentified Genes

9.6.2.1 *inaE*

However, if the microarray strategy were to be generally useful, we need to show that it can be successfully applied to previously unidentified genes. Moreover, most mutants are not null mutants and their phenotypes are often relatively mild. We therefore need to show that this strategy can be used in the isolation of genes in which only mild mutants have been isolated. We chose the *inaE* gene for the first "live test" of the strategy. The mutant *inaE*[N125] is one of the two classically generated mutant alleles of the *inaE* gene (41), which we had been attempting to clone for some time with limited success. From the mutant phenotype, the *inaE*-encoded protein was expected to be involved in phototransduction. We felt *inaE*[N125] was a very good test of the strategy because its phenotype is rather mild and, for technical reasons, only one of the two allelic mutants could be used for microarray analysis. Thus, it was expected to tax the efficacy of the strategy.

The *inaE* gene was mapped to the 12C5-6;12C8-D1 region of the X chromosome by deficiency mapping. Two critical deficiencies for this mapping were *Df(1)AR10* (12B1-2;12C8-D1) and *Df(1)ben*[CO2] (12C5-6;12E), both of which uncovered the mutation; i.e., they did not complement the mutation. Again, we searched a much broader region than that determined by mapping: 12B4 to 12D4, a ~350 kb region with 41 genes (table 9.2). In this region, probe sets for the following three genes detected log-fold expression changes greater than 0.3 in absolute value in the *inaE* mutant compared to wild type: *Yp3* at 12C1, *CG33174* at 12C4-5, and *CG32626* at 12C6-7. *CG33174* was of particular interest because this was the only gene in this region for which the changes in expression detected by its probe sets were assigned a P value of 0 (table 9.2). Southern blots of deficiency heterozygotes, *Df(1)AR10*/+ and *Df(1)ben*[CO2]/+, probed with *CG32626* probes showed that *CG32626* is outside the mapped limits of *inaE*. The *Yolk protein 3* gene (*Yp3*) had been previously characterized both genetically and molecularly (e.g., 42–44). We obtained two *Yp3* mutants from Dr. Mary Bownes of the University of Edinburgh and carried out complementation tests between them and *inaE*. A *Yp3* mutant allele when placed in trans with *inaE* complemented the *inaE* phenotype, definitively ruling out *Yp3* as the *inaE* gene.

Thus, the preceding analysis left *CG33174* as the only remaining candidate gene for *inaE*. To determine whether *CG33174* indeed is *inaE*, we carried out the following three independent lines of validation experiments (41): (a) sequencing the *CG33174* gene in the *inaE*[N125] mutant, (b) complementation tests between *inaE* and existing P-insertion mutations in *CG33174*, and (c) in vivo rescue of *inaE* by germ-line transformation of *inaE* with a wild-type copy of *CG33174*. Sequencing showed

TABLE 9.2

Genes Showing Expression Changes in *inaE* (12B4-D4)

Probe Set ID	Log-Fold Change	P Value	Alignments	Gene Symbol	Cytol Position
1641025_at	−0.086208	0.010916	X:13468801-13469838	CG11134	12B6
1631419_at	−0.64749	7.84E-05	X:13490203-13491835	Yp3	12C1
1639574_a_at	0.17604	6.75E-05	X:13493428-13507399	rdgB	12C1-2
1637006_at	0.13555	0.0061398	X:13514407-13536302	CG33174	12C4-5
1627706_a_at	0.657	0	X:13514407-13536302	CG33174	12C4-5
1625176_at	0.56423	0	X:13514407-13542160	CG33174	12C4-5
1640287_s_at	0.49505	0	X:13540023-13541832	CG33174	12C5
1631073_at	0.06994	0.0079756	X:13544100-13552983	l(1)G0053	12C5-6
1636030_s_at	0.13365	0.0015452	X:13568261-13568744	CG32626	12C6-7
1632081_s_at	0.40171	1.56E-09	X:13574612-13576134	CG32626	12C6-7
1638749_at	0.13435	0.0012134	X:13600040-13616171	CG32611	12C7
1632900_at	0.12029	0.002592	X:13632111-13632282	CG11072	12C8
1625285_at	0.10004	0.0057516	X:13640807-13642515		12D1
1630614_s_at	−0.18765	0.0003122	X:13737316-13737562		12D2

Note: Again, only those probe sets satisfying the significance level of 0.05 (9.5 statistical analysis) are included in the table. The search covered the region from 12B4 to 12D4. However, the table shows probe sets in a slightly narrower region because the other probe sets did not satisfy the preceding criterion. The same comments as in table 9.1 regarding alignments apply.

that the *inaE* mutant carries a mutation at the 5′ splice site of the 11th intron. As a consequence, this intron fails to be spliced out, and translation proceeds into the intron sequence until a stop codon is encountered within the intron.

The *CG33174* gene had been characterized only electronically. There have been no classically generated mutants in this gene. However, there were four P insertion alleles listed in Flybase. These were obtained, and complementation tests were carried out between each of these and *inaE^{N125}*. *inaE^{N125}* is characterized by two distinct ERG phenotypes. One is that its light-evoked response starts decaying during illumination, and the other is that it enhances (makes the phenotype stronger) the ERG phenotype of *Trp^{P365}*/+. If the mutation is relatively mild, the first of these phenotypes can be troublesome to detect in a red-eye background, as is the case of P-insertion lines. In fact, none of the four P-insertion lines displayed this phenotype. The second of these phenotypes is an almost infallible diagnostic test for *inaE^{N125}*. *Trp^{P365}* is a semidominant mutation in the *trp* gene, which encodes subunits of the TRP channel, the major carrier of the light-evoked current in *Drosophila* photoreceptors (45). The *Trp^{P365}* mutation makes the TRP channel constitutively open and kills the photoreceptor cells by allowing excessive Ca^{2+} entry. In *Trp^{P365}* homozygotes, the ERG amplitude is nearly zero even at 1 day post eclosion because of extensive photoreceptor degeneration. In *Trp^{P365}* heterozygotes (*Trp^{P365}*/+), however, the time course of degeneration is much slower and a substantial ERG (10–15 mV) is present at 1 day post eclosion (table 9.3, row 1) and disappears with further ageing (hence, the mutation is said to be semidominant). If *inaE* is combined with *Trp^{P365}* (i.e., *inaE*/

TABLE 9.3

Complementation Test between *inaE^{N125}* and *14959* P Insertion Mutants

Genotype	ERG	Amplitudes (mV)
+/+;;P365/+	14.0 ± 2.3	(n = 9)
N125/+;;P365/+	12.1 ± 1.5	(n = 6)
14959/+;;P365/+	13.5 ± 1.2	(n = 9)
N125/N125;;P365/+	0.4 ± 0.3	(n = 10)
N125/14959;;P365/+	3.0 ± 0.3	(n = 6)
14959/14959;;P365/+	4.8 ± 0.7	(n = 6)

Note: All are white-eyed flies marked with either white (*w*) or cinnabar brown (*cn bw*) mutations. All ERGs were performed on 1 d.o. flies. *Trp^{P365}* and *inaE^{N125}* are abbreviated as *P365* and *N125*, respectively.

inaE;;*P365*/+ for female), it enhances the aforementioned phenotype of *Trp^{P365}*/+ so that only a very small ERG is present at 1 day post eclosion (table 9.3, row 4). We found that one of the P insertion mutants of the *CG33147* gene, *14959*, also had this property (table 9.3, row 6). Moreover, *14959* in heterozygous combination with *inaE* also displayed the same property; that is, *inaE/inaE*;;*P365*/+, *inaE/14959*;;*P365*/+ and *14959/14959*;;*P365*/+ all had only a very small ERG at 1 day post eclosion (table 9.3, rows 4, 5, and 6) (41). Thus, *14959* P insertion allele and *inaE* are mutations in the same gene, or *inaE* and *CG33174* are the same gene.

The most definitive proof that a correct gene has been cloned is generally considered to be the restoration of the wild-type phenotype in the mutant in vivo through the introduction of the cloned gene sequence by germ-line transformation. We took advantage of the ability of the *inaE* mutation to enhance the *Trp*/+ phenotype, described earlier, to assess whether or not the introduced sequence rescues the mutant phenotype. Thus, *Trp^{P365}* heterozygotes (*Trp^{P365}*/+) have 10–15 mV ERG responses at 1 day post eclosion. In the double mutant of *inaE* and *Trp^{P365}*/+ (*inaE/inaE*;;*Trp^{P365}*/+), however, the responses are nearly zero at 1 day post eclosion because of the enhancement of the *Trp^{P365}*/+ phenotype by *inaE*. When the rescue construct, consisting of the wild-type *inaE* cDNA sequence driven by a photoreceptor specific promoter, was introduced into the *inaE/inaE*;;*Trp^{P365}*/+ double mutant, the transformant behaved like the *Trp^{P365}* heterozygote. That is, the rescue construct nullified the effects of *inaE* mutation in the double mutant, or rescued the *inaE* phenotype. Thus, all three lines of validation experiments unequivocally demonstrated that *inaE* and *CG33174* are the same gene.

9.6.2.2 *P222* and *P255*

We have identified two other new genes by the microarray approach, *P222* and *P255*. Mutants in both these genes are impaired in synaptic transmission. Validation of identification is not as far advanced for these genes as for *inaE*. Nevertheless, the data obtained to date are sufficiently strong to lead one to believe that the identification is secure.

There are four mutants falling into the *P255* complementation group. Microarray experiments were carried out on two of these, *JK1285* and *JK1503*. Deficiency mapping showed that *P255* is uncovered by (included within) both the deficiencies *Df(2L)VA6* and *Df(2L)DS6*, which have reported breakpoints at (37D2;38C6-E3) and (38E2;39E7), respectively. Thus, this mapping localized the *P255* gene to the region of overlap between these two deficiencies, 38E2-3. However, the breakpoints of these deficiencies had not been determined at the molecular level. Therefore, we needed to allow for possible mismatches between the reported breakpoints of deficiencies and the locations of annotated genes. Again, we searched a much wider region than that determined by mapping—from 38D1 to 38F1, an approximately 220 kb region with 40 plus genes. When we examined the statistically analyzed microarray data for both *JK1285* and *JK1503* in this region, we found that the data from the two mutants were very different in quality. The data from the *JK1503* mutant were noisy, with many genes of similar log-fold changes in expression and similar P values. Although these results suggested possible problems somewhere in the experimental steps, we made no attempts to repeat these data because the data from the other mutant were so clear-cut. In the data from the other mutant, *JK1285*, only two genes stood out in this entire region: *CG9317* at 38E4 with a log-fold change of −1.31 and a P value of 0 and *diaphanous* (*CG1768*) at 38E7-8 with a log-fold change of −2.02 and a P value of 0. These were the only genes with log-fold changes >1.0 in absolute value and P values of 0 (table 9.4). The only other gene displaying a log-fold change >0.4 in absolute value was *CG17470* at 38D2 (log-fold change: 0.65; P value: 1.98×10^{-2}). However, its P value was relatively large and its location at 38D2 was clearly outside the mapped limits of *P255*.

To choose between *CG9317* and *diaphanous*, we sequenced both cDNA and genomic sequences of both genes in *JK1285* and *JK1505* mutants as well as a wild-type control. There were no mutations in either the coding region or exon–intron boundaries of the *diaphanous* gene of either mutant, whereas there were mutations in the coding region of the *CG9317* gene of both *JK1285* and *JK1503*. *JK1503* harbored a mutation that would introduce a stop codon near the 5′ end of the coding sequence. *JK1285*, on the other hand, harbored a mutation that is predicted to cause a Gly-Asp amino acid change within one of the transmembrane segments of the encoded protein. This mutation is likely to drastically alter the hydrophobicity of the transmembrane segment. These results strongly suggested that *CG9317* is the *P255* gene.

As for *P222*, there are three mutants falling into the *P222* complementation group: *P222*, *P223*, and *JK1669*. Microarray experiments were carried out on all three alleles and the corresponding wild-type controls. Earlier deficiency mapping had placed the *P222* gene at 25D2-4;25D6-7 of the left arm of the second chromosome. This region happened to be well covered by deficiencies whose breakpoints have been determined at the molecular level. Remapping with some of these deficiencies placed the *P222* gene between the right breakpoint of the deficiency *Df(2L) Exel6011* and the left breakpoint of the deficiency *Df(2L)cl7* (figure 9.2). The right breakpoint of *Df(2L)Exel6011* at 25D5 has a molecular coordinate of 2L:5320589 (Flybase). The left breakpoint of *Df(2L)cl7* at 25D7 is defined at the molecular level

TABLE 9.4
Genes Showing Expression Changes in *JK1285* (38D1-F1)

Probe Set ID	Log-Fold Change	P Value	SD	Alignments	Gene Symbol	Cytol Position
1627780_at	0.162140079	0.003849598	0.24638223	2L:20592770-20602014	—	38D2
1641727_at	0.651517765	0.019801799	1.244615801	2L:20612044-20612986	*phr6-4*	38D2
1635020_s_at	0.274181393	0.00014085	0.307431775	2L:20613089-20615531	*CG17470*	38D2
1640158_at	-0.1297622	0.021034361	0.250476964	2L:20615565-20617123	*CG2608*	38D2
1639142_s_at	-0.257087611	0.00104006	0.340491415	2L:20617274-20618592	*pncr012:12L*	38D2
1639951_at	-0.141575284	0.004522731	0.219294899	2L:20619935-20620642	*CG2611*	38D2
1629579_at	0.304217116	5.94E-09	0.203277575	2L:20620496-20625918	*CG2478*	38D2
1640130_at	0.136468743	0.018082278	0.256781459	2L:20628871-20645683	*CG31678*	38D2-4
1639807_s_at	-0.112077832	0.001094611	0.149160382	2L:20648662-20651586	*CG1962*	38D5
1627392_at	0.18375385	0.015396777	0.336901807	2L:20678158-20678516	—	
1630152_at	0.144351724	0.008338214	0.241865569	2L:2068145-2069086	—	
1625635_at	0.150378928	0.001755247	0.209691477	2L:20693946-20696293	*CG9316*	E4
1634836_a_at	-1.3142205	0	0.38681972	2L:20696884-20700051	*CG9317*	E4
1636977_at	0.128201465	0.008035779	0.213727089	2L:20700386-20704052	*CG9318*	38E4-5
1635766_at	0.210830067	0.003112032	0.312635319	2L:20704656-20709521	*Fs(2)Ket*	
1635645_at	0.166335291	4.68E-05	0.172689394	2L:20720675-20722082	*CG9319*	E6
1625942_s_at	-2.0230015	0	0.29921406	2L:20727888-20737265	*dia*	38E7-8
1627598_at	0.07940608	0.021928796	0.154364841	2L:207398-210506 (+)	*CG11490*	
1636545_at	0.290662014	7.37E-06	0.269521319	2L:20789892-20790609	*CheB38c*	38F1
1623649_at	0.242106415	0.000633153	0.306428811	2L:20791470-20793622	—	
1635390_s_at	0.09671275	0.001372121	0.13156216	2L:20794177-20797486	*CG9331*	38F1
1623291_at	0.351549306	4.52E-10	0.213810033	2L:20799411-20800805	*CG31673*	38F1
1638368_at	0.364121531	2.05E-06	0.315055654	2L:20801219-20802525	*CG31674*	38F1

Note: "Alignments" are as explained in table 9.1.

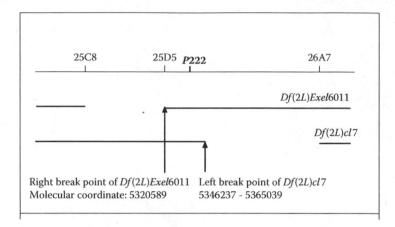

FIGURE 9.2 Deficiency mapping of *P222*. *P222* was mapped between the right break-point of *Df(2L)Exel6011* and the left breakpoint of *Df(2L)cl7*. The right breakpoint of *Df(2L) Exel6011* is at 25D5 cytologically, and its molecular coordinate is 2L:5320589. The left break-point of *Df(2L)cl7* is at 25D7 and is defined molecularly by the deficiency's ability to disrupt or delete the nompC gene, located at 2L:5346237 to 5365039. Both the foregoing deficiencies complement *P222*. Another deficiency, *Df(2L)Exel6012*, with breakpoints at 25D5 and 25E6, uncovers the mutation; i.e., the mutation is within the deficiency. Deficiencies are shown as breaks in solid lines.

by the fact that the deficiency disrupts or deletes the *nompC* gene, which occupies the nucleotide coordinates 2L:5346237-5365039 (Flybase) (figure 9.2).

The microarray data were somewhat noisier than those obtained with other mutants. Nevertheless, the data were clear in identifying the candidate gene in the mapped region. Table 9.5 displays all probe sets in the 25D4 to 25D7 region found to detect expression changes that are significant at a 0.05 level (see section 9.5) in the three mutants of the *P222* complementation group. The probe set consistently detecting the largest log-fold change accompanied by a very low P value in each of the three mutants was that corresponding to *CG14021*. In addition, another probe set corresponding to *cype* showed moderately high log-fold changes and P values of 0 in the three mutants. A third one corresponding to *vri* had a relatively high log-fold change with a P value of 0 but only in one mutant, *P223*. Results of mapping, because they were defined at the molecular level, helped eliminate several of the probe sets shown in table 9.5. The last entry corresponding to the *nompC* gene was eliminated because *nompC* is disrupted by the *Df(2L)cl7* deficiency (figure 9.2). The first four entries, from *vri* to *CG14023*, were eliminated because they lie at or to the left of the right breakpoint of the *Df(2L)Excel6011* deficiency (figure 9.2). The right break-point of *Df(2L)Excel6011* at 2L:5320589 runs through the *CG14023* gene, located at 2L:5316240-5324927, and the other three genes, *vri* to *CG14024* lie to the left of this breakpoint. Thus, the only two genes in the mapped region showing changes in expression of any significance in all three mutants were *cype* and *CG14021* (table 9.5). We sequenced both these genes in all three mutants. There were no sequence altera-tions in the *cype* gene in any of the three mutants. By contrast, stop codons were found

TABLE 9.5

Genes Showing Changes in Expression in *P222*, *P223*, and *JK1669* (25D4-7)

Probe Set ID	P222		P223		JK1669		Gene Symbol	Cytol Position
	Log-Fold	P Value	Log-Fold	P Value	Log-Fold	P Value		
1639273_s_at	−0.281933617	2.62E-13	−0.680404453	0	0.261083184	1.53E-06	vri	25D4-5
1623674_at			−0.1920818	0.004021573	0.223023798	0.000136228	—	25D5
1636447_at			−0.279272584	3.72E-05	0.220514924	0.003405732	CG14024	25D5-6
1640735_at			−0.36860693	1.99E-06	0.429930948	4.32E-07	CG14023	25D6
1638416_at	−0.351956818	0	−0.45925195	0	0.208014883	0	cype	25D6
1625430_at	−0.197400154	4.18E-07	−0.363730714	7.77E-15	0.252175926	3.98E-11	CG14022	25D6
1623635_at	−0.336016225	5.41E-06	−0.539940969	3.38E-13	0.313795286	0.00136493	TotM	25D6
1640365_s_at	−0.593472788	0	−0.895430605	0	−0.55860993	3.77E-15	CG14021	25D6
1638636_at	−0.230068711	2.40E-05	−0.447498813	2.22E-16	0.233696	0.0008696	CG12512	25D6-7
1626936_at			−0.209368132	0.000649058			nompC	25D6-7

Note: Blank spaces in *P222* columns correspond to probe sets that did not detect expression changes at the 0.05 significance level.

in the coding region of *CG14021* in all three mutants. These results strongly suggest that *CG14021* is the *P222* gene.

9.7 CONCLUDING REMARKS

In this chapter, we have reviewed a novel technique involving the use of DNA microarrays for cloning genes that are identified only by chemically induced ERG-defect-causing mutations. We have presented five applications of this approach, two to previously identified genes and three to new ones that had not been isolated previously. In all cases, the approach rapidly and reliably identified one or two candidate genes within the mapped interval, even though the quality of the microarray data varied with mutants. The main advantage of this approach is the speed with which the identification of a very small number of candidate genes can be achieved. It took, on the average, 3½ to 4 months from the time of fly head collection for RNA isolation to the time of identification of candidate genes. More than 50% of this time was taken up by the expansion of fly stocks, collection of heads, and RNA isolation. Had we used the two-cycle protocol for cRNA amplification (see section 9.4), which allows the use of as little as 10 ng amount of total RNA for microarray target preparation, the time of 3½ to 4 month could have been slashed by another month or so. To our knowledge, no currently available technique can match this feat.

The speed is achieved by allowing one to skip the labor-intensive intermediate steps of fine mapping and proceed directly from deficiency mapping to the identification of a very small number of candidate genes. The final steps of gene identification and validation, on the other hand, are similar to those of other strategies. From the small number of candidate genes identified, the final identification of the single most likely candidate gene must be made, and the identification needs to be validated. These are achieved, as in other gene cloning strategies, by various combinations of gene sequencing, complementation with existing mutations, if available, and germ-line transformation. However, the present strategy does have one other advantage over other methods in that it provides information as to which of the candidate genes is likely to be the correct gene, allowing one to concentrate on it. By contrast, in other methods, all genes falling within the mapped interval must be considered equally.

Microarray studies are not inexpensive. If we were to examine two allelic mutants in a gene, we need to prepare nine independent RNA samples, three each for the two mutants and a wild-type control. The cost of the nine chips required for these samples is $300 × 9 = $2700, and the charges of the Genomic Facility are $250 × 9 = $2250. In addition, an allowance needs to be made for the statistician's time. However, these costs must be weighed against the time saved. If the target mutation happens to fall in a region rich in molecularly defined deficiencies and can be mapped within a region containing less than 5–6 genes by deficiency mapping alone, one could attempt to identify the gene directly by either sequencing or some other methods of mutation detection. However, it is still rare to achieve that kind of accuracy and resolution with deficiency mapping alone. One would need to map the mutant more accurately using the procedures we have outlined earlier (section 9.2.2). These procedures unfortunately can go on for months or even years. Our experience with the *inaE* gene is instructive in this regard. We had been attempting to clone this gene using the traditional methods for about 2 years with limited success. Had we applied the microarray strategy to this gene earlier, we estimate that we would have saved approximately 2 years of graduate student or postdoctoral time for this gene alone. The microarray costs, as exorbitant as they may seem, represent only a fraction of the money we spent on salaries. An even more important consideration than money is the invaluable time saved by using this approach.

ACKNOWLEDGMENTS

We thank Young Seok Hong for suggesting the use of the microarray approach for gene cloning and Kim Gilbert for help in the preparation of the manuscript. Some of the mutants mentioned in the paper were generated in other laboratories and generously given to us many years ago. The mutant *inaE*[N125] was generated by Martin Heisenberg, and *JK1285*, *JK1503*, and *JK1669* were generated by Jane Koenig and John Merriam. The first microarray experiment was done at the Center for Medical Genomics, Indiana University School of Medicine, Indianapolis, and all subsequent experiments were done at the Purdue Genomics Core Facility. We thank Phillip San Miguel, Ann Feil, and Fred Rakhshan of PGCF. This work was supported by grants from the National Eye Institute (EY00033) and National Institute of Mental Health (MH075041) to WLP.

NOTE

1. These investigators used the EP element for the generation of P insertion lines. The EP element is a P element containing a Gal4-responsive enhancer at one end (Rørth, 1996). If flies carrying an EP element are crossed to flies carrying Gal4 fused to a suitable promoter (46), in the progeny carrying both elements, an endogenous gene immediately adjacent to the EP element is induced. Thus those lines with EP insertions immediately upstream of a gene can be used for overexpression of the gene, and those with EP insertions in the antisense strand immediately downstream of a gene can be used for antisense knockdown of a gene.

REFERENCES

1. Pak, W.L., Mutations affecting the vision of *Drosophila melanogaster*. In *Handbook of genetics* 3rd ed. King, R.C., ed., New York: Plenum, 703–733, 1975.
2. Pak, W.L., Study of photoreceptor function using *Drosophila mutant*. In *Neurogenetics: Genetic approaches to the nervous system*, ed. Breakefield, X., Amsterdam: Elsevier North, 67–99, 1979.
3. Pak, W.L., *Drosophila* in Vision Research: The Friedenwald Lecture, *Invest Ophthalmol Vis Sci*, 36: 2340–2357, 1995.
4. Pak, W.L. and Leung, H.T., Genetic approaches to visual transduction in *Drosophila melanogaster*. *Receptor Channels*, 9: 149–167, 2003.
5. Pak, W.L., Grossfield, J., and White, N.V., Nonphototactic mutants in a study of vision of *Drosophila*. *Nature*, 222(191): 351–354, 1969.
6. Hotta, Y. and Benzer, S., Abnormal electroretinograms in visual mutants of *Drosophila*. *Nature*, 222, 354–356 (26 April 1969); doi: 10.1038/222354a0,1969.

7. Hotta, Y. and Benzer, S., Genetic dissection of the *Drosophila* nervous system by means of mosaics. *Proc Natl Acad Sci U.S.A.*, 67(3): 1156–1163, 1970.

8. Heisenberg, M., Isolation of mutants lacking the optomotor response, *Dros Inf Svc*, 46: 68, 1971.

9. Heisenberg, M., Behavioral diagnostics: A way to analyze visual mutants of *Drosphila*. In *Information processing in the visual systems of arthropods*, ed. Wehner, R., Berlin: Springer-Verlag, 265–268, 1972.

10. Koenig, J. and Merriam, J., Autosomal ERG mutants, *Dros Inf Serv* 52: 50–51, 1977.

11. Zhai, R.G. et al., Mapping *Drosophila* mutations with molecularly defined P element insertions. *Proc Natl Acad Sci U.S.A.*, 16;100(19): 10860–10865, 2003.

12. Parks, A.L. et al., Systematic generation of high-resolution deletion coverage of the *Drosophila melanogaster* genome. *Nat Genet*, 36, 288–292, 2004.

13. Ryder, E. et al., The DrosDel Collection: a set of *P*-element in *Drosophila melanogaster. Genetics* 167: 797–813, 2004.

14. Venken, K.J. and Bellen, H.J., Emerging technologies for gene manipulation in *Drosophila melanogaster. Nat Rev Genet*, 6(4): 340, Apr 2005.

15. Preston, C.R., Sved, J.A., and Engels, W.R., Flanking duplications and deletions associated with P-induced male recombination in *Drosophila. Genetics*, 144(4): 1623–1638, 1996.

16. Chen, B. et al., Mapping of *Drosophila* mutations using site-specific male recombination, *Genetics*, 149: 157–163, 1998.

17. Berger, J. et al., Genetic mapping with SNP markers in *Drosophila, Nat Genet*, 29: 475–481, 2001.

18. Schena, M. et al., Quantitative monitoring of gene expression patterns with a complementary DNA microarray. *Science,* 270: 467–470, October 1995.

19. Lockhart, D.J. et al., Expression monitoring by hybridization to high-density oligonucleotide arrays. *Nat Biotechnol*, 14(13): 1675–1680, December 1996.

20. Adams, M.D. et al., The genome sequence of *Drosophila melanogaster, Science,* 287: 2185–2195, 2000.

21. Bloomquist, B.T. et al., Isolation of a putative phospholipase C gene of *Drosophila*, norpA, and its role in phototransduction. *Cell*, 54(5): 723–733, August 26, 1988.

22. Montell, C. and Rubin, G.M. Molecular characterization of the *Drosophila trp* locus: A putative integral membrane protein required for phototransduction, *Neuron* 2(2): 1313–1323, April 1989.

23. Schneuwly, S. et al., *Drosophila* ninaA gene encodes an eye-specific cyclophilin (cyclosporine A binding protein), *Proc Natl Acad Sci U.S.A.*, 86(14): 5390–5394, July 1989.

24. Li, C. et al., INAF, a protein required for transient receptor potential Ca^{2+} channel function. *Proc Natl Acad Sci U.S.A.*, 96: 13474–13479, 1999.

25. Burg, M.G. et al., Genetic and molecular identification of a *Drosophila* histidine decarboxylase gene required in photoreceptor transmitter synthesis, *EMBO J*, 12: 911–919, 1993.

26. Burg, M.G. et al., *Drosophila rosA* gene, which causes aberrant photoreceptor oscillation, encodes a novel neurotransmitter transporter homologue, *J Neurogenet*, 11: 59–79, 1996.

27. Geng, C. et al, Target of *Drosophila* photoreceptor synaptic transmission is histamine-gated chloride channel encoded by *ort (hclA)*. *J Biol Chem*, 277, 42113–42120, 2002.

28. Lee, H.S. Personal communication. 2004.

29. Kornberg, T.B. and Krasnow, M.A., The *Drosophila* genome sequence: Implications for biology and medicine. *Science*, 287(5461): 2218–2220, 2000.

30. Ishikawa, H., Evolution of ribosomal DNA, *Comp Biochem Physiol*, 58(1): 1–7, 1977.

31. Craig, B.A., Black, M.A., and Doerge, R.W., Gene expression data: The technology and statistical analysis. *J Agric Biol Environ Statistics (JABES)* 8(1): 1–28, 2003.

32. Benjamini, Y. and Hochberg, Y., Controlling the false discovery rate: a practical and powerful approach to multiple testing, *J R Stat Soc*, Series B, 57: 289–300, 1995.

33. Holm, S., A simple sequentially rejective multiple test procedure, *Scand J Statistics*, 6: 65–70, 1979.

34. Cosens, D.J. and Manning, A., Abnormal electroretinogram from a *Drosophila* mutant, *Nature*, 285–287; doi: 10.1038/224285a0, Oct 18, 1969.

35. Minke, B., Wu, C., and Pak, W.L., Induction of photoreceptor voltage noise in the dark in *Drosophila* mutant. *Nature*, 258(5530): 84–87, 1975.

36. Hardie, R.C. and Minke, B., The *trp* gene is essential for a light-activated Ca^{2+} channel in *Drosophila* photoreceptors. *Neuron*, 8(4): 643–651, 1992.

37. Wong, F. et al., Proper function of the *Drosophila trp* gene product during pupal development is important for normal visual transduction in the adult, *Neuron*, 3(1): 81–94, 1989.

38. Montell, C., The TRP superfamily of cation channels. *Sci STKE*, 2005(272): re3. Review, February 22 2005.

39. Minke, B., TRP channels and Ca^{2+} signaling. *Cell Calcium*, 40(3): 261–275. Review, September 2006.

40. Hardie, R.C., TRP channels and lipids: From *Drosophila* to mammalian physiology. *J Physiol*. 578: 9–24, 2007.

41. Leung, H-T et al., unpublished results, n.d.

42. Postlethwait, J.H. and Handler, A.M., The roles of juvenile hormone and ecdysone during vitellogenesis in isolated abdomens of *Drosophila melanogaster*. *J Insect Physiol*, 25: 455–460, 1979.

43. Barnett, T.C. et al., The isolation and characterization of *Drosophila* yolk protein genes. *Cell*, 21: 729–738, 1980.

44. Hovemann, B. et al., In *Drosophila* melanogaster: Sequence of the yolk protein I gene and its flanking regions. *Nucl Acids Res*, 9(18): 4721–4734, September 25 1981.

45. Yoon, J. et al., Novel mechanism of massive photoreceptor degeneration caused by mutations in the *trp* gene of *Drosophila*. *J Neurosci*, 20(2): 649–659, 2000.

46. Brant, A.H. and Perrimon, N., Targeted gene expression as a means of altering cell fates and generating dominant phenotypes. *Development* 118(2): 401–415, 1993.

4

Insulin Receptor-Based Signaling in the Vertebrate Retina

10 Insulin Receptor-Based Signaling in the Retina

Patrice E. Fort, Ravi S.J. Singh,
Mandy K. Losiewicz, and Thomas W. Gardner

CONTENTS

10.1   INTRODUCTION

The retina is one of the numerous insulin-sensitive tissues. Studies conducted in mice, rats, and rabbits have shown that the insulin receptor (IR) has a high basal kinase activity in retina that does not fluctuate with feeding or fasting. By contrast, IR activity in the liver varies markedly with fasting (low plasma insulin levels, low IR activity) and feeding (high plasma insulin levels, high IR activity). The IR, along with insulin-like growth factor-1 receptor (IGF-1R) and insulin receptor-related receptor (IRR), belongs to a family of heterotetrameric ($\alpha2\beta2$) transmembrane glycoprotein receptors that are widely expressed in mammalian tissues. Although little is known about the IRR, the IR and IGF-1R are the products of different genes that lead to the expression of proreceptors that display more than 50% amino acid sequence identity. The IR family ligands include three structurally related peptides that also present a high amino acid sequence identity: insulin, IGF-1, and IGF-2. Posttranslational processing results in the dimerization and disulfide linkage of proreceptors followed by proteolytic cleavage that generates α and β subunits (1). The extracellular α subunit contains the ligand-specific binding site, and the transmembrane β subunit possesses the tyrosine kinase activity. The kinase activity is induced by the interaction of the ligand with the α subunit that induces the transautophosphorylation of the β subunits.

Despite their high degree of identity, insulin and IGFs can trigger very different functions. Insulin is mostly involved in the regulation of the metabolic functions of the classically insulin-responsive tissues, liver, adipose, and skeletal muscle. Evidence also suggests that insulin acts on neural tissue and can modulate neural metabolism, synapse activity, and feeding behaviors (2). Insulin receptors are expressed on both the vasculature and neurons of the retina, but their functions are not completely defined. IGFs seem to play essential roles in the regulation of growth and development in different tissues as shown by different transgenic models. However, in the central nervous system, including the retina, evidence suggests that insulin action stimulates neuronal development, differentiation, growth, and survival, rather than acutely stimulating nutrient metabolism; e.g., glucose uptake as in skeletal muscle (3–6). IR from retinal neurons and blood vessels share similar properties with IR from other peripheral tissues, and retinal neurons express numerous proteins that are attributed to the insulin signaling cascade, as in other tissues. However, undefined neuron-specific signals downstream of the insulin receptor are also likely to exist.

The retina and brain are unique in containing a high number of different cell types and being heterogeneous compared to other insulin-sensitive tissues such as liver and skeletal muscles. Although the distinct physiological functions of insulin and IGFs depend on differences in the distribution and/or signaling potential of their respective receptors, the various retinal cell types can differentially express these receptors and thus have varying sensitivity to these ligands. Moreover, IR and IGF-1R receptors are able to assemble not only as homodimers but also as heterodimeric hybrid structures containing an ($\alpha\beta$) half of the IR disulfide-linked to an ($\alpha\beta$) half of the IGF-1R in different tissues and cell lines (7, 8). This feature adds another level of complexity to the function of these pathways and their sensitivity to the different ligands.

In this chapter, we will detail how the expression, and more importantly, the activity of the different proteins of these pathways, can be assessed in the retinal tissue. We will particularly discuss the various parameters of these assays that can be modified. We will finally discuss two different methods that can be used to characterize more deeply the signaling cascades involved in these pathways.

10.2 EXPRESSION AND PHOSPHORYLATION OF THE INSULIN RECEPTOR

10.2.1 RECEPTOR AUTOPHOSPHORYLATION

The first step of the activation of both insulin and IGF pathways is the trans-auto-phosphorylation of the receptors induced by ligand binding. This step is necessary for the activation of the kinase activity of the β subunits of both receptors. Note that the IGF-2 receptor does not possess tyrosine kinase activity, and IGF-2 stimulates the IGF-1R and IR. Characterization of the phosphorylation state can be assessed using specific antibodies directed against either IRβ or IGF1-Rβ (Biotechnology, Santa Cruz, California) to immunoprecipitate the receptors and purify them from original retinal lysates (9). Lysates are prepared in IP buffer (50 mM HEPES, 137 mM NaCl, 1 mM $MgCl_2$, 1 mM $CaCl_2$, 2 mM EDTA, 2 mM $NaVO_4$, 10 mM NaF, 10 mM sodium pyrophosphate, 10 mM benzamidine, 2 mM phenylmethylsulfonyl fluoride, 10% glycerol, 1% NP-40, and protease inhibitor tablet; Roche, Mannheim, Germany).

1. Preincubate 5 μL of antibody with 50 μL of 50% slurry protein A equilibrated in PBS containing 1% of BSA in 500 μL of IP buffer, for 4 h at 4°C to allow the interaction of the antibody with the beads.
2. Centrifuge for 1 min at 500 × g, and discard the supernatant.
3. Wash the beads pellet with 250 μL of IP buffer. (Repeat steps 2–3, twice.)
4. Add 500 μg of protein lysate in 1 mL of IP buffer final, and incubate overnight at 4°C rocking to allow the interaction of the prebound antibody with its epitope.
5. Centrifuge for 1 min at 500 × g, and discard the supernatant.
6. Wash the beads pellet with 500 μL of IP buffer. (Repeat steps 5–6, twice.)
7. Add 30 μL of 2× sample buffer (0.13 M Tris, pH 6.8, 25% glycerol, 2.5% β-mercaptoethanol, 2% SDS, 2.3 mg/mL bromophenol blue) and boil for 5 min before proceeding with SDS-polyacrylamide gel electrophoresis.
8. Centrifuge for 2 min at 2000 × g to pellet the beads, and run the supernatant on 4%–12% Bis–Tris gel in MOPS buffer at 200 V for 55 min.
9. Proteins are then electrotransferred to nitrocellulose membrane for 2 h at 400 mA at 4°C.
10. Block the membrane with TBST containing 3% of BSA for at least 1 h.
11. Incubate primary antibody antiphosphotyrosine (Upstate, Charlottesville, Virginia) in blocking solution overnight at 4°C.
12. Wash thrice for 5 min each in TBST.
13. Incubate horseradish peroxidase-conjugated secondary antibody 1 h at room temperature in the blocking solution.
14. Wash thrice for 5 min each in TBST.
15. Add detection reagents (ECL+, Amersham, Piscataway, New Jersey) onto the membrane, and incubate for 5 min at room temperature.
16. Expose autoradiography film on top of the membrane and develop.

10.2.2 Receptor Ligand Affinity

Insulin stimulates the type 1 IGF receptor homodimer as well as the IR–IGF-1R heterodimer in different cell types, including retinal cells during development, although it has a much lower affinity for the IGF-1R than for IR (10). This has been shown using binding assays that measure the displacement of radiolabeled ligand by addition of unlabeled ligand. This technique can be used on retinal explants or retinal cells in culture.

1. Cells are incubated for 4 h at 10°C in HEPES binding buffer, pH 7.8 (100 mM HEPES, 120 mM NaCl, 5 mM KCl, 1.2 mM $MgSO_4$, 8 mM glucose, and 0.1% bovine serum albumin) with the addition of 12.5×10^{-12} M of ^{125}I-insulin or ^{125}I-IGF-1 and increasing concentrations of unlabeled insulin or IGF-1.
2. The cells are then washed 4 times with ice-cold PBS.
3. The cells are then homogenized in 0.1% SDS, and the radioactivity is measured in a gamma counter.
4. Unspecific binding is defined as the binding of radiolabeled ligand in the presence of excess unlabeled ligand.

This method allowed Nitert et al. (8) to show that human endothelial cells have a larger amount of IGF-1R than IR at their surface because the specific binding of IGF-1 is more than twofold higher compared to insulin.

The same technique is used on coimmunoprecipitated dimers formed by IR and IGFR to characterize their respective ligand affinity. As mentioned earlier, IR and IGFR can form heterodimers in several tissues and cell types in addition to homodimers, which have different properties that add another level of regulation of the IR and IGF-1R pathways.

10.3 FUNCTIONAL ASSAYS

The insulin and insulin-like growth factor pathways are composed of a series of phosphorylation-activated kinases following the receptor–ligand interaction.

10.3.1 IR Kinase Assay

The first step in the insulin signaling pathway is tyrosine autophosphorylation of the receptor inducing the activation of the kinase domain of the IR. Insulin receptor kinase activity depends on autophosphorylation, as well as serine phosphorylation and adaptor proteins, such as Grb14 and Src. The β subunit of the IR is immunoprecipitated using the same antibody mentioned in section 10.2.2.

1. Preincubate 5 μL of antibody with 50 μL of slurry protein A equilibrated in PBS containing 1% of BSA in 500 μL of IP buffer for 4 h at 4°C to allow the interaction of the antibody with the beads.
2. Centrifuge for 1 min at 500 × g and discard the supernatant.
3. Wash the beads pellet with 250 μL of IP buffer. (Repeat steps 2–3, twice.)

4. Add 500 µg of protein lysate in a final volume of 1 mL of IP buffer and incubate overnight at 4°C rocking to allow the interaction of the antibody with its epitope.
5. Centrifuge at 500 × g for 1 min, and discard the supernatant.
6. Add 500 µL of kinase buffer (50 mM HEPES, pH 7.3, 150 mM NaCl, 20 mM $MgCl_2$, 2 mM $MnCl_2$, 0.1% Triton X-100, and 0.05% bovine serum albumin). (Repeat steps 5–6, twice.)
7. Add 200 µL of kinase mix (112.5 µM ATP, 4.5 mg/mL polyGlu:Tyr, 0.25 µCi/mL ^{32}P–ATP in kinase buffer), and rock for 60 min at room temperature.
8. Pulse spin the samples to stop the reaction, and transfer 40 µL on filter paper. Let it dry for 15 s, and then place the filter paper in 0.75% phosphoric acid.
9. Wash filters 5 times for 5 min each in 0.75% phosphoric acid and once in acetone. Air-dry filters, and transfer in tubes with 5 mL of scintillation liquid.

Using this technique, our group was able to show that retinal IR has a high constitutive kinase activity, equivalent to that of postprandial liver, but it does not fluctuate with feeding or fasting (11). Comparison of kinase activities from control and diabetic rats retinas also enabled our group to show that diabetes induces a 25% decrease in the IR activity as early as 4 weeks after the onset of diabetes (12), whereas the tyrosine phosphorylation decrease was not yet detectable (figure 10.1a) before 12 weeks of streptozotocin-induced diabetes (figure 10.1b).

10.3.2 KINASE ASSAYS FOR OTHER PROTEINS OF THE IR SIGNALING PATHWAY

The activity of several other downstream kinases of the insulin signaling pathways can be tested, including Akt–1, Akt–3, and PI3–kinase, which are activated following IR activation. Here, we will describe two different kinase assays.

10.3.2.1 Akt-1 Kinase Assay

1. Preincubate 5 µL of antibody with 50 µL of slurry protein G in 50 µL of buffer A (2 M Tris, pH 8, 0.5 M EDTA, 1 M EGTA, 0.5 M sodium fluoride, 0.5 M sodium β-glycerophosphate, 0.5 M sodium pyrophosphate, 0.5 M sodium orthovanadate, 10% Triton X-100, 0.1% β-mercaptoethanol, 1 mM LR-microcystin, and proteinase inhibitors cocktail), standing for 1 h at 4°C to allow the interaction of the antibody with the beads.
2. Centrifuge for 1 min at 500 × g, and discard the supernatant.
3. Wash the beads pellet with 500 µL of buffer A. (Repeat steps 2–3, twice.)
4. Add 500 µg of protein lysate in a final volume of 500 µL of buffer A and incubate for 1 h at 4°C, rocking to allow the interaction of the antibody with its epitope.
5. Centrifuge at 500 × g for 1 min, and discard the supernatant.
6. Wash the beads pellet with 500 µL of buffer A (thrice), then twice with buffer B (2 M Tris, pH 7.5, 30% Brij 35, 1 M EGTA, 0.1% β-mercaptoethanol) and once with ADBI buffer (0.5 M MOPS, 0.5 M sodium β-glycerophosphate, 1M EGTA, 0.5 M sodium orthovanadate, 0.5 M DTT).

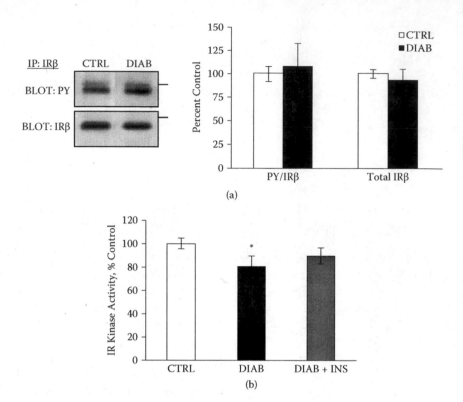

FIGURE 10.1   Reduced retinal insulin receptor phosphorylation and kinase activity in diabetes. The insulin receptor (IR) from retina was immunoprecipitated (IP) and analyzed for
tyrosine phosphorylation (PY) or kinase activity. (a) Analysis of immunoblots for tyrosine
phosphorylation of the insulin receptor-β (IRβ) and total insulin receptor-β reveals equivalent
phosphorylation and expression in the retinal tissue of control (CTRL) and 4-week-diabetic
(DIAB) rats (representative immunoblots shown, $n = 8$ control and 5 diabetic rats). (b) Four
weeks of diabetes ($n = 26$) reduces insulin receptor kinase activity compared with controls
($n = 27$) ($P < 0.001$). Retinal insulin receptor kinase activity was normal in diabetic rats
treated with insulin pellets (INS) ($n = 14$) when compared with controls ($P = 0.12$). Retinal insulin receptor kinase activity was not significantly reduced in 4-week-old diabetic rats
when compared with diabetic rats treated with insulin pellets ($P = 0.23$); however, treatment
with insulin pellets trended toward improved insulin receptor kinase activity.

7. Add 40 μL of kinase mix (40 μM PKA, 0.4 mM Akt/SGK Substrate Peptide, 112.5 μM ATP, 0.25 μCi/μL ^{32}P-ATP) and rock for 60 min at 30°C.
8. Pulse spin the samples to stop the reaction, and transfer 40 μL on filter paper.
   Let it dry for 15 s, and then place the filter paper in 0.75% phosphoric acid.
9. Wash filters 5 times for 5 min each in 0.75% phosphoric acid and once in acetone. Air-dry filters, and transfer in tubes with 5 mL of scintillation liquid.

Using this method, we have shown that the Akt-1 and Akt-3 activities were reduced in
rat retina as early as 4 weeks after the onset of diabetes, whereas the expression and
phosphorylation level were decreased after 12 weeks of diabetes (figure 10.2) (12).

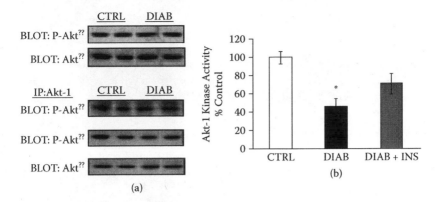

(a)

(b)

FIGURE 10.2 Reduced Akt kinase activity by diabetes in rat retina. (a) Retinal lysates from control (CTRL) and diabetic (DIAB) rats were analyzed by immunoblot analysis. In retina, Akt phosphorylation of threonine (Thr) 308 and serine (Ser) 473 and total expression were unaltered in the diabetic state. (b) After 4 weeks of diabetes, Akt-1 kinase activity in retina was reduced by 54% (*$P < 0.01$ by ANOVA and Tukey–Kramer multiple comparisons post hoc test, $n = 7$/group).

10.3.2.2 PI3-Kinase Assay

Phosphatidylinositol 3-phosphate (PI3-K) is involved in numerous different mechanisms in cell activity. Hence, the assay for PI3-K activity specifically involved in the insulin signaling pathway is based on its interaction with the specific docking proteins called IRS-1 and IRS-2 for insulin receptor substrate.

1. Preincubate with 2 µg each of anti–IRS-1 and anti–IRS-2 (Santa Cruz Biotechnology, Santa Cruz, California) with 30 µL of 50% protein G Sepharose slurry (Amersham) in 500 µL of IP buffer for 4 h.
2. Centrifuge for 1 min at 500 × g, and discard the supernatant.
3. Wash the beads pellet with 500 µL of IP buffer. (Repeat steps 2–3, twice.)
4. Add 75 µg of protein lysate in 1 mL of IP buffer final, and incubate overnight at 4°C rocking.
5. The immune complexes are washed once with buffer A containing 0.5 mol/L NaCl, once with buffer B (50 mmol/L Tris-HCl, pH 7.5, 0.03% Brij-35 [vol/vol], 0.1 mmol/L EGTA, and 0.1% β-mercaptoethanol [vol/vol]), and once with TNE buffer consisting of 20 mmol/L Tris-HCl, pH 7.5, 100 mmol/L NaCl, 0.5 mmol/L EGTA, and 0.1 mmol/L sodium orthovanadate.
6. The immune complexes are then incubated at 35°C for 10 min in 50 µL TNE buffer, pH 7.4, in the presence of ^{32}P-ATP (10 µCi/assay) and the substrate phosphatidylinositol (20 µg/assay).
7. The reaction is stopped by adding 20 µL of 6 N HCl and 160 µL of $CHCl_3$/CH_3OH (1:1).
8. The organic phase is spotted on a thin-layer chromatography plate and subjected to ascending chromatography using the solvent $CHCl_3$/CH_3OH/H_2O/NH_4OH (60:47:11.3:2).

9. Phosphatidylinositol 3-phosphate, thus resolved, is quantified by Phosphor-Imager analysis (Molecular Dynamics, Sunnyvale, California).

With this assay, our group was able to show that IRS2-dependent PI3-K activity was reduced in the retina after 4 weeks of diabetes, whereas the level of expression and phosphorylation were unchanged (figure 10.3) (12).

10.3.3 POTENTIAL PITFALLS

The efficiency of IP-based kinase assays depends on multiple parameters that can be optimized according to the nature of the targeted kinase. This section will concentrate on these diverse parameters.

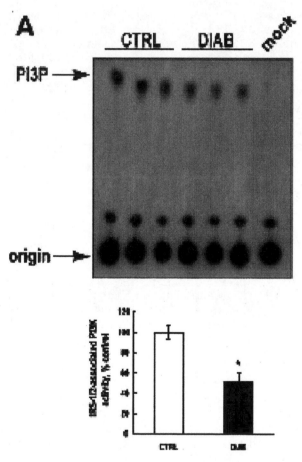

FIGURE 10.3 Reduced IRS-1/2–PI3-K activity in retina of diabetic rats. Retinal lysates were dual-immunoprecipitated for IRS-1/2, and PI3-K activity was measured. A representative TLC plate image is shown demonstrating reduced phosphatidylinositol 3-phosphate (PI3P) formation in lysates of 4-week-old diabetic rats ($n = 14$ control [CTRL] and 15 diabetic [DIAB] rats), with corresponding graphic representation.

The first parameter to be optimized is the protein amount used for the IP, which will affect the intensity of the final signal. The amount of protein used to perform the IP must be adapted to the relative concentration of the kinase in the original lysate and the affinity of the antibody. Although IR and Akt-1 proteins are expressed at modest levels in the retina, we used a high amount of total retinal lysate to perform the IPs and found that 500 μg of protein leads to a good signal-to-noise ratio. This ratio also depends on the efficiency of the washes performed during the IP and after the substrate phosphorylation. The step of prebinding the antibody with the beads, and the washes following are especially important with weakly expressed proteins, because they reduce the loss of protein due to binding to free antibody. The series of washes performed after the lysate incubation is necessary to decrease the background activity induced by kinases that nonspecifically interact with the beads. During the optimization process, we determined that the number of washes was even more crucial than the volume of buffer used; three washes of 500 μL of IP buffer were more efficient than two washes of 1 mL. Finally, we also determined that at least five washes were required following the substrate incubation to decrease nonspecific binding of the free radioactive ATP to the filter paper.

Another parameter to be optimized is the duration of substrate incubation, which determines the intensity of the final signal obtained. The duration of the reaction is correlated to the abundance and the activity of the kinase in the lysate tested. In the retina, we have determined that a 1 h reaction is the optimum duration to obtain the best signal-to-noise ratio for IR and Akt1 kinases.

Key components of the immunoprecipitation-based kinase assays, especially those using nonspecific substrates, are the negative controls without antibody and, to a lesser extent, without lysate to prove the specificity of the approach. The so-called no-lysate control is necessary to show that the signal obtained for the IPs of interest is due to a specific interaction of the specific antibody used and not due to any nonspecific binding to the beads.

10.3.4 BIOLUMINESCENT ASSAY (13)

Another way to test the activity of both IR and IGF-1R uses a technology called bioluminescence resonance energy transfer (BRET). Contrary to assays previously mentioned that allow measurement from tissue samples, this assay can only be used in cells and allows the measurement of the IR or IGF-1R activity in living cells in real time. Moreover, this assay does not measure the kinase activity but the conformational change following substrate activation of the receptors necessary for trans-autophosphorylation. Because transfecting primary retinal cells remains particularly challenging, this technique is of more interest for use in retinal cell lines such as R28 or RGC-5. The principle of the method is to transfect cells with two different constructs: one is fused to the 35 kDa Renilla luciferase (RLuc, the donor), and the other is fused to the 27 kDa enhanced yellow fluorescent protein (EYFP, the acceptor). The synthetic RLuc substrate, coelenterazine h, is able to permeate the cell membrane and is converted by RLuc under the emission of blue light of 480 nm. When EYFP is in close proximity (<100 Å), a nonradioactive transfer of energy causes its excitation, and yellow/green light of 527 nm is emitted. In our case, both constructs could be either IR or IGFR monomers or a combination of both, allowing study of homo- or heterodimer characteristics.

10.4 RETINAL SPECIFICITY COMPARED TO OTHER INSULIN-SENSITIVE TISSUES

As we previously mentioned, one of the unique retinal features is its complex cellular composition compared to the other insulin-sensitive tissues. To study the relative importance of the insulin and IGFs signaling pathways in the different retinal cells, several approaches can be used, among them in situ hybridization or laser capture dissection of the retinal layers followed by RNA analysis for specific proteins of these pathways such as IR or Akt.

10.4.1 In Situ Hybridization

Using this method, we have shown that the Akt isoforms are differentially localized in the retina, with high expression of Akt-1 and Akt-3 mRNAs in the inner nuclear and ganglion cell layers (figure 10.4).

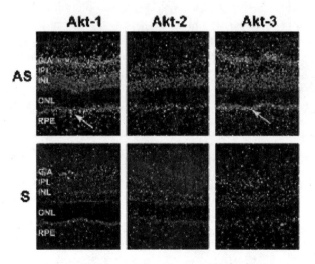

FIGURE 10.4 Rat retina expresses Akt 1–3 mRNA species. cRNA in situ hybridization was performed for all three Akt isoforms in intact eyes with antisense (AS) and sense (S) probes and counterstained with hematoxylin. Representative dark-field images are shown. RPE, retinal pigmented epithelium; ONL, outer nuclear layer; INL, inner nuclear layer; IPL, inner plexiform layer; G/A, ganglion and astrocyte layer. Akt-1 and Akt-3 mRNA expression has similar profiles among all retinal layers, whereas Akt-2 mRNA is observed primarily in the INL and G/A layers. Akt 1–3 mRNA is also expressed in the outer segments of the photoreceptors (arrows). Results are representative of duplicate experiments.

10.4.1.1 Akt-1, -2, and -3 In Situ Hybridization

1. The synthesis of ^{35}S-labeled riboprobes and RNA in situ hybridization was performed essentially as described (14,15) on optimal cutting tissue (OCT)-embedded cryostat sections, 16 μm thick, of whole eyes from normal rats. Vectors for the sense and antisense riboprobes of Akt-1 (bp 1696–2306), -2 (bp 1632–2229), and -3 (bp 1476–1724) were kindly provided by Dr. Morris Birnbaum, University of Pennsylvania.
2. Hybridization stringency was optimized by applying probes (107 cpm/mL) in formamide hybridization buffer to sections, coverslipping, and incubating in a humidified chamber overnight at 55°C.
3. Slides were washed in 4× SSC to remove the coverslip and buffer, dehydrated, and immersed in 0.3 M NaCl, 50% formamide, 20 mM Tris-HCl, and 1 mM EDTA at 60°C for 15 min, followed by RNase A (20 μg/mL) treatment for 30 min at 37°C.
4. Slides were passed through graded salt solutions and washed in 0.1× SSC for 15 min and 0.05× SSC for 30 min at 60°C.
5. Air-dried slides were exposed to Hyperfilm-max (Amersham) for 5–7 days, dipped in Kodak NTB3 photographic emulsion, stored with desiccant at 4°C for 10–14 days, and developed and stained with hematoxylin for evaluation.
6. All sections were processed together to facilitate signal comparisons among the groups.

10.4.2 LASER CAPTURE MICROSCOPY

Another method to assess the cellular localization of the different elements of the IR and IGF-1R signaling pathway in the retina involves the isolation of the different layers of the retina using laser capture microscopy. The different cells of the retina are organized in successive layers that can be isolated using this method. Once isolated, relative quantification of any specific message, including IR, IGF-1R, Akt, or any other protein, can be done on the different layers using real-time RT-PCR.

1. Whole eyes from control rats are harvested and fixed overnight in 4% paraformaldehyde prepared in ultrapure water.
2. Whole eyes are then incubated in successive baths of 10, 20, and 30% sucrose in ultrapure water respectively for 1 h, 1 h, and overnight.
3. Whole eyes are embedded in OCT, and sections of 10 μm are placed on polyethylene naphthalate (PEN) membrane-covered glass slides from P.A.L.M. (Bernried, Germany).
4. The retinal nuclear layers are stained with a labeling solution consisting of a 500 nM solution of RNASelect—green fluorescent cell stain in PBS (Molecular Probes, Eugene, Oregon) that is incubated for 20 min at room temperature.
5. Sections are washed for 5 min in PBS.

6. LCM is then performed using a PALM MicroLaser system (PALM–Zeiss, Bernreid, Germany) containing a PALM MicroBeam (driven by PALM MicroBeam software), and PALM RoboStage and for high-throughput sample collection a PALM RoboMover (driven by PALM RoboSoftware v2.2).
7. The different retinal layers visualized with the cytoRNA staining are cut under a 5× ocular lens and catapulted directly into RNase-free microtube for RNA extraction followed by real-time RT-PCR.

These two methods allow the study of specific proteins known to be involved in the IR and IGF-1R signaling pathways and to determine their regional or cellular localization in the retina. As we will see now, other methods exist to determine what other proteins are involved in these pathways.

10.5 GENERAL APPROACHES

Although the insulin and IGF signaling pathways are involved in numerous cellular processes, broader approaches can be used to study them. We will mention two of them in this chapter. The first approach is used in many topics to determine the broad influence of any compound at the protein level by determining the modification of the proteome of cells or tissues. The second approach is more specifically dedicated to the study of pathways such as the insulin signaling pathways because it consists of the study of the kinome of a cell or tissue.

10.5.1 Proteomic Analysis

Characterization of the insulin signaling pathway can be done using several proteomic approaches, including DIGE technology, on which we will focus in this paragraph. The basic idea is similar in any "proteomic" approach to compare the level of expression of proteins in different conditions. In the specific case of the IR and IGF-1R signaling pathways in the retina, it can be used to compare the proteome of: (1) ex vivo retinal tissue exposed to insulin or IGF stimulation and (2) retinal tissue from animals in which IR or IGF-1R signaling as been disrupted, i.e., siRNA, tissue-specific knockout.

1. Tissues are homogenized in standard cell lysis buffer by intermittent sonication.
2. Centrifuge the cell lysate at 4°C for 10 min at 12,000 × g, and transfer the supernatant to another tube (the pH of the cell lysate must be 8.0–9.0).
3. Add 1 μL of working dye solution to a volume of protein sample equivalent to 50 μg to a microfuge tube. Mix dye and protein sample thoroughly by vortexing.
4. Centrifuge briefly in a microcentrifuge to collect the solution at the bottom of the tube. Leave on ice for 30 min in the dark.
5. Add 1 μL of 10 mM lysine to stop the reaction. Mix and spin briefly in a microcentrifuge. Leave for 10 min on ice, in the dark.
6. A different dye is used for each of the conditions tested. Combine the two differentially labeled samples into a single microfuge tube and mix.

7. Add an equal volume of 2× sample buffer to the labeled protein samples, and leave on ice for at least 10 min.
8. Proceed with the 2-D focusing using the Ettan IPGphor isoelectric focusing system.
9. Proceed with the 2-D electrophoresis using the Ettan DALT electrophoresis system.
10. Image the gels using Typhoon Variable Mode Imager with the Ettan DIGE system.
11. Analyze the images using the DeCyder Differential Analysis Software to determine the spots that present an increase or decrease intensity between the two different conditions.
12. A picking gel run with unlabeled pooled samples in parallel to the analysis gel is used to pick the spots of interest after labeling with SyproRuby.
13. Identification of the proteins is done using MS/MS.

10.5.2 KINOME ANALYSIS

The insulin and IGF signaling pathways involve a number of proteins with kinase activity. One of the approaches to study the effect of a drug, treatment, agonist, or antagonist is to determine the modifications of the cells or tissues "kinome" induced by those treatments. This can be done using a new technology called Pepchip® kinase, which consists of testing protein extracts of interest in in vitro kinase reactions on a Pepchip microarray to analyze the phosphorylation profiles of kinase substrates in these different samples. In our case, the samples could be retinal lysate from control and any conditions modifying the IR or IGF-1R signaling pathways, such as conditional knockout or insulin administration.

1. The retinal tissues are lysed in 25 mM HEPES (pH 7.6), 50 mM NaCl, 0.1 mM EDTA, 0.5% Triton X-100, 0.1 mM sodium orthovanadate, and 1 mM phenylmethylsulfonyl fluoride.
2. Centrifuge the lysate at 4°C for 10 min at 12,000 × g and transfer the supernatant to another tube.
3. Use the supernatants for the screening according to the instructions of the manufacturer.
4. The microarrays are scanned by PhosphorImager (Molecular Dynamics, Sunnyvale, California) with a resolution of 50 μm.
5. The data are evaluated and compared using ImageQuant software.
6. The signal intensity is normalized according to the instructions of the manufacturer.

10.6 CONCLUSIONS

An interesting variety of complementary approaches are available nowadays to study cellular signaling pathways, ranging from study of the ligand affinity to the activation of downstream effectors. As previously mentioned, these techniques can be protein specific to identify the proteins involved, or less specifically, to characterize the proteins

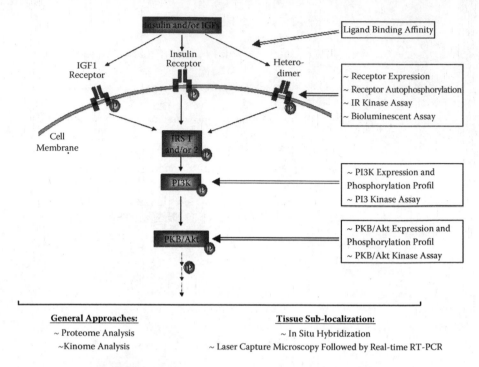

General Approaches:
~ Proteome Analysis
~Kinome Analysis

Tissue Sub-localization:
~ In Situ Hybridization
~ Laser Capture Microscopy Followed by Real-time RT-PCR

FIGURE 10.5   (See accompanying color CD.) Schematic representation of the first steps of the insulin/IGFs pathways and the different methods allowing their assessments.

involved in the downstream cascade (figure 10.5). This variety of approaches is necessary in many scientific areas, but even more so in cell signaling studies, where the answers are so wide-ranging.

ACKNOWLEDGMENTS

Juvenile Diabetes Research Foundation (JDRF), American Diabetes Association (ADA), The Pennsylvania Lions Sight Conservation and Eye Research Foundation, and the Turner Professorship.

REFERENCES

1. Olson, T.S., Bamberger, M.J., and Lane, M.D.: Post-translational changes in tertiary and quaternary structure of the insulin proreceptor. Correlation with acquisition of function. *J Biol Chem* 263: 7342–7351, 1988.
2. Abbott, M.A., Wells, D.G., and Fallon, J.R.: The insulin receptor tyrosine kinase substrate p58/53 and the insulin receptor are components of CNS synapses. *J Neurosci* 19: 7300–7308, 1999.
3. Barber, A.J., Lieth, E., Khin, S.A., Antonetti, D.A., Buchanan, A.G., and Gardner, T.W.: Neural apoptosis in the retina during experimental and human diabetes. Early onset and effect of insulin. *J Clin Invest* 102: 783–791, 1998.

4. Diaz, B., Serna, J., De Pablo, F., and de la Rosa, E.J.: In vivo regulation of cell death by embryonic (pro)insulin and the insulin receptor during early retinal neurogenesis. *Development* 127: 1641–1649, 2000.
5. Hernandez-Sanchez, C., Lopez-Carranza, A., Alarcon, C., de La Rosa, E.J., and de Pablo, F.: Autocrine/paracrine role of insulin-related growth factors in neurogenesis: local expression and effects on cell proliferation and differentiation in retina. *Proc Natl Acad Sci U.S.A.* 92: 9834–9838, 1995.
6. Waldbillig, R.J., Arnold, D.R., Fletcher, R.T., and Chader, G.J.: Insulin and IGF-I binding in developing chick neural retina and pigment epithelium: a characterization of binding and structural differences. *Exp Eye Res* 53: 13–22, 1991.
7. Bailyes, E.M., Nave, B.T., Soos, M.A., Orr, S.R., Hayward, A.C., and Siddle, K.: Insulin receptor/IGF-I receptor hybrids are widely distributed in mammalian tissues: quantification of individual receptor species by selective immunoprecipitation and immunoblotting. *Biochem J* 327 (Pt 1): 209–215, 1997.
8. Nitert, M.D., Chisalita, S.I., Olsson, K., Bornfeldt, K.E., and Arnqvist, H.J.: IGF-I/insulin hybrid receptors in human endothelial cells. *Mol Cell Endocrinol* 229: 31–37, 2005.
9. Barber, A.J., Nakamura, M., Wolpert, E.B., Reiter, C.E., Seigel, G.M., Antonetti, D.A., and Gardner, T.W.: Insulin rescues retinal neurons from apoptosis by a phosphatidylinositol 3-kinase/Akt-mediated mechanism that reduces the activation of caspase-3. *J Biol Chem* 276: 32814–32821, 2001.
10. Garcia-de Lacoba, M., Alarcon, C., de la Rosa, E.J., and de Pablo, F.: Insulin/insulin-like growth factor-I hybrid receptors with high affinity for insulin are developmentally regulated during neurogenesis. *Endocrinology* 140: 233–243, 1999.
11. Reiter, C.E., Sandirasegarane, L., Wolpert, E.B., Klinger, M., Simpson, I.A., Barber, A.J., Antonetti, D.A., Kester, M., and Gardner, T.W.: Characterization of insulin signaling in rat retina in vivo and ex vivo. *Am J Physiol Endocrinol Metab* 285: E763–774, 2003.
12. Reiter, C.E., Wu, X., Sandirasegarane, L., Nakamura, M., Gilbert, K.A., Singh, R.S., Fort, P.E., Antonetti, D.A., and Gardner, T.W.: Diabetes reduces basal retinal insulin receptor signaling: reversal with systemic and local insulin. *Diabetes* 55: 1148–1156, 2006.
13. Laursen, L.S. and Oxvig, C.: Real-time measurement in living cells of insulin-like growth factor activity using bioluminescence resonance energy transfer. *Biochem Pharmacol* 69: 1723–1732, 2005.
14. Koehler-Stec, E.M., Simpson, I.A., Vannucci, S.J., Landschulz, K.T., and Landschulz, W.H.: Monocarboxylate transporter expression in mouse brain. *Am J Physiol* 275: E516–524, 1998.
15. Vannucci, S.J., Maher, F., and Simpson, I.A.: Glucose transporter proteins in brain: delivery of glucose to neurons and glia. *Glia* 21: 2–21, 1997.

11 Probing the Interactions between the Retinal Insulin Receptor and Its Downstream Effectors

Raju V.S. Rajala and Robert E. Anderson

CONTENTS

11.1 INTRODUCTION

The insulin receptor (IR) is a transmembrane receptor that is activated by insulin. It belongs to the large class of tyrosine kinase receptors. It induces cellular responses by phosphorylating proteins on their tyrosine residues. The IR and IR signaling proteins are widely distributed throughout the central nervous system (1). IR signaling is shown to regulate food intake (2,3) and neuronal growth and differentiation (4,5).

IR signaling provides a trophic signal for transformed retinal neurons in culture (6), but the in vivo role of the IR is unknown. Light induces tyrosine phosphorylation of the retinal IR, and this activation leads to the binding of phosphoinositide 3-kinase (PI3K) to rod outer segment (ROS) membranes (7). IR signaling is also involved in 17β-estradiol-mediated neuroprotection in the retina (8). Recent evidence suggests a downregulation of IR kinase activity in diabetic retinopathy that is associated with the deregulation of downstream signaling molecules (9). Deletion of several downstream effector molecules of the IR signaling pathway such as IRS-2 (10), Akt2 (11), and bcl-xl (12) in the retina results in a photoreceptor degeneration phenotype. These studies clearly indicate the importance of the IR signaling pathway in the retina. Most of these IR signaling events are mediated through PI3K, the first step in the IR signaling pathway. This chapter focuses on the interaction between the IR and PI3K, and IR-induced activation of downstream effectors, and describes methods that are useful for the study of such interactions.

11.2 INSULIN RECEPTOR

11.2.1 STRUCTURE

The IR is present in virtually all vertebrate tissues, although the concentration varies from as few as 40 receptors on circulating erythrocytes to more than 200,000 on adipocytes and hepatocytes (13). The receptor gene is located on the short arm of human chromosome 19 and is more than 150 kb in length with 22 exons that encode a 4.2 kb cDNA (14). The IR is composed of two α-subunits that are each linked to a β-subunit and to each other by disulfide bonds (13). Both subunits are derived from a single proreceptor by proteolytic processing at a cleavage site that consists of four basic amino acids.

The mature heterotetramer (α2β2) contains complex N-linked carbohydrate side chains capped by terminal sialic acid residues and migrates with a molecular mass of 300–400 kDa by sodium dodecyl sulfate polyacrylamide gel electrophoresis (SDS-PAGE). The α-subunits are located entirely outside of the cell and contain the insulin-binding sites; the intracellular portion of the β-subunit contains the insulin-regulated tyrosine kinase (13).

11.2.2 SPECIES HOMOLOGY

The IR is highly conserved across evolution. It regulates neuronal survival in *C. elegans* (15). It serves an important function to guide retinal photoreceptor axons from the retina to the brain during development in *Drosophila* (16). In addition, the IR influences the size and number of photoreceptors (17). Mutations in either the IR (16) or its binding partner Dock (18) in *Drosophila* result in a severe photoreceptor axonal misguidance phenotype. There is a high degree of IR signaling homology between *C. elegans*, *Drosophila,* and humans, suggesting a functional conservation in the mammalian retina.

11.2.3 STIMULATION

The actions of insulin are initiated by its binding to the IR (19–21). Insulin binds to two asymmetric sites on the extracellular α-subunits, causing conformational changes that lead to autophosphorylation of the membrane-spanning β-subunits and activation of the receptor's intrinsic tyrosine kinase (22,23). IR activation is the first step in the IR signaling pathway, and the methods to study the IR activation in the retina are described in the following text. IR activation can be studied using both radioactive and nonradioactive methods.

PROTOCOL 1: DISSECTION OF RETINAS

Dark-adapt animals (rats or mice) overnight, and next morning kill them by CO_2 asphyxiation, followed by cervical dislocation. Quickly remove retinas by a technique we call "winkleing" (24). A deep cut is made across the corneal surface, and a pair of curved forceps is placed behind the eyeball on either side of the optic nerve head. The forceps are squeezed and brought forward, forcing the contents of the eye to be extruded through the hole in the cornea. The retina can be recovered relatively intact and placed immediately into ice-cold buffer or snap-frozen. This procedure takes about 5–10 s in experienced hands and is recommended when it is necessary to produce retinas quickly (i.e., RNA preparation) or when a large number of retinas need to be isolated. It is not recommended if the retinas will be used for proteomic studies, because there will inevitably be some contamination with lens proteins.

PROTOCOL 2: INSULIN-INDUCED ACTIVATION OF THE RETINAL IR EX VIVO

1. Obtain retinas as described in protocol 1.
2. Incubate retinas in Dulbecco's Modified Eagle's Medium (DMEM, Gibco BRL) with or without 1 µM human insulin [Human insulin R (rDNA origin, Eli Lilly & Co)] at 37°C for 5 min.
3. Flash-freeze retinas in liquid nitrogen.
4. Add 400 µL lysis buffer [1% NP 40, 20 mM HEPES (pH 7.4), 2 mM EDTA, phosphatase inhibitors (100 mM NaF, 10 mM $Na_4P_2O_7$, 1 mM $NaVO_3$, and 1 mM molybdate), and protease inhibitors (10 µM leupeptin, 10 µg/mL aprotinin, and 1 mM PMSF)] to each retina.
5. Sonicate retinal suspension for 10 s, and cool the tube on ice for 1 min. Repeat this twice more.
6. Keep retina suspensions on ice for 10 min, then centrifuge at 16,100 × g for 20 min at 4°C.
7. Aspirate supernatant into a clean Eppendorf tube, and determine its protein concentration by BCA reagent (Pierce) following the manufacturer's instructions.
8. Add 50 µL Protein A Sepharose beads to 500 µg protein, and incubate at 4°C for 60 min with continuous shaking.

9. Spin at high speed (16,100 × g) using a table-top centrifuge for 2 min at 4°C.
10. Transfer the supernatant into a clean Eppendorf tube, add 4 μg β-subunit of the IR (IRβ) antibody, and incubate the reaction at 4°C for either 2 h or overnight.
11. Add 50 μL Protein A Sepharose beads to the antibody reaction, and continue to shake the reaction for another 60 min at 4°C.
12. Wash the immune complexes thrice with wash buffer [50 mM HEPES (pH 7.4), 118 mM NaCl, 100 mM NaF, 2 mM NaVO$_3$, 0.1% (w/v) SDS, and 1% (v/v) Triton X-100]; then spin the reaction at high speed using a table-top centrifuge for 2 min at 4°C.
13. Add 20 μL 2× SDS sample buffer [0.125 M Tris-HCl (pH 6.8), 4% SDS, 5% 2-mercaptoethanol, 20% glycerol, and 0.01% bromophenol blue] to washed IR immunoprecipitates (IP), and incubate the tubes at 100°C for 5 min. Then, centrifuge at 16,100 × g for 1 min at 4°C.
14. Electrophorese the proteins on a 10% Tris-glycine SDS precast gel (Invitrogen). Transfer the SDS gel onto nitrocellulose membrane (either BioRad or Invitrogen) using 100 V for 90 min at 4°C.
15. Wash the membrane in 1× TTBS [20 mM Tris-HCl (pH 7.4), 100 mM NaCl, and 0.1% Tween-20] for 5 min, and block the membrane with 5% nonfat milk for 60 min at room temperature (RT).
16. Rinse the membrane with 1× TTBS for 5 min, then add antiphosphotyrosine antibody, and gently shake the blot on a rocker at 4°C, overnight.
17. Wash the blot twice with 1× TTBS for 5 min each, then add antimouse secondary antibody, and incubate the blot at RT for 60 min. Wash the blot thrice with 1× TTBS for 10 min each. Develop the blot with ECL reagent.
18. Verify equal protein loading by incubating the blot with 20 mL stripping reagent (Pierce) for 15 min at 37°C, and wash the membrane 5 times with 1× TTBS. Block the membrane with 5% nonfat milk, and probe the blot with anti-IRβ antibody. Develop the blot with ECL Reagent.
19. Wash the blot twice with 1× TTBS for 5 min each, then add anti-rabbit secondary antibody and incubate the blot at RT for 60 min. Wash the blot thrice with 1× TTBS for 10 min each. Develop the blot with ECL reagent.

Figure 11.1A shows the autophosphorylation of the retinal IR as detected using protocol 1. Figure 11.1B is a typical result, showing equal amounts of the IR in each immunoprecipitate obtained from rat retinas.

11.2.4 ACTIVITY

The IR is known to phosphorylate several proteins in the cytoplasm, including insulin receptor substrates (IRS) and Shc (25). Both IRS and Shc function as docking proteins for Src homology 2 (SH2) domain-containing signaling molecules, which perform the next step in the signaling cascade (26). This kinase activity of the activated IR can be quantitated by using exogenous substrates, such as polyGlu:Tyr peptide (27) or IRS -1 peptide (28).

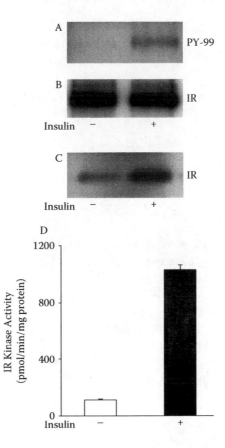

FIGURE 11.1 Insulin-induced activation of the rat retinal IR. Retinas were cultured in DMEM in the absence or presence of 1 μM insulin. The IR was immunoprecipitated with anti-IRβ antibody, (B) followed by Western blot analysis with anti-PY-99 antibody (A). Autophosphorylation (C) and receptor kinase activity (D) were measured in the presence or absence of 1 μM insulin. Receptor kinase activity was measured by in vitro phosphorylation of a poly (Glu:Tyr) substrate in the presence of [γ³²P]ATP. IR kinase activity was expressed as pmol per min per mg of protein. Data are expressed as mean ± SD.

PROTOCOL 3: INSULIN RECEPTOR KINASE ACTIVITY

1. Prepare IR immunoprecipitates as described in steps 1–11 of protocol 2.
2. Wash the immunoprecipitates twice with kinase buffer [50 mM Tris-HCl (pH 7.0), 50 mM $MgCl_2$, 5 mM $MnCl_2$, 50 mM $NaVO_3$, and 7 μg/mL p-nitrophenyl phosphate].
3. Add 5 μL kinase buffer, ATP (to a final concentration of 100 μM), insulin (to a final concentration of 1 μM), polyGlu:Tyr peptide (to a final concentration of 3 mg/mL, Sigma), and ~10 μCi [γ-³²P]ATP to the IR immunoprecipitates to a final volume of 25 μL, and incubate at RT for 15–60 min.
4. Terminate the reaction by adding 10 μL of 50% (vol/vol) acetic acid followed by centrifugation at 16,100 × g, for 2 min at RT. Remove the supernatant

to a clean Eppendorf tube. The supernatant contains the phosphorylated peptides, and the precipitate contains the autophosphorylated IR.

5. Add 2× sample buffer to the precipitate, and incubate the reaction at 100°C for 5 min.

6. Load sample on a 10% SDS gel, and resolve proteins. Wash the gel thrice in distilled water, and dry the gel for 60 min in a gel dryer. Follow with autoradiography.

7. For kinase activity calculations, spot 25 μL of the supernatant (from step 4) on to Whatman p81 phosphocellulose paper; wash the filter paper thrice for 5 min in 0.75% phosphoric acid to remove radioactive ATP and once for 5 min in acetone to remove the water from the p81 filter papers. Place the filter papers in scintillation cocktail and count the radioactivity of the phosphorylated peptides.

8. Dilute 1 μL [γ-^{32}P]ATP solution in 1 mL of 0.1 N HCl, add 1 μL diluted ATP solution to scintillation cocktail, and count. Knowing the amount of ATP in the incubation solution, the specific activity of the ATP [disintegrations per minute (DPM)/pmol] can be calculated.

Knowing the DPM in the product and the specific activity of the precursor, the IR kinase activity (pmol/min/mg protein) can be calculated. This calculation is based on the amount of protein used for immunoprecipitations and therefore is only a relative value useful for comparisons between treatments (i.e., light versus dark). It also assumes uniform immunoprecipitation, which can be determined by using anti-IR antibodies with Western blots of the retina IP (figure 11.1B).

Figure 11.1C shows the autophosphorylation of the IR in the presence or absence of insulin. Figure 11.1D shows the IR kinase activity toward the synthetic substrate. This method is very useful to measure both autophosphorylation and kinase activity in the same sample. This technique has been used to study the differences between the IR autophosphorylation and kinase activity in a Streptozotocin-induced experimental diabetes model in rats (9). We have also reported that light induced the tyrosine phosphorylation of the IR in rod photoreceptor cells of rats (7); this technique can also be used to study the autophosphorylation and kinase activity between light- and dark-adapted states. This assay is well applicable to screen for regulators of the IR in vitro (29).

11.3 DOWNSTREAM EFFECTORS

Many cell proliferation and cell survival pathways are initiated upon activation of tyrosine kinase receptors, which transduce their signals by recruiting a variety of cytoplasmic signaling proteins (30,31). Many of the signaling proteins contain phosphotyrosine-binding (PTB) domains, SH2 domains, and Src Homology 3 (SH3) domains, which are involved in mediating protein–protein interactions (32,33). The phosphotyrosine-dependent interaction between different PTB and SH2 domain-containing proteins with activated receptors initiates cellular changes that take place in response to a wide range of extracellular signals (31).

11.3.1 PHOSPHOINOSITIDE 3-KINASE

PI3K constitutes a family of enzymes that phosphorylate the D-3 position of the inositol ring of phosphoinositides (PI) and its derivatives (figure 11.2). The lipid products of PI3K serve as second messengers to recruit specific phospholipid-binding proteins to the plasma membrane and control the activity and subcellular localization of a diverse array of signal transduction molecules (30). The PI3K activity is also involved in various cellular responses, which include cell growth, inhibition of apoptosis, actin cytoskeletal reorganization, and protein trafficking (30,34). Multiple forms of PI3K exist in higher eukaryotes; the class Ia enzymes are responsible for the generation of D-3 PI in response to growth factors (35). PI3K consists of an ~85 kDa regulatory subunit (p85) and a ~110 kDa catalytic subunit (p110), which are responsible for the phosphorylation at the D-3 position of phosphatidylinositol lipids and serine phosphorylation of proteins (35–37). The p85 subunit contains an SH3 domain that is capable of binding to proline-rich sequences, a region of homology to the breakpoint cluster region (BCR) gene product, a p110-binding domain, and an SH2 domain on both the N and C terminuses. PI3K activity increases in response to receptor activation by the direct binding of the p85 SH2 domain to phosphorylated tyrosine sites on the receptor (38,39).

Several targets of the PI3K signal pathway use this regulatory cascade to promote survival (34). PI3K activity is a measure of the cell fate, and understanding the

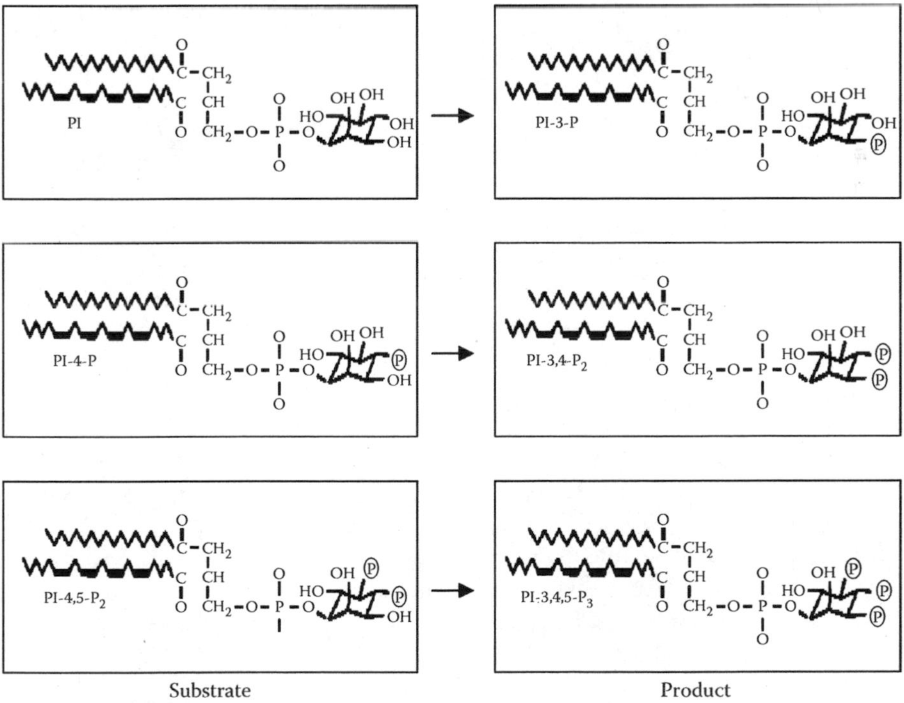

Substrate	Product

FIGURE 11.2 Substrates and phosphorylation products of phosphoinositide 3-kinase.

complexity of the PI3K pathway may provide new avenues for therapeutic intervention in a variety of diseases.

PROTOCOL 4: PHOSPHOINOSITIDE 3-KINASE ASSAY

1. Prepare the following reagents:
    a. Lysis buffer: 1% NP 40, 20 mM HEPES (pH 7.4), 2 mM EDTA, phosphatase inhibitors (100 mM NaF, 10 mM $Na_4P_2O_7$, 1 mM $NaVO_3$, and 1 mM molybdate), and protease inhibitors (10 µM leupeptin, 10 µg/mL aprotinin, and 1 mM PMSF).
    b. PI3K buffer: 20 mM Tris-HCl (pH 7.4), 100 mM NaCl, and 0.5 mM EGTA.
    c. 5× reaction buffer: 100 mM Tris (pH 7.4), 500 mM NaCl, and 2.5 mM EGTA.
    d. Lipid substrate: Dissolve 1 mg $PI-4,5-P_2$ in 2 mL chloroform/methanol (1:1) in a glass tube; vortex and dry the organic solvents under nitrogen gas. Add 1 mL PI3K buffer, vortex, and sonicate for 10 min.
    e. Preparation of ATP: Prepare fresh ATP in deionized H_2O. For each reaction, add 2.5 µL cold ATP (final concentration 20 µM) plus ~10 µCi [γ-^{32}P]ATP.
2. Prepare thin-layer chromatography (TLC) plates by spraying with 1% (w/v) potassium oxalate in 50% methanol (v/v), air-dry for 5 min, and bake in an oven at 100°C for 1 h.
3. Immunoprecipitate PI3K (250–500 µg protein) from retinas using the anti-p85 subunit of PI3K antibody overnight, and wash the immunoprecipitates thrice in 1× lysis buffer and twice in PI3K buffer.
4. Add 10 µL 5× PI3K reaction buffer and 10 µL lipid substrate (0.2 mg/mL $PI-4,5-P_2$) to either retina lysates or immunoprecipitates for a final volume of 50 µL.
5. Initiate the reaction by the addition of 3.5 µL ATP, and incubate the reaction for 15–60 min at RT. Terminate the reaction by the addition of 100 µL of 1 N HCl.
6. Add 200 µL of chloroform/methanol (1:1, v/v), and vortex, followed by 2 min centrifugation on a table-top centrifuge.
7. Remove the chloroform layer to a clean Eppendorf tube. Apply the entire chloroform layer to a TLC plate by spotting 20 µL at a time on oxalate-coated plates (silica gel 60). Air-dry the plate between each application.
8. Resolve lipids in a 1-D solvent system of 2-propanol:2 M acetic acid (63:35, v/v).
9. Expose TLC plate to x-ray film overnight at −70°C. Develop the film, and place over the TLC plate to mark the radioactive lipid areas.
10. Scrape the radioactive lipids into liquid scintillation fluid, acidify with 100 µL 1 N HCl, and determine DPM.

Determine the specific activity of the ATP as described in protocol 2, and use the DPM in each lipid spot from the TLC to calculate the pmol of lipid formed per

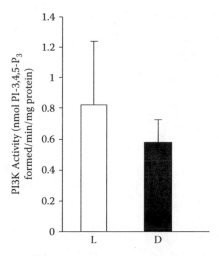

FIGURE 11.3 Measurement of PI3K activity. Lysates from dark (D)- and light (L)-adapted retinas were immunoprecipitated with anti-p85 antibody. PI3K activity was measured using PI-4,5-P_2 and $[\gamma^{32}P]$ATP as substrates. The radioactive spots of PI-3,4,5-P_3 were scraped from the TLC plate and counted. Data are mean ± SE (n = 6). (From Rajala, R.V., McClellan, M.E., Ash, J.D., and Anderson, R.E., In vivo regulation of phosphoinositide 3-kinase in retina through light-induced tyrosine phosphorylation of the insulin receptor beta-subunit, *J. Biol. Chem.*, 277, 43319, 2002. With permission.)

min per mg protein. Figure 11.3 shows the PI3K activity from the p85 immunoprecipitates recovered from light- and dark-adapted retinas. With this assay, we have demonstrated that light does not activate the PI3K enzyme activity but does facilitate PI3K binding to the ROS (7).

11.3.1.1 Insulin-Induced Activation of PI3K

PI3K was initially found to be associated with middle-T/pp60c-Src (40), pp60v-Src, and platelet-derived growth factor (PDGF) receptors in both normal and transformed NIH3T3 fibroblast cells (41,42). PI3K activity increases in response to PDGF binding to its receptor, in large part because the p85–p110 complex is translocated from the cytosol to the plasma membrane by the direct binding of the p85 SH2 domain to tyrosine-phosphorylated sites on the receptor (38,39). Rudermen et al. (43) first demonstrated the activation of PI3K by insulin, either by stimulating CHO cells with insulin or transfecting the CHO cells with the human IR. PI3K activity is also regulated by activated-receptor tyrosine kinase substrates, such as IRS-1 (33). PI3K is a major signaling molecule that is activated by the binding to IR directly (44,45) or via IRS (46) and is important in coupling the IR to glucose uptake processes (25). In the retina, insulin stimulates the activation of PI3K and also rescues the death of transformed retinal neurons in culture through PI3K activation (6). These studies clearly suggest a neuroprotective role for the IR through its activation of PI3K in the retina. Studying the interaction between the IR and PI3K is important to understanding how this neuroprotective signal transduction pathway is activated.

PROTOCOL 5: INSULIN-INDUCED ACTIVATION OF PI3K

1. Incubate rat retinas at 37°C in DMEM medium (Gibco BRL) in the presence or absence of 1 μM insulin for 0–60 min.
2. Remove retinas from the medium, snap-freeze in liquid nitrogen, and store at −80°C until analysis.
3. Solubilize the retinas in lysis buffer [1% NP 40, 20 mM HEPES (pH 7.4) and 2 mM EDTA containing phosphatase inhibitors (100 mM NaF, 10 mM $Na_4P_2O_7$, 1 mM $NaVO_3$, and 1 mM molybdate), and protease inhibitors (10 μM leupeptin, 10 μg/mL aprotinin, and 1 mM PMSF)]. Sonicate retinal suspension for 10 s and cool the tubes on ice for 1 min. Repeat twice more. Centrifuge the homogenate at 17,000 × g for 20 min at 4°C to remove insoluble material.
4. Measure the protein concentration of the lysates using BCA assay.
5. Preclear lysates of insulin-stimulated or control retinas (250 μg protein) with 50 μL of Protein A-Sepharose for 1 h at 4°C with gentle shaking.
6. Centrifuge the sample at 4°C for 1 min, and transfer the supernatant into a clean Eppendorf tube.
7. Add 2 μg anti-IRβ antibody, and incubate overnight at 4°C with continuous shaking.
8. Add 50 μL of protein A-Sepharose beads and incubate at 4°C for 60 min.
9. Wash IR immunoprecipitates thrice in 1× lysis buffer and twice in PI3K buffer.
10. Measure PI3K activity associated with anti-IRβ immunoprecipitates as described in protocol 4.

The TLC autoradiogram in figure 11.4A shows an increase in labeled PI-3,4,5-P_3 in lysates of rat retinas stimulated with insulin, indicating that PI3K was bound to the activated IR. This technique has been used to study the differences in the PI3K activity between light- and dark-adapted retinas (7).

PROTOCOL 6: INTERACTION BETWEEN THE IR AND THE P85 SUBUNIT OF PI3K USING A GST PULL-DOWN ASSAY

The p85 subunit of PI3K binds specifically to phosphorylated Y1322 present on the C-terminus of the IR (47,48). Substitution of phenylalanine for tyrosine at this position abolished p85 binding to the phosphorylated IR (28). Glutathione S-transferase (GST)-fusion proteins containing the N-terminal SH2 domain of p85 have been shown to bind the phosphorylated IR (45). So, GST pull-down experiments followed by Western blot analysis with anti-IRβ antibody can be used to directly measure the interaction between the phosphorylated IR and PI3K. This technique is also suitable to measure direct phosphorylation of the IR when the sample volume is a limiting factor.

1. Transform GST-p85 N (SH2) fusion vectors (45) into BL21 (DE3) maximum efficiency *E. coli* competent cells.
2. Pick individual clones, and grow overnight in 25 mL LB/Amp (100 μg/mL) medium.

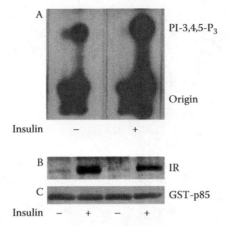

FIGURE 11.4 Insulin-induced activation of PI3K and the interaction between IR and PI3K. TLC autoradiogram of PI3K activity (from Rajala, R.V., McClellan, M.E., Ash, J.D., and Anderson, R.E., In vivo regulation of phosphoinositide 3-kinase in retina through light-induced tyrosine phosphorylation of the insulin receptor beta-subunit, *J. Biol. Chem.*, 277, 43319, 2002. With permission) measured in anti-IRβ immunoprecipitates of insulin-stimulated and unstimulated retinas using PI-4,5-P_2 and [γ^{32}P]ATP as substrates. (A). Insulin-stimulated and unstimulated retinas were lysed and subjected to GST pull-down assays (125 μg protein) with GST fusion proteins of N-SH2 domains of p85. Western blot analysis of the bound proteins was probed with anti-IRβ antibody. (B) The blot was stripped and reprobed with anti-GST p85 antibody as a positive control (C). (From Rajala, R.V., McClellan, M.E., Chan, M.D., Tsiokas, L., and Anderson, R.E., Interaction of the retinal insulin receptor beta-subunit with the p85 subunit of phosphoinositide 3-kinase, *Biochemistry*, 43, 5637, 2004. With permission.)

3. Dilute culture with LB/Amp (1:10) medium, and grow for 1 h at 30°C.
4. Induce GST protein synthesis with the addition of Isopropyl-β-D-thiogalactopyranoside (IPTG) to 0.1 mM final concentration for 1 h at 30°C.
5. Spin the bacteria at 2987 × g for 10 min at 4°C, and decant the medium.
6. Add 10 mL ice-cold PBS [137 mM NaCl, 2.7 mM KCl, 4.3 mM $Na_2HPO_4\cdot7H_2O$, and 1.4 mM KH_2PO_4], and spin the cells at 11,951 × g for 10 min at 4°C.
7. Remove the PBS, and store the bacterial pellets at −80°C until use.
8. Add 1 mL lysis buffer [50 mM HEPES (pH 7.5), 50 mM NaCl, 10% glycerol, 1% Triton X-100, 1 mM EDTA, 1 mM EGTA, 10 μg/mL aprotinin, 10 μg/mL leupeptin, and 1 mM phenylmethylsulfonyl fluoride (PMSF)] to bacterial pellets and sonicate the suspension for 10 s on ice. Repeat this step twice more.
9. Spin the suspension at 16,100 × g for 10 min at 4°C and collect the supernatant.
10. Add 100 μL glutathione Sepharose (50% v/w) beads to the supernatant from step 9, and continue to shake for 30 min at 4°C.
11. Wash the bound GST-fusion proteins with 500 μL of 1× PBS, and repeat this step twice more. Resuspend the beads in 500 μL lysis buffer.

12. Add 50 µL of glutathione Sepharose 4B beads to 100–250 µg of tissue extract, and shake the reaction at 4°C for 1 h.
13. Spin the reaction on a table-top centrifuge, and remove the supernatant to a clean Eppendorf tube.
14. Add 5 µg of GST-p85 N (SH2) fusion protein and 20 µL of glutathione Sepharose beads, and incubate the reaction at 4°C for 2 h.
15. Wash the GST fusions thrice with wash buffer 50 mM HEPES (pH 7.4) containing 118 mM NaCl, 100 mM NaF, 2 mM NaVO$_3$, 0.1% (w/v) SDS, and 1% (v/v) Triton X-100.
16. Add 25 µL 2× sample buffer, and incubate the sample for 5 min at 100°C.
17. Resolve the proteins on 10% SDS-PAGE followed by Western blot analysis with anti-IRβ antibody.

The interaction of the IR with the p85 subunit of PI3K is shown in figures 11.4B and 11.4C. Activation by insulin leads to phosphorylation of tyrosines on the C-terminus (not shown) and subsequent binding to the p85 subunit of PI3K through the N-terminal SH2 domain. We have used this technique and identified the PI3K as the binding partner of the IR in the ROS (45). This technique is a very reliable and sensitive method to directly study IR phosphorylation. This method is also useful as an alternative approach to conventional immunoprecipitation when sample concentration is low. With this technique, one can study IR autophosphorylation with 50 µg of protein.

11.3.2 IRS PROTEINS

Insulin binding to its receptor leads to the autophosphorylation of its β-subunits and to the subsequent tyrosine phosphorylation of IRS by the activated IR (49). The phosphorylated IRS proteins provide a docking site for binding of the p85 subunit of PI3K.

PROTOCOL 7: INTERACTION OF IRS PROTEINS AND THE p85 SUBUNIT OF PI3K

1. Stimulate the retinas in culture with insulin, and prepare immunoprecipitates as described in protocol 2.
2. Add anti-IRS-1 and anti-IRS-2 antibodies, and perform Western blot analysis with the anti-p85 antibody.

The binding of the p85 subunit of PI3K to anti-IRS-1 and anti-IRS-2 antibodies to the retinal immunoprecipitates is shown in figure 11.5. With this technique, we have demonstrated that IRS-1 does not interact with the p85 subunit of PI3K under insulin-stimulated conditions (figure 11.5A). Control experiments carried out with 3T3 adipocytes (28) show the enhanced interaction between IRS-1 and p85 under insulin-stimulated conditions (figure 11.5A). Using this technique, we also demonstrated an enhanced interaction between IRS-2 and p85 in the retina in response to insulin stimulation (figure 11.5B).

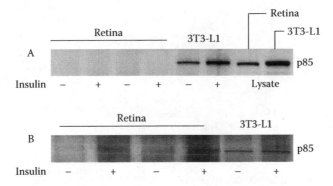

FIGURE 11.5   Interaction of IRS proteins with the p85 subunit of PI3K. 3T3-L1 adipocytes and rat retinas were stimulated in culture with or without 1 μM insulin. After insulin stimulation, retinal lysates were immunoprecipitated with either anti-IRS-1 (A) or anti-IRS-2 (B) agarose conjugates and immunoblotted with the anti-p85 antibody. Rat retina and 3T3-L1 adipocyte lysates were used as control. IRS-2 was activated in response to insulin in the retina, but not IRS-1. (From Rajala, R.V., McClellan, M.E., Chan, M.D., Tsiokas, L., and Anderson, R.E., Interaction of the retinal insulin receptor beta-subunit with the p85 subunit of phosphoinositide 3-kinase, *Biochemistry.*, 43, 5637, 2004. With permission.)

11.3.3   AKT

Akt is a downstream effector molecule of PI3K that is activated through the activation of the IR (6). This protocol describes a method to study PI3K-dependent activation of Akt.

PROTOCOL 8: INSULIN-INDUCED ACTIVATION OF AKT

1. Stimulate the retinas in culture with insulin, and prepare retina lysates as described in protocol 2.
2. Perform Western blot analysis with 20 μg of retina lysate, and probe the membrane with antiphosphoAkt (Ser473) antibody. Strip the blot and reprobe with the total Akt antibody to ensure that equal amounts of protein were loaded.
3. Preincubate the retinal organ cultures with or without 50 μM of the PI3K inhibitor, LY294002, for 60 min at 37°C.
4. Incubate the retinas with 1 μM insulin for 5 min. Lyse the retinas, and extract the proteins.
5. Perform Western blot analysis with 20 μg of retinal lysate proteins with antiphosphoAkt (Ser 473) and total Akt antibodies.

The phosphorylation of Akt following insulin stimulation of intact rat retinas is shown in figures 11.6A and 11.6B. Reduction of Akt phosphorylation following incubation with a PI3K inhibitor indicates PI3K-dependent activation of Akt (figures 11.6C and 11.6D). The retinal organ cultures may be useful to study various intricate signaling events through the addition of either activators or inhibitors.

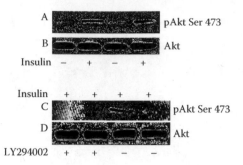

FIGURE 11.6 Insulin-induced activation of Akt. Rat retinas were dissected and incubated in DMEM medium without or with 100 nM insulin. Twenty micrograms of retina lysate was subjected to Western blot analysis with antiphospho-Akt (Ser 473) (A) or anti-Akt (B) antibodies. Rat retinas were preincubated in DMEM medium with or without the PI3K inhibitor LY294002 (50 µM) for 60 min prior to the addition of 100 nM insulin. Twenty micrograms of retina lysate were subjected to Western blot analysis with antiphospho-Akt (Ser 473) (C) or total Akt (D) antibodies. (From Rajala, R.V., McClellan, M.E., Chan, M.D., Tsiokas, L., and Anderson, R.E., Interaction of the retinal insulin receptor beta-subunit with the p85 subunit of phosphoinositide 3-kinase, *Biochemistry.*, 43, 5637, 2004. With permission.)

11.3.4   GROWTH FACTOR RECEPTOR BOUND PROTEIN 14 (GRB14)

Grb14 is a member of an emerging family of noncatalytic adapter proteins that also includes Grb7 and Grb10 (50,51). Characteristic features of the Grb7 family members include a central pleckstrin homology (PH) domain, a C-terminal SH2 domain, an N-terminal proline-rich motif (which is similar to a SH3-binding site), and a homology to Ras-associating (RA) domain. The presence of these functional domains indicates that Grb7 family members potentially interact with a variety of proteins. Grb14 has been shown to bind to Tek/Tie2, PDGF, EGFR, FGFR, and IR (52–55). The large number of binding partners for the Grb7 family members suggests that they play key roles as cytoplasmic signaling intermediates. We screened a bovine retinal cDNA library against the cytoplasmic domain of the retinal IR, and identified Grb14 as one of the binding partners to the IR (29). Grb14 binds to the IR on tyrosine residues Y1146, Y1150, and Y1151, and the binding is mediated through a PIR (phosphorylated IR interacting) domain (56).

PROTOCOL 9: INTERACTION BETWEEN THE IR AND GRB14

1. Stimulate retinal organ cultures with 1 µM insulin for various times (0, 2, 5, 15, 30, and 60 min) (see protocol 2).
2. Lyse the retinas in lysis buffer, and incubate with GST-PIR-SH2 fusion protein (56), which specifically binds to the phosphorylated IR. Perform GST pull-down assay as described in protocol 6.
3. Wash the GST fusions thrice in wash buffer, add 25 µL sample buffer, and incubate for 5 min at 100°C.
4. Resolve the proteins on 10% SDS-PAGE followed by Western blot analysis with anti-IR antibody.

Time
(min)   0   2   5   15   30   60

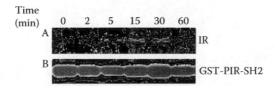

FIGURE 11.7 Interaction between the retinal IR and Grb14. Insulin-stimulated retina lysates were incubated with GST-PIR-SH2 fusion protein followed by Western blot analysis with anti-IR IRβ antibody (A) and reprobed with anti-GST antibody (B). (From Rajala, R.V., Chan, M.D., and Rajala, A. Lipid-protein interactions of growth factor receptor-bound protein 14 in insulin receptor signaling, *Biochemistry*, 44, 15461, 2005. With permission.)

The binding of Grb14 to the phosphorylated IR as a function of time of insulin stimulation is shown in figure 11.7A. The relative amount of GST-PIR-SH2 in each pull-down is shown in figure 11.7B. Unlike the p85 subunit of PI3K, which binds through the SH2 domain, Grb14 binds to the IR through PIR domain. The SH2 domain in the Grb14 does not bind to the phosphorylated IR (56). PIR-SH2 expression and solubility are much better than PIR alone and therefore is the form we use most often.

11.4 LIGHT-INDUCED ACTIVATION OF THE RETINAL INSULIN RECEPTOR

In ROS, the IR is activated in a light-dependent manner (7). The light-induced activation of the IR is independent of insulin secretion, and the light effect is localized to photoreceptor neurons (7). The light-activated IR results in the subsequent activation of PI3K in ROS (7). Methods are described to study the light-induced tyrosine phosphorylation of the IR and also to study the interaction between the IR and PI3K in a light-dependent manner.

PROTOCOL 10: LIGHT-INDUCED ACTIVATION OF PI3K THROUGH THE RETINAL IR

We have demonstrated that PI3K can be activated by light in intact animals as well as by insulin in retinal explant cultures (28). Recent studies have allowed us to show that activation in vivo is through the IR and is carried out in an insulin-independent manner (7). Albino Sprague Dawley rats (150–200 g) were used for these studies. Ideally, we prefer to breed and raise rats in our vivarium to control for light history, which we and others have shown is important is some aspects of retinal function (57–59). If adult albino rats are purchased from a supplier, care must be taken to ensure that the animals have been raised in cyclic light that does not exceed 150–250 lux. Once received into the vivarium, we recommend that these rats be acclimated for at least 2 weeks in dim cyclic light (12 h ON; 12 h OFF; 5–10 lux).

1. Rats are dark-adapted overnight and exposed to normal room light (~300 lux) for 30 min. Dark-adapted control animals should remain in the dark for 30 min.
2. Asphyxiate rats with CO_2 followed by cervical dislocation.
3. Remove retinas and place in lysis buffer [1% NP 40, 20 mM HEPES (pH 7.4), and 2 mM EDTA] containing phosphatase inhibitors (100 mM NaF, 10 mM $Na_4P_2O_7$, and 1 mM molybdate) and protease inhibitors (10 μM leupeptin, 10 μg/mL aprotinin, and 1 mM PMSF).
4. Immunoprecipitate IRs from the light-exposed and dark-adapted retinas as described in protocol 2.
5. Determine PI3K activity in the IR immunoprecipitates as described in protocol 4.

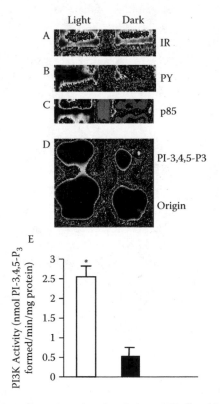

FIGURE 11.8 In vivo light-dependent phosphorylation of IR. Rats were either dark- or light-adapted, and retinas from each rat were pooled, homogenized, immunoprecipitated with anti-IRβ antibodies, and immunoblotted with anti-IRβ (A) anti-PY (B), or anti-p85 (C) antibodies. PI3K activity of light-stimulated or dark-adapted retinas was measured using PI-4,5-P_2 and [γ^{32}P]ATP as substrates (D). The radioactive spots of PI-3,4,5-P_3 were scraped from the TLC plate, counted, and expressed as nmol PI-3,4,5-P_3 formed/min/mg protein (E). Data are mean + SE (n = 6). The difference between light-stimulated and dark-adapted retinal PI3K activity was significant at P < 0.01. (From Rajala, R.V., McClellan, M.E., Ash, J.D., and Anderson, R.E., In vivo regulation of phosphoinositide 3-kinase in retina through light-induced tyrosine phosphorylation of the insulin receptor beta-subunit, *J. Biol. Chem.*, 277, 43319, 2002. With permission.)

6. Immunoprecipitate the IR from light-exposed and dark-adapted retinal lysates and subject to Western blot analysis with anti-PY99 antibody to determine tyrosine phosphorylation of the IR or Western blot analysis with anti-p85 antibody to examine the interaction between IR and p85 subunit of PI3K.
7. Strip the blot with stripping reagent and reprobe with anti-IRβ antibody to ensure equal amounts of protein were loaded.

Results of a typical experiment using whole rat retinas are shown in figure 11.8. Immunoprecipitates of light-exposed and dark-adapted retinas have the same amount of IR (figure 11.8A), but the light-exposed retinas have greater tyrosine phosphorylation (figure 11.8B). PI3K activity is also higher in the light-exposed retinal immunoprecipitates, as seen in the autoradiogram of the TLC plate (figure 11.8D) and in the radioactivity determined from the scraped TLC spots (figure 11.8E). Probing the gels shown in figures 11.8A and 11.8B with an anti-p85 antibody also revealed a greater binding of p85 in the immunoprecipitates from the light-exposed retinas (figure 11.8C). With this technique, we were the first to show that light induced tyrosine phosphorylation of the retinal IR (7).

PROTOCOL 11: PREPARATION OF RETINAL ROD OUTER SEGMENTS

ROSs are among the most abundant cellular organelles whose membranes can be prepared in relatively large quantities. Unlike mitochondria or endoplasmic reticulum, the ROSs offer a huge advantage in that they are primarily derived from a single class of neurons (rod photoreceptor cells) about which a large amount of information is available. There are two convenient methods for preparing ROS: continuous and discontinuous sucrose differential gradient centrifugation. ROS prepared on continuous gradients contain soluble and membrane-bound proteins and may have some inner segment blebs attached at the base, whereas ROS prepared on discontinuous gradients have only membrane-bound proteins and are less likely to be contaminated by membranes from other organelles. For example, ROS prepared on discontinuous gradients have no immunoreactivity toward α1- or α3-Na-K-ATPases (7), which are found in retinal pigment epithelium (RPE) and inner segments, respectively. For the dark-adapted retinas (unexposed), all procedures must be done in the dark or under dim red light.

1. Obtain retinas as described in protocol 1.
2. Homogenize ~10 retinas in 4.0 mL of ice-cold 47% sucrose solution containing 100 mM NaCl, 1 mM EDTA, 1 mM NaVO₃, 1 mM PMSF, and 10 mM Tris-HCl (pH 7.4) (buffer A).
3. Transfer the homogenate into 15 mL centrifuge tubes and subsequently overlay with 4.0 mL of 42%, 4.0 mL of 37%, and 4.0 mL of 32% sucrose dissolved in buffer A. For mouse or single rat retinas, a 5 mL tube can be used and volumes proportionately reduced. On the other hand, the procedure can be adapted for 35 mL tubes to prepare bovine ROS.
4. Centrifuge the gradients at 82,000 × g for 1.5 h at 4°C in a swinging bucket rotor.
5. Collect the ROS band at the interface between 32 and 37% into clean 40 mL centrifuge tubes.

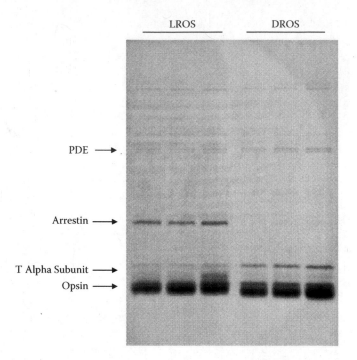

FIGURE 11.9 SDS gel electrophoresis of light- and dark-adapted ROS. Five micrograms of light-stimulated or dark-adapted ROS fractions were subjected to electrophoresis followed by staining the gel with GelCode blue. Arrestin and Talpha subunit indicates the positive controls for light stimulation and dark adaptation, respectively. LROS, light-stimulated ROS; DROS, dark-adapted ROS; PDE, phosphodiesterase.

6. Dilute the ROS with 10 mM Tris-HCl (pH 7.4) containing 100 mM NaCl, 1 mM EDTA, and 1 mM orthovanadate (buffer B), and centrifuge at 27,000 × g for 30 min.

7. Resuspend the ROS in buffer B, wrap the dark-adapted ROS in foil, and store the samples at –70°C until used.

8. For quality control, check the ROS purity by running 5 µg of unboiled ROS proteins on 10% SDS-PAGE. Wash the gel for 5 min in distilled water and repeat the wash step twice more. Stain the gel with GelCode Blue (Pierce) for 1 h. Decant the GelCode blue, and wash the gel with distilled water.

A polyacrylamide gel of a typical ROS preparation is shown in figure 11.9. The major band is rhodopsin. Other identifiable bands are transducin alpha, rhodopsin dimer and trimer, and the doublet of phosphodiesterase. Arrestin is not seen in this preparation of dark-adapted ROS.

PROTOCOL 12: PI3K ACTIVITY ASSOCIATED WITH LIGHT- AND DARK-ADAPTED ROS

ROS membranes contain the visual pigment rhodopsin, which absorbs photons and initiates visual transduction. We have shown that PI3K and the IR are present in the ROS and that both are activated by light (7). The previous protocol can be modified for analysis of light-stimulation of PI3K activity in the ROS.

1. Add 5 µg of either light-exposed or dark-adapted ROS, 10 µL 5× PI3K assay buffer (see protocol 3), and 10 µL lipid substrate (0.2 mg/mL PI-4, 5-P$_2$) to a total volume of 50 µL.
2. Initiate the reaction by the addition of 3.5 µL ATP (cold and [γ-^{32}P]ATP) to a final concentration of 20 µM, and incubate the reaction for 15–60 min at RT.
3. Terminate the reaction by the addition of 100 µL of 1 N HCl. Add 200 µL of chloroform/methanol (1:1, v/v), and vortex the reaction followed by 2 min centrifugation on a table-top centrifuge.
4. Separate lipids by TLC and determine PI3K activity as described in protocol 3.

Figure 11.10A is an autoradiogram of a TLC plate showing the labeling of PI-3,4,5-P$_3$ in the ROS isolated from light-exposed or dark-adapted retinas. As the membranes were prepared from homogenized retinas on a discontinuous sucrose gradient, the reaction mixture contains only membrane-bound proteins. There is clearly greater PI3K activity (figure 11.10A) and greater p85 immunoreactivity (figure 11.10B) from

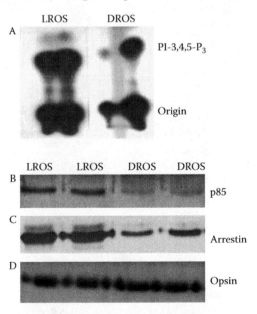

FIGURE 11.10 p85 binding and PI3K activity associated with ROS membranes. PI3K activity was measured from light- and dark-adapted ROS using PI-4,5-P$_2$ and [γ^{32}P] ATP as substrates (A). Specific antibodies were used to show the expression of p85 (B), arrestin (C), and opsin (D) in light- and dark-adapted rat ROS. LROS, light-stimulated ROS; DROS, dark-adapted ROS. (From Rajala, R.V., McClellan, M.E., Ash, J.D., and Anderson, R.E., In vivo regulation of phosphoinositide 3-kinase in retina through light-induced tyrosine phosphorylation of the insulin receptor beta-subunit, *J. Biol. Chem.*, 277, 43319, 2002. With permission.)

the light-exposed ROS. The light-exposed state was confirmed by the presence of greater amounts of arrestin in the light-exposed ROS (figure 11.10C). Opsin immunoreactivity was used as loading control (figure 11.10D).

11.5   YEAST TWO-HYBRID GENETIC ASSAY

In the yeast two-hybrid system (60), fusion proteins to the DNA-binding domain of *E. coli* LexA protein and to the activation domain of the herpes simplex virus VP16 protein are expressed on 2μ high-copy number plasmids in the *Saccharomyces cerevisiae* L40 reporter strain (61). The genotype of L40 reporter strain is *MATα HIS3Δ200 trp1-901 leu2-3,112 ade2 LYS2:: (lexop)4-HIS3 URA3:: (lexAop)8-lacZ GAL4)*. The L40 strain contains two integrated reporters, the yeast *HIS3* gene and the bacterial *lacZ* gene. The first reporter places the *HIS3* gene under control of four LexA operators, whereas the second reporter places the *lacZ* gene under control of eight LexA operators. Association of the two fusion proteins activates transcription of both the *HIS3* gene and *lacZ* gene. As a result, the yeast grow in the absence of histidine and turn blue when β-galacotosidase activity is assayed, using the chromogenic substrate 5-bromo-4-chloro-3-indolyl- β-D-galactoside (X-Gal).

PROTOCOL 13: YEAST TWO-HYBRID ASSAY TO STUDY THE INTERACTION BETWEEN THE IR AND P85

1. The following media will be used:
    a. Grow yeast in supplemented SD minimal medium to maintain selection for plasmids introduced into the strains and for the integrated reporter constructs.
    b. Omit tryptophan (W) from the medium to maintain selection for the LexA-fusion (bait) plasmid.
    c. Omit leucine (L) from the medium to maintain selection for the prey or library plasmid.
    d. Omit histidine (H) to select for an interaction event between the bait and prey or library plasmid.
    e. 1× YPD medium: Place in a 2 L Erlenmyer flask 10 g Bacto yeast extract, 20 g Bacto peptone, 20 g dextrose, and 0.3 g L-tryptophan, and add 1 L water (use water from the Millipore Milli Q machine). Add a stir bar and stir until in solution. Autoclave for 30 min on liquid cycle in a shallow tray of water.
    f. NZY medium: 10 gm NZ amine (casein hydrolysate, 5 gm yeast extract, and 5 g NaCl). Adjust the pH to 7.5 using NaOH, make the volume to 1 L, and then autoclave. Add the following supplements prior to use: 12.5 mL 1 M MgCl$_2$, 12.5 mL 1M MgSO$_4$, and 10 mL of 2 M filter sterilized glucose solution.
2. The following plasmids are used to express the interacting proteins in the yeast:
    a. The bait plasmid, pBTM116, contains the entire coding region of the *E. coli* LexA protein, expressed from the yeast alcohol dehydrogenase I

(*ADH1*) promoter, followed by a polylinker for inserting cDNA to generate in-frame fusion to LexA. In addition, the bait plasmid contains a *TRP1* gene that allows yeast containing the plasmid to grow in minimal media lacking tryptophan, the 2μ origin of replication for maintenance of the plasmid in yeast, a bacterial origin of replication, and the β-lactimase gene, which confers ampicillin resistance to *E. coli*.

b. The prey or library plasmid, VP16, contains the VP16 acidic activation domain, expressed from the yeast *ADH1* promoter, followed by a polylinker for inserting cDNAs to generate in-frame fusion to VP16. Upstream of and in-frame to the VP16 coding region is the simian virus 40 (SV40) large T antigen nuclear localization sequence. In addition, the plasmid contains the *LEU2* gene, which allows yeast containing this plasmid to grow in the absence of leucine, the 2μ origin of replication, a bacterial origin of replication, and the β-lactimase gene.

3. Isolate RNA from rat retina using any standard RNA isolation kit, and synthesize first-strand cDNA using a cDNA synthesis kit.

4. Synthesize PCR primers for cytoplasmic domain of IR (amino acids 941–1343) and added *Eco*RI site at the 5′ end (sense: CCGTCGACGAATTCAGGAAGAGGCAGCCAGAT) and *Sal*I site at the 3′ end (antisense: CGTCGACTTAGGAAGGGTTCGACCTCGGCGAGG).

5. Amplify the cytoplasmic domain of IR using sense and antisense primers.

6. Resolve the PCR product on 0.8% agarose gel to examine the amplification of correction molecular size fragment.

7. Clone the PCR products into TOPO cloning vector (Invitrogen), and sequence the PCR product using M13 forward and reverse primers.

8. Excise the fragment as *Eco*RI/*Sal*1 from TOPO vector, and clone into pBTM-116 yeast two-hybrid DNA binding vector.

9. Amplify the cDNA fragments encoding the N-SH2 domain of p85 and its respective SH2 mutant (R358A from the full-length p85 cDNA), and subclone into pVP16 activation domain.

10. Perform site-directed mutagenesis (SDM) by adding the reaction mixture containing SDM buffer [200 mM Tris-HCl (pH 8.8), 100 mM KCl, 100 mM $(NH_4)_2SO_4$, 20 mM $MgSO_4$, 1% Triton X-100, 1 mg/mL nuclease-free bovine serum albumin, and 1 mM deoxynucleotide mix (dATP, dCTP, dTTP, and dGTP)] to 50 ng of vector, and 125 ng of sense and antisense primers with mutations in a total volume of 50 μL, followed by the addition of 2.5 units of *pfu* DNA polymerase.

a. K1018A primer (sense: GAGACCCGTGTTGCGGTGGCGACGGTCAATGAG; antisense: CTCATTGACCGTCGCCACCGCAACACG).

b. p85 N-SH2 (R358A) primer (sense: ACCTTTTTGGTAGCAGACGCATCTACTAAA; antisense: TTTAGTAGATGCGTCTGCTACCAAAAAGGT).

11. Perform PCR with initial denaturation at 95°C for 30 s, followed by 16 cycles at 95°C for 30 s, 55°C for 1 min, and at 68°C for 12 min (2 min/kb of plasmid length).

12. Place the reaction on ice for 2 min, then add 10 units of Dpn1 restriction enzyme, mix, and incubate at 37°C for 60 min.
13. Transform 1 μL of the Dpn1-treated reaction into Epicurean XL-blue supercompetent cells, and keep the reaction on ice for 30 min. Heat-shock the reaction at 42°C for 45 s, and immediately place the reaction on ice for 2 min.
14. Add 1 mL of NZY medium, and incubate the reaction at 37°C for 60 min.
15. Spread 200 μL reaction onto LB/Amp (100 μg/mL) plates, and incubate the plates overnight at 37°C.
16. Pick up at least three independent clones, and grow 5 mL of liquid culture LB/Amp (100 μg/mL) overnight at 37°C.
17. Isolate plasmid DNA, and sequence the mutants.
18. Excise the IR-K1018 as *Eco*RI/*Sal*1 and p85 (R358A) as *Bam*HI/*Eco*RI.
19. Clone the IR-K1018A into pBTM-116 and p85 (R358A) into pVP16 vector.
20. Streak L40 yeast strain onto YPD agar plate.
21. Inoculate a single yeast colony of L40 into 5 mL of liquid YPD medium and grow overnight at 30°C with shaking to an OD_{546} of 0.6–0.8.
22. Centrifuge 2 mL of yeast culture at 700 g for 5 min, and resuspend the pellet in 100 μL H_2O.
23. Add 2–3 μL mini-prep DNA (300–500 ng/mL) to an Eppendorf tube.
24. Prepare polyethylene glycol (PEG)/lithium acetate (LIOAc) master mix: 240 μL 50% PEG, 36 μL LIOAc, and 25 μL salmon sperm single-stranded DNA (2mg/mL).
25. Add 300 μL PEG/LIOAc mix to each tube, and vortex briefly. Add 100 μL yeast cells to each tube, and vortex 1 min to thoroughly mix all components. Incubate in a 42°C water bath for 45 min.
26. Pellet yeast at 700 g for 5 min at 30°C. Dissolve pellet in 100 μL 0.9% NaCl, and plate each transformation on to plater containing appropriate dropout media; incubate for 2–3 days at 30°C.

The yeast grown in this manner will express functional fusion proteins and can be used for studying the interactions between the IR and several downstream effectors.

11.5.1 Interaction Mating

Yeast mating is a convenient method of introducing two different plasmids into the same host cell and, in some applications, can be used as a convenient alternative to yeast cotransformation (62,63).

Protocol 14: Interaction Mating to Determine the Interaction between IR and PI3K

1. Transform LexA-IR plasmid into L40 yeast strain [*MATα HIS3Δ200 trp1-901 leu2-3,112 ade2 LYS2:: (lexop)4-HIS3 URA3:: (lexAop)8-lacZ GAL4)*] and p85 subunit of P3K to AMR-70 [*MATα HIS3Δ200 trp1-901 leu2-3,112 ade2 LYS2:: (lexop)4-HIS3 URA3:: (lexAop)8-lacZ GAL4* yeast strain.
2. Pick each colony of each type to use in the mating. Use only large (2–3 mm), fresh colonies from the working stock plate.

3. Place both colonies in the same 1.5 mL microcentrifuge tube containing 0.5 mL of YPD medium. Vortex tubes to completely resuspend the cells.

4. Incubate the microcentrifuge tubes overnight at 30°C (20–24 h) with shaking at 200 rpm.

5. Spread 100 μL aliquots of the mating culture on to plates containing the double dropout medium (SD-Trp/Leu) and triple dropout medium (SD-Trp/Leu/His). Growth on the double dropout medium indicates coexpression. Growth on the triple dropout medium indicates interaction that drives expression of the *HIS3* gene.

6. To detect protein–protein interactions, assay the fresh diploid colonies from the SD selection plates for β-galctosidase assay using the colony-lift filter assay or liquid for β-galactosidase assay.

PROTOCOL 15: COLONY-LIFT FILTER ASSAY

1. Prepare the following reagents:
   a. Z buffer (16.1 g/L $Na_2HPO_4 \cdot 7H_2O$, 5.5 g/L $NaH_2PO_4 \cdot H_2O$, 0.75 g/L KCl, and 0.246 g/L $MgSO_4 \cdot 7H_2O$). Adjust to pH 7.0, and autoclave. Can be stored at RT for up to 1 year.
   b. X-gal stock solution: Dissolve 5-bromo-4-chloro-3-indolyl-β-D-galactopyranoside (X-GAL) in N,N-dimethylformamide (DMF) at a concentration of 20 mg/mL. Store in the dark at −20°C.
   c. β-Mercaptoethanol (β-ME and Sigma #M-6250).
   d. Z buffer/X-gal solution (100 mL Z buffer, 0.27 mL β-ME, and 1.67 mL X-gal stock solution).

2. Soak sterile Whatman #5 filters in 2.5 mL Z buffer/X-gal solution in a clean 100 mm tissue culture plate.

3. With a forceps, place a clean, dry filter over the surface of the plate of colonies to be assayed. Gently rub the filter with the side of the forceps to help colonies cling to the filter. Evenly wet the filter.

4. Carefully lift the filter off the agar plate with forceps, and transfer it (colonies facing up) to a pool of liquid nitrogen. Using the forceps, completely submerge the filters for 10 s. (Liquid nitrogen should be handled with care; always wear thick gloves and goggles.)

5. Remove the filter from the liquid nitrogen and allow it to thaw at RT. Carefully place the filter, colony side up, on another presoaked filter. Avoid trapping air bubbles under or between the filters.

6. Incubate the filters at 30°C, and check periodically for the appearance of blue colonies.

Expression of the cytoplasmic domain of the IR in the absence of α-subunits has been shown to be constitutively active and the protein autophosphorylates (47). The p85 subunit forms a specific complex with the cytoplasmic domain of the IR when both are expressed as hybrid proteins in yeast cells (figure 11.11A). This interaction is strictly dependent on receptor tyrosine kinase activity, because p85 shows no interaction with a kinase-inactive receptor hybrid containing a mutated ATP-binding

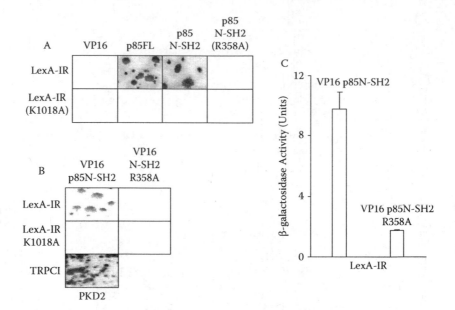

FIGURE 11.11   Interactions of the IR and the p85 subunit of PI3K. The entire cytoplasmic
domains of the retinal insulin receptor (IR) or K1018A mutant retinal IR were fused to the
LexA DNA-binding domain (DBD). The p85 full-length (p85FL), p85 (N-SH2 domain), and
p85 (N-SH2 R353A mutant) proteins were coexpressed along with either wild-type (LexA-
IR) or mutant IR (K1018A). Transformants were assayed for β-galactosidase assay by colony
color (A). Tyrosine phosphorylation and SH2-dependent binding of the retinal IR and the p85
subunit of PI3K were studied through interacting mating. Yeast diploids were obtained by
mating L40 that were transformed with the entire cytoplasmic domain (wild-type or K1018A
mutant) of retinal IR (LexA DNA-binding domain) with AMR-70 transformed with constructs carrying VP16-p85 N-SH2 or its mutant (R358A). For a positive control, we used
TRPCI (L40) and PKD2 (AMR-70), and carried out the interaction mating experiments.
Transformants were assayed for β-galactosidase assay by colony color (B). Yeast cells were
grown overnight and transformants were isolated on selection medium. Transformants were
assayed for β-galactosidase activity using the solution assay (C). Data are mean ± SD, $n = 6$.
(From Rajala, R.V., McClellan, M.E., Chan, M.D., Tsiokas, L., and Anderson, R.E., Interaction of the retinal insulin receptor beta-subunit with the p85 subunit of phosphoinositide
3-kinase, *Biochemistry*, 43, 5637, 2004. With permission.)

site (K1018A) (figure 11.11A). These experiments demonstrate the role of phosphorylation in the binding between p85 and IR. The data suggest that the interaction
between the p85 subunit and receptor is direct and provide evidence that tyrosine
kinase activity is necessary. Substitution of arginine with alanine (R358A) in the
N-SH2 domain of p85 prevented any interaction with wild-type IR (figure 11.11A),
further supporting the specific interaction between the N-SH2 domain and phosphorylated tyrosine residues in the IR.

The p85 subunit of PI3K and the IR were expressed individually in haploid
strains, which were subsequently assayed for β-galactosidase activity after interaction mating (diploids). There was a positive read-out phenotype in the wild type and
absence of β-galactosidase expression when K1018A mutant IR and p85 N-SH2

subunit of PI3K were used (figure 11.11B). These interaction-mating experiments further confirm the direct association between p85 and IR in diploid strains. To confirm the interactions, we used a positive control, which is the interactions between TRPCI and PKD2 (64) (figure 11.11B).

Protocol 16: Liquid Culture Assay Using ONPG Substrate

1. Prepare the following reagents:
   a. Z buffer (as described in protocol 15).
   b. Z buffer with β-ME: 100 mL Z buffer and 0.27 mL of β-ME.
   c. Z buffer/X-gal solution: 100 mL Z buffer, 0.27 mL β-ME, and 1.67 mL X-gal stock solution.
   d. ONPG (O-nitrophenyl β-D-galactopyranoside). Dissolve 4 mg ONPG in 1 mL Z buffer, adjust to pH 7.0, and mix well. ONPG requires 1–2 h to dissolve, and always prepare fresh before each use.
   e. 1 M Na_2CO_3.
2. Prepare 5 mL of overnight cultures in SD-Trp/Leu/His medium.
3. On the day of the experiment, dissolve ONPG at 4 mg/mL in Z buffer with shaking for 1–2 h.
4. Vortex the culture tube for 0.5–1 min to disperse cell clumps. Immediately transfer 2 mL of the overnight cultures to 8 mL of YPD.
5. Incubate the fresh culture at 30°C for 3–5 h with shaking (230–250 rpm) until the cells are in midlog phase (OD_{600} of 1 mL = 0.5–08). Record the exact OD_{600} when you harvest the cells.
6. Place 1.5 mL of culture into each of three 1.5 mL microcentifuge tubes. Centrifuge at 10,000 × g for 30 s.
7. Carefully remove supernatants. Add 1.5 mL of Z buffer to each tube, and vortex until cells are resuspended.
8. Centrifuge cells again and remove supernatants. Resuspend each pellet in 300 μL of Z buffer. (Thus, the concentration factor is 1.5/0.3 = 5-fold.)
9. Transfer 0.1 mL of the cell suspension to a fresh microcentrifuge tube. Place tubes in liquid nitrogen until the cells are frozen (0.5–1 min). Place frozen tubes in a 37°C water bath for 0.5–1 min to thaw. Repeat the freeze/thaw cycle twice more to ensure that the cells are broken up.
10. Set up a blank tube with 100 μL of Z buffer.
11. Add 0.7 mL of Z buffer + β-ME to the reaction and blank tubes. Do not add Z buffer prior to freezing samples.
12. Start timer. Immediately add 160 μL of ONPG in Z buffer to the reaction and blank tubes, and place tubes in a 30°C incubator.
13. After the yellow color develops (usually 15–20 min), add 0.4 mL of 1 M Na_2CO_3 to the reaction and blank tubes. Record elapsed time in minutes.
14. The time will vary (3–15 min for a single-plasmid β-gal positive control, ~30 min for a two-hybrid positive control, and up to 24 h for weaker interaction).
15. The yellow color is not stable and will become more intense with time. A new blank must be run with every batch.
16. Centrifuge reaction tubes for 10 min at 16,100 × g to pellet cell debris.

17. Carefully transfer supernatants to clean cuvettes. Calibrate the spectrophotometer against the blank at A420 and measure the OD_{420} of the sample relative to the blank. The ODs should be between 0.02 and 1.0 to be within the linear range of the assay.

18. Calculate β-galactosidase units: 1 unit of β-galactosidase is defined as the amount that hydrolyzes 1 μmol of ONPG to O-nitrophenol and D-galactose per min per cell (65).

$$\text{β-galactosidase units} = 1000 \, OD_{420}/(T_*V_*OD_{600}),$$

where

$\quad\quad$ T = elapsed time (in min) of incubation,

$\quad\quad$ V = 0.1 mL X concentration factor*, and

$\quad$ OD_{600} = A_{600} of 1 mL culture.

* The concentration factor is 5. However, it may be necessary to do several dilutions of cells at this step (hence, may have different concentration factors) to remain within the linear range of the assay.

The quantitative β-galactosidase assay can be used to study the extent of interaction between two signaling proteins. This assay would also allow quantitation of the percentage of interaction to study domain function, contribution of individual amino acids, and importance of posttranslational modifications in the interaction between two proteins. Figure 11.11C shows that SH2 domain of p85 is required for the binding to the IR. Substitution of arginine with alanine (R358A) in the N-SH2 domain of p85 prevented the interaction with wild-type IR (figure 11.11C).

ACKNOWLEDGMENTS

Raju V.S. Rajala is a recipient of Career Development Award from Research to Prevent Blindness, Inc. This work was supported by grants from the National Institutes of Health (EY016507, EY00871, EY12190, and RR17703); Research to Prevent Blindness, Inc.

REFERENCES

1. Havrankova, J., Roth, J., and Brownstein, M., Insulin receptors are widely distributed in the central nervous system of the rat, *Nature*, 272, 827, 1978.

2. Baskin, D.G., Figlewicz, L.D., Seeley, R.J., Woods, S.C., Porte, D., Jr., and Schwartz, M. W., Insulin and leptin: dual adiposity signals to the brain for the regulation of food intake and body weight, *Brain Res.*, 848, 114, 1999.

3. Schwartz, M.W., Sipols, A.J., Marks, J.L., Sanacora, G., White, J.D., Scheurink, A., Kahn, S.E., Baskin, D.G., Woods, S.C., and Figlewicz, D.P., Inhibition of hypothalamic neuropeptide Y gene expression by insulin, *Endocrinology*, 130, 3608, 1992.

4. Heidenreich, K.A., Insulin and IGF-I receptor signaling in cultured neurons, *Ann. N.Y. Acad. Sci.*, 692, 72, 1993.

5. Robinson, L.J., Leitner, W., Draznin, B., and Heidenreich, K.A., Evidence that p21ras mediates the neurotrophic effects of insulin and insulinlike growth factor I in chick forebrain neurons, *Endocrinology*, 135, 2568, 1994.
6. Barber, A.J., Nakamura, M., Wolpert, E.B., Reiter, C.E., Seigel, G.M., Antonetti, D.A., and Gardner, T.W., Insulin rescues retinal neurons from apoptosis by a phosphatidylinositol 3-kinase/Akt-mediated mechanism that reduces the activation of caspase-3, *J. Biol. Chem.*, 276, 32814, 2001.
7. Rajala, R.V., Chan, M.D., and Rajala, A., Lipid-protein interactions of growth factor receptor-bound protein 14 in insulin receptor signaling. *Biochemistry*, 44, 15461, 2005.
8. Yu, X., Rajala, R.V., McGinnis, J.F., Li, F., Anderson, R.E., Yan, X., Li, S., Elias, R.V., Knapp, R.R., Zhou, X., and Cao, W., Involvement of insulin/phosphoinositide 3-kinase/Akt signal pathway in 17 beta-estradiol-mediated neuroprotection, *J. Biol. Chem.*, 279, 13086, 2004.
9. Reiter, C.E., Wu, X., Sandirasegarane, L., Nakamura, M., Gilbert, K.A., Singh, R.S., Fort, P.E., Antonetti, D.A., and Gardner, T.W., Diabetes reduces basal retinal insulin receptor signaling: reversal with systemic and local insulin, *Diabetes*, 55, 1148, 2006.
10. Yi, X., Schubert, M., Peachey, N.S., Suzuma, K., Burks, D.J., Kushner, J.A., Suzuma, I., Cahill, C., Flint, C.L., Dow, M.A., Leshan, R.L., King, G.L., and White, M.F., Insulin receptor substrate 2 is essential for maturation and survival of photoreceptor cells, *J. Neurosci.*, 25, 1240, 2005.
11. Li, G., Anderson, R.E., Tomita, H., Adler, R., Liu, X., Zack, D.J., and Rajala, R.V., Nonredundant role of Akt2 for neuroprotection of rod photoreceptor cells from light-induced cell death, *J. Neurosci.*, 27, 203, 2007.
12. Zheng, L., Anderson, R.E., Agbaga, M.P., Rucker, I.E.B., and Le, Y.Z., Loss of Bcl-XL causes increased rod photoreceptor susceptibility to bright light damage, *Invest.Ophthalmol.Vis.Sci.*, 47, 5583, 2006.
13. White, M.F. and Kahn, C.R., The insulin signaling system, *J. Biol. Chem.*, 269, 1, 1994.
14. Seino, S. and Bell, G.I., Alternative splicing of human insulin receptor messenger RNA, *Biochem. Biophys. Res. Commun.*, 159, 312, 1989.
15. Wolkow, C.A., Kimura, K.D., Lee, M.S., and Ruvkun, G., Regulation of C. elegans life-span by insulinlike signaling in the nervous system, *Science*, 290, 147, 2000.
16. Song, J., Wu, L., Chen, Z., Kohanski, R.A., and Pick, L., Axons guided by insulin receptor in Drosophila visual system, *Science*, 300, 502, 2003.
17. Brogiolo, W., Stocker, H., Ikeya, T., Rintelen, F., Fernandez, R., and Hafen, E., An evolutionarily conserved function of the Drosophila insulin receptor and insulin-like peptides in growth control, *Curr. Biol.*, 11, 213, 2001.
18. Garrity, P.A., Rao, Y., Salecker, I., McGlade, J., Pawson, T., and Zipursky, S.L., Drosophila photoreceptor axon guidance and targeting requires the dreadlocks SH2/SH3 adapter protein, *Cell*, 85, 639, 1996.
19. Rosen, O.M., Banting lecture 1989. Structure and function of insulin receptors, *Diabetes*, 38, 1508, 1989.
20. Ebina, Y., Ellis, L., Jarnagin, K., Edery, M., Graf, L., Clauser, E., Ou, J.H., Masiarz, F., Kan, Y.W., and Goldfine, I.D., The human insulin receptor cDNA: the structural basis for hormone-activated transmembrane signalling, *Cell*, 40, 747, 1985.
21. Ullrich, A., Bell, J.R., Chen, E.Y., Herrera, R., Petruzzelli, L.M., Dull, T.J., Gray, A., Coussens, L., Liao, Y.C., and Tsubokawa, M., Human insulin receptor and its relationship to the tyrosine kinase family of oncogenes, *Nature*, 313, 756, 1985.
22. Hubbard, S.R., Wei, L., Ellis, L., and Hendrickson, W.A., Crystal structure of the tyrosine kinase domain of the human insulin receptor, *Nature*, 372, 746, 1994.
23. Hubbard, S.R., Crystal structure of the activated insulin receptor tyrosine kinase in complex with peptide substrate and ATP analog, *EMBO J.*, 16, 5572, 1997.

24. Winkler, B.S., Dependence of fast components of the electroretinogram of the isolated rat retina on the ionic environment, *Vision Res.*, 13, 457, 1973.
25. Saltiel, A.R. and Pessin, J.E., Insulin signaling pathways in time and space, *Trends Cell Biol.*, 12, 65, 2002.
26. Craparo, A., O'Neill, T.J., and Gustafson, T.A., Non-SH2 domains within insulin receptor substrate-1 and SHC mediate their phosphotyrosine-dependent interaction with the NPEY motif of the insulin-like growth factor I receptor, *J. Biol. Chem.*, 270, 15639, 1995.
27. Reiter, C.E., Sandirasegarane, L., Wolpert, E.B., Klinger, M., Simpson, I.A., Barber, A.J., Antonetti, D.A., Kester, M., and Gardner, T.W., Characterization of insulin signaling in rat retina in vivo and ex vivo, *Am. J. Physiol. Endocrinol. Metab.*, 285, E763-E774, 2003.
28. Rajala, R.V., McClellan, M.E., Chan, M.D., Tsiokas, L., and Anderson, R.E., Interaction of the retinal insulin receptor beta-subunit with the p85 subunit of phosphoinositide 3-kinase, *Biochemistry*, 43, 5637, 2004.
29. Rajala, R.V. and Chan, M.D., Identification of a NPXY motif in growth factor receptor-bound protein 14 (Grb14) and its interaction with the phosphotyrosine-binding (PTB) domain of IRS-1, *Biochemistry*, 44, 7929, 2005.
30. Schlessinger, J., Cell signaling by receptor tyrosine kinases, *Cell*, 103, 211, 2000.
31. Pawson, T. and Nash, P., Protein-protein interactions define specificity in signal transduction, *Genes Dev.*, 14, 1027, 2000.
32. Koch, C.A., Moran, M.F., Anderson, D., Liu, X.Q., Mbamalu, G., and Pawson, T., Multiple SH2-mediated interactions in v-src-transformed cells, *Mol. Cell Biol.*, 12, 1366, 1992.
33. McGlade, C.J., Ellis, C., Reedijk, M., Anderson, D., Mbamalu, G., Reith, A.D., Panayotou, G., End, P., Bernstein, A., and Kazlauskas, A., SH2 domains of the p85 alpha subunit of phosphatidylinositol 3-kinase regulate binding to growth factor receptors, *Mol. Cell Biol.*, 12, 991, 1992.
34. Cantley, L.C., The phosphoinositide 3-kinase pathway, *Science*, 296, 1655, 2002.
35. Skolnik, E.Y., Margolis, B., Mohammadi, M., Lowenstein, E., Fischer, R., Drepps, A., Ullrich, A., and Schlessinger, J., Cloning of PI3 kinase-associated p85 utilizing a novel method for expression/cloning of target proteins for receptor tyrosine kinases, *Cell*, 65, 83, 1991.
36. Carpenter, C.L., Duckworth, B.C., Auger, K.R., Cohen, B., Schaffhausen, B.S., and Cantley, L.C., Purification and characterization of phosphoinositide 3-kinase from rat liver, *J. Biol. Chem.*, 265, 19704, 1990.
37. Dhand, R., Hiles, I., Panayotou, G., Roche, S., Fry, M.J., Gout, I., Totty, N.F., Truong, O., Vicendo, P., and Yonezawa, K., PI 3-kinase is a dual specificity enzyme: autoregulation by an intrinsic protein-serine kinase activity, *EMBO J.*, 13, 522, 1994.
38. Hu, P., Mondino, A., Skolnik, E.Y., and Schlessinger, J., Cloning of a novel, ubiquitously expressed human phosphatidylinositol 3-kinase and identification of its binding site on p85, *Mol. Cell Biol.*, 13, 7677, 1993.
39. Klippel, A., Escobedo, J.A., Fantl, W.J., and Williams, L.T., The C-terminal SH2 domain of p85 accounts for the high affinity and specificity of the binding of phosphatidylinositol 3-kinase to phosphorylated platelet-derived growth factor beta receptor, *Mol. Cell Biol.*, 12, 1451, 1992.
40. Yoakim, M., Hou, W., Liu, Y., Carpenter, C.L., Kapeller, R., and Schaffhausen, B.S., Interactions of polyomavirus middle T with the SH2 domains of the pp85 subunit of phosphatidylinositol-3-kinase, *J. Virol.*, 66, 5485, 1992.
41. Whitman, M., Downes, C.P., Keeler, M., Keller, T., and Cantley, L., Type I phosphatidylinositol kinase makes a novel inositol phospholipid, phosphatidylinositol-3-phosphate, *Nature*, 332, 644, 1988.
42. Williams, L.T., Signal transduction by the platelet-derived growth factor receptor, *Science*, 243, 1564, 1989.

43. Ruderman, N.B., Kapeller, R., White, M.F., and Cantley, L.C., Activation of phosphatidylinositol 3-kinase by insulin, *Proc. Natl. Acad. Sci. U.S.A.*, 87, 1411, 1990.
44. Van Horn, D.J., Myers, M.G., Jr., and Backer, J.M., Direct activation of the phosphatidylinositol 3'-kinase by the insulin receptor, *J. Biol. Chem.*, 269, 29, 1994.
45. Rajala, R.V. and Anderson, R.E., Interaction of the insulin receptor beta-subunit with phosphatidylinositol 3-kinase in bovine ROS, *Invest. Ophthalmol. Vis. Sci.*, 42, 3110, 2001.
46. Kahn, C.R., White, M.F., Shoelson, S.E., Backer, J.M., Araki, E., Cheatham, B., Csermely, P., Folli, F., Goldstein, B.J., and Huertas, P., The insulin receptor and its substrate: molecular determinants of early events in insulin action, *Recent Prog. Horm. Res.*, 48, 291, 1993.
47. O'Neill, T.J., Craparo, A., and Gustafson, T.A., Characterization of an interaction between insulin receptor substrate 1 and the insulin receptor by using the two-hybrid system, *Mol. Cell Biol.*, 14, 6433, 1994.
48. Gustafson, T.A., He, W., Craparo, A., Schaub, C.D., and O'Neill, T.J., Phosphotyrosine-dependent interaction of SHC and insulin receptor substrate 1 with the NPEY motif of the insulin receptor via a novel non-SH2 domain, *Mol. Cell Biol.*, 15, 2500, 1995.
49. White, M.F., The IRS-signalling system: a network of docking proteins that mediate insulin action, *Mol.Cell Biochem.*, 182, 3, 1998.
50. Han, D.C., Shen, T.L., and Guan, J.L., The Grb7 family proteins: structure, interactions with other signaling molecules and potential cellular functions, *Oncogene*, 20, 6315, 2001.
51. He, W., Rose, D.W., Olefsky, J.M., and Gustafson, T.A., Grb10 interacts differentially with the insulin receptor, insulin-like growth factor I receptor, and epidermal growth factor receptor via the Grb10 Src homology 2 (SH2) domain and a second novel domain located between the pleckstrin homology and SH2 domains, *J. Biol. Chem.*, 273, 6860, 1998.
52. Jones, N., Master, Z., Jones, J., Bouchard, D., Gunji, Y., Sasaki, H., Daly, R., Alitalo, K., and Dumont, D.J., Identification of Tek/Tie2 binding partners. Binding to a multifunctional docking site mediates cell survival and migration, *J. Biol. Chem.*, 274, 30896, 1999.
53. Daly, R.J., Sanderson, G.M., Janes, P.W., and Sutherland, R.L., Cloning and characterization of GRB14, a novel member of the GRB7 gene family, *J. Biol. Chem.*, 271, 12502, 1996.
54. Reilly, J.F., Mickey, G., and Maher, P.A., Association of fibroblast growth factor receptor 1 with the adaptor protein Grb14. Characterization of a new receptor binding partner, *J. Biol. Chem.*, 275, 7771, 2000.
55. Kasus-Jacobi, A., Perdereau, D., Auzan, C., Clauser, E., Van Obberghen, E., Mauvais-Jarvis, F., Girard, J., and Burnol, A.F., Identification of the rat adapter Grb14 as an inhibitor of insulin actions, *J. Biol. Chem.*, 273, 26026, 1998.
56. Rajala, R.V., Chan, M.D., and Rajala, A., Lipid-protein interactions of growth factor receptor-bound protein 14 in insulin receptor signaling, *Biochemistry*, 44, 15461, 2005.
57. Organisciak, D.T., Darrow, R.M., Barsalou, L., Darrow, R.A., Kutty, R.K., Kutty, G., and Wiggert, B., Light history and age-related changes in retinal light damage, *Invest. Ophthalmol. Vis. Sci.*, 39, 1107, 1998.
58. Penn, J.S. and Thum, L.A., A comparison of the retinal effects of light damage and high illuminance light history, *Prog. Clin. Biol. Res.*, 247, 425, 1987.
59. Li, F., Cao, W., and Anderson, R.E., Protection of photoreceptor cells in adult rats from light-induced degeneration by adaptation to bright cyclic light, *Exp. Eye Res.*, 73, 569, 2001.
60. Fields, S. and Song, O., A novel genetic system to detect protein-protein interactions, *Nature*, 340, 245, 1989.
61. Fields, S. and Sternglanz, R., The two-hybrid system: an assay for protein-protein interactions, *Trends Genet.*, 10, 286, 1994.

62. Finley, R.L., Jr. and Brent, R., Interaction mating reveals binary and ternary connections between Drosophila cell cycle regulators, *Proc. Natl. Acad. Sci. U.S.A.*, 91, 12980, 1994.

63. Bendixen, C., Gangloff, S., and Rothstein, R., A yeast mating-selection scheme for detection of protein-protein interactions, *Nucl. Acids Res.*, 22, 1778, 1994.

64. Tsiokas, L., Arnould, T., Zhu, C., Kim, E., Walz, G., and Sukhatme, V.P., Specific association of the gene product of PKD2 with the TRPC1 channel, *Proc. Natl. Acad. Sci. U.S.A.*, 96, 3934, 1999.

65. Breeden, L. and Nasmyth, K., Cell cycle control of the yeast HO gene: cis- and trans-acting regulators, *Cell*, 48, 389, 1987.

5

Signal Transduction in Vertebrate Retinal Development and Vascular Homeostasis

12 Signal Transduction in Retinal Progenitor Cells

Branden R. Nelson, Susan Hayes,
Byron H. Hartman, and Thomas A. Reh

CONTENTS

12.1   INTRODUCTION

The neural retina is a complex neural tissue and, like the rest of the central nervous system, the neurons and glia of the retina arise from stem cells and progenitor cells of the neural tube. The neural tube itself is derived from a region of ectodermal cells, which are induced to become neural tissue through interactions with subjacent, prospective mesodermal cells in a process known as *neural induction* (1). Following neural induction and neural tube formation, cells that line the diencephalic vesicle of the tube evaginate to form the optic vesicles. The optic vesicle undergoes further morphogenesis to form a two-layered optic cup. The cells of the optic cup (now called retinal progenitors or multipotent progenitors) proliferate extensively; some of their progeny withdraw from the mitotic cycle to develop as neurons, whereas others remain in the mitotic cell cycle and continue to produce additional cells.

We are only beginning to understand the extraordinarily complicated process of making the mature retina; however, some key concepts have emerged. First, the progenitor cells generate different types of retinal neurons at different times in development. Ganglion cells, cone photoreceptors, and some horizontal cells are generated by the progenitors as an initial cohort, followed by genesis of most amacrine cells and rod photoreceptors; finally, bipolar cells and Müller glia differentiate. Second, tracing the progeny of single progenitor cells has revealed that they are multipotent and can generate different cell types, even at their final mitotic division. Third, cell-to-cell interactions are critical for the maintenance of the proliferation of progenitor cells, and for the fates of their progeny. Fourth, progenitor cells express a diverse set of transcription factors, many of which are critical for their ability to generate the large number of different retinal cell types. One challenge for those interested in retinal histogenesis is to define how the complex network of cell signals controls the expression of these key transcription factors, and some upcoming insights may come from studies of signal transduction in these cells.

The fact that the proliferation, differentiation, and fates of the progeny of retinal progenitor cells are all regulated by intercellular interactions (2–5) led to investigations into the effects of various signaling molecules on these cells (6–8). These studies were done initially using cell culture approaches, but more recently, sophisticated genetic tools are allowing these studies to be confirmed and extended in vivo. To date, a large number of signaling molecules have been shown to affect retinal development in general, and, specifically, retinal progenitor cells (9). However, relatively few studies have focused on the signal transduction cascades that use these signaling molecules in the progenitor cells. Even fewer studies have analyzed interactions between signal transduction pathways in retinal progenitors, although these are undoubtedly of key importance in understanding how these multiple signals are integrated.

In this review, we will first describe some of the assays that have been used to assess the effects of signaling factors on retinal progenitor cells, and those studies that have addressed signal transduction in these cells. We also detail the methods for assaying Notch signaling, because this is a critical pathway for the maintenance of retinal progenitor cells in their undifferentiated state, and therefore allows the other factors to regulate the proliferation of these cells and to direct them to specific fates.

12.2   RECEPTOR TYROSINE KINASES AND THEIR LIGANDS

Among the first factors that were demonstrated to stimulate proliferation of retinal progenitor cells in vitro were members of the epidermal growth factor/transforming growth factor-alpha (EGF/TGFα) family of proteins (6,7,10). EGF was among the first secreted proteins shown to have mitogenic properties, when added to cells in tissue culture (11). EGF and related ligands bind to one of four related receptor tyrosine kinases (ErbB1-4), activating the receptors and initiating a complex signal transduction network. The main downstream pathways that are activated by binding of EGF to the ErbB1 are the Ras/MAPK and PI3K/Akt pathways (see reference 12 for recent review), although the STAT pathway is also activated by EGF in some cells. The activation of these pathways has effects on various aspects of a cell's physiology, including proliferation, migration, and survival (13).

In the developing retina, the activation of the ErbB1 stimulates mitotic activity in progenitor cells. In the rat retina, this effect is observed soon after embryonic day 16, although before that time the cells are relatively unresponsive to the factor (6,7). Several other features distinguish progenitors during the different stages of their development (14,15), and the EGF-responsive progenitors may represent a subset of the progenitors (i.e., those competent to generate Müller glia). In vivo evidence for a requirement for ErbB1 comes from the analysis of mice with loss-of-function mutations in this receptor. Although the early stages of progenitor proliferation are not significantly affected by the loss of the *EbrB1*, there is significantly less proliferation in the retinas of *ErbB1⁻/⁻* mice at later stages of histogenesis when compared with littermate control animals (16). The mitogenic effects of EGF are presumably due to the activation of the MAPK pathway, although other pathways activated by ErbB1 may also be involved. Rhee et al. (17) found that EGF stimulated ERK1/2 phosphorylation in neonatal mouse retina. In addition to its effects on retinal progenitor proliferation, addition of EGF to cultures of retinal cells, or overexpression of the ErbB1 in progenitors from late embryonic or postnatal stages of retinal development biases the cells toward the Müller glial fate (18). EGF treatment of primary retinal cultures also suppresses the differentiation of rhodopsin in the cells, though it is not clear whether this effect occurs in the progenitor cells, or in the newly postmitotic rods (7,10).

Fibroblast growth factors (FGFs) comprise a large family of related soluble signaling proteins that signal through receptor tyrosine kinases, FGFRs (for review see Reference 19). FGFs are key regulators of various tissues during embryonic development. Many FGFs are expressed in very specific regions of the developing nervous system and are both necessary and sufficient for their development. For example, FGF8 is expressed at the midbrain/hindbrain border in the embryonic brain, and is a key organizer of this region of the CNS (20). Several members of the FGF family are expressed in the developing retina, including FGF1, FGF2, FGF3, FGF8, FGF15/19, FGF18, and others (21–27), and the retinal progenitors have at least one receptor for FGFs: FGFR1. The effect of FGF on retinal progenitor cells depends in part on the species and the particular assay. FGFs act as mitogens for rat retinal progenitors (7,10). The addition of FGF1 or FGF2 to dissociated cultures of retinal progenitors stimulates their proliferation, although not as strongly as EGF or TGFα. Another difference between these two groups of growth factors is that whereas EGF is mitogenic only at relatively late stages of retinal histogenesis, FGFs are mitogenic even in progenitors isolated from retinas at very early stages of development.

In addition to their role as mitogenic factors, FGFs stimulate the production of retinal ganglion cells from the progenitor cells (23,28). This effect may be due in part to the upregulation of Pea3 (29), a transcription factor downstream of FGF signaling in several developing tissues, and ultimately an upregulation in Ath5 (28–30), a transcription factor necessary for ganglion cell development (31). Similar to EGF, FGF is known to function primarily through the MAPK pathway, although it is not clear how the two receptor tyrosine kinase families (EGFR and FGFRs) generate qualitatively different downstream responses in the progenitor cells.

Vascular endothelial growth factors (VEGFs) and their receptors (Flk1/Flt1,4) are best known for their role in the regulation of endothelial cell proliferation (32). However, in a screen for receptor tyrosine kinases expressed in the developing retina, Yang and Cepko (33) reported that Flk1 is also expressed by retinal progenitors. Yourey et al. (34) found that VEGF-2 stimulated proliferation of retinal progenitors in P1 rat retinal cultures. Similar to the response to EGF and other growth factors, this response was developmentally regulated, with maximal stimulation in the newborn retinal progenitors and almost an order of magnitude reduced response at embryonic day 15. The response was different from the response to EGF in that there was no inhibition of rhodopsin expression with VEGF-2, whereas in cultures treated with either EGF or CNTF, the number of rhodopsin-positive cells was greatly reduced. Thus, the response of retinal progenitors to VEGF-2 is most similar to that of FGF. Hashimoto et al. (35) recently reported similar effects for VEGF in embryonic chick retina. VEGF can promote proliferation of retinal progenitors, whereas addition of a VEGF receptor (Flk-1) inhibitor reduces their proliferation. The mitogenic effect of VEGF is likely mediated through the MAPK pathway, because it can be blocked by treatment with UO126. However, studies with mice have failed to confirm a critical role for VEGF or Flk1 in retinal progenitor proliferation. Cerebral cortical and retinal defects caused by neural progenitor-specific VEGF-A deletion are not reproduced by neural progenitor-specific deletion of its receptor, leading Haigh et al. (36) to argue that VEGF-A/Flk1 signaling does not have a significant autocrine role in CNS development; rather, its effects are mediated through the regulation of endothelial cell growth.

There is also evidence that neurotrophins (NTs) regulate retinal progenitor proliferation. In chick embryos, intraocular injection of a blocking antibody to NT-3 causes an increase in proliferation of progenitors; this led these researchers to propose that NT-3 is an antiproliferative signal. By contrast, viral misexpression of truncated TrkC in chick embryo retina, which should interfere with neurotrophin signaling, also causes a reduction in the number of BrdU-labeled cells (37). Deletion of the p75NTR also results in a small, but significant reduction in BrdU-labeled cells in the embryonic mouse retina (38); although this effect is likely mediated through the retinal ganglion cells, because p75NTR does not appear to be expressed by the progenitors themselves (38).

12.3 HEDGEHOG SIGNALING

Another signaling molecule important at several stages of retinal development is Sonic hedgehog (Shh). Shh signaling in the retina is the subject of another chapter in this volume, and so will be only briefly described here. Shh is a member of a small family of secreted proteins, known as Hedgehogs, that are involved in regulating pattern formation of many tissues during embryonic development. Hedgehog proteins have been shown to function in the regulation of cell proliferation, differentiation, and survival. Hedgehogs function by binding to a cell surface receptor complex that includes two proteins: Patched (Ptch) and Smoothened (Smo). When there is no Hedgehog bound to Ptch, it inhibits the activity of Smo and key downstream transcription factors, and Gli 1 is proteolytically cleaved into a transcriptional repressor. When Hedgehog binds to Ptch, Smo repression is relieved, and Gli1 is no longer targeted for proteolysis; the full-length form can act as a transcriptional activator (see reference 39 for review).

The Shh pathway is important at several distinct periods during eye development. In the earliest stages of eye development, Shh secreted from cells of the presomitic mesoderm suppresses eye formation at the ventral midline of the neural tube, effectively splitting the presumptive eye field into the two eyes (40). Thus, loss of the Shh signal at this early stage of embryogenesis results in cyclopia (41). Later in eye development, Shh stimulates proliferation of the retinal progenitors in vitro and in vivo (42–45). In addition to promoting cell proliferation, Shh appears to have effects on cell fate, and studies have shown some role for Shh in development of retinal ganglion cells, (46) rod photoreceptors, and Müller glia (42,43,47,48). Targeted deletion of Shh in the retinal ganglion cells (the primary source of Shh in the retina) results in mice with significant defects in retinal growth (44). In addition, Spence et al. (49) and Moshiri and Reh (50,51) have shown that Shh stimulates retinal progenitor proliferation and regeneration from the ciliary marginal zone of embryonic and posthatch chicks. In all these functions of Shh during eye development, the Ptc and Smo receptors are likely to mediate the responses, via the Gli family of transcription factors (43,51–53).

12.4 TGFβ/ACTIVINS/BMPS AND ALKS

The TGFβ signaling pathway regulates several aspects of retinal patterning, proliferation, differentiation, and apoptosis. TGFβ signaling is initiated when a secreted ligand binds to the type II receptor. The type II receptor then binds to and phophorylates the type I receptor. The type I receptor subsequently phosphorylates and activates Smad proteins. The activated Smad then binds to the co-Smad, Smad 4. The activated Smad complex then translocates to the nucleus, where it activates transcription of downstream targets.

There are several structurally related TGFβ ligands in the TGFβ superfamily, including TGFβ, BMP (bone morphogenic protein), and activins. These receptors can be activated by more than one ligand but with variable affinity. In the human genome, there are seven type I TGFβ receptors (ALK 1-7, ALK-2 is also known as ActRI, ALK-4 as ActRIB, ALK-5 as TGFβRI, ALK-3 as Bmpr1a, and ALK 6 as Bmpr1b) and five type II TGFβ receptors (ActRIIA, ActRIIB, BmprII, AMHRII, and TbrII) (54). There is evidence for the expression of all these TGFβ receptors, excluding ALK-1 and ALK-7, in the developing retina. The TGFβ and activin receptors activate Smad 2 and 3. The BMP receptors activate Smad 1, 5, and 8 (54).

Before differentiation, retinal precursor cells exit the cell cycle at G1. The TGFβ signaling pathway acts as a stop signal to proliferation. When the Dpp (*Drosophila* BMP) pathway is mutated (either by mutating the type I receptors, tkv or sax, or the Smad, MAD), there is a delay in G1 arrest (55–57). Likewise, in the rat retina, activin inhibits retinal progenitor proliferation and, interestingly, also promotes differentiation into rod photoreceptors (58). Activin βA is expressed in the rat pigment epithelium at E15. When E18 rat retinal progenitor cells are cultured and treated with activin βA, there is a significant decrease in proliferation followed by a differentiation. A large fraction of these cells differentiate into rod photoreceptors. The activin receptor, ActRIB, is detected in the rat retina as early as E17; the other activin type I receptor, ActRI, is not detected until P6. The type II activin receptor mRNA are detectable by in situ hybridization; ActRII mRNA is detected as early as E16, and ActRIIB is detected as early as E12 in rat eyes (59).

Another member of the TGFβ family, TGFβ2, may also work as a stop signal for proliferation in postnatal rat retinal progenitors (60,61). The proliferation of retinal progenitors normally continues through the first postnatal week in rodents. However, between postnatal day 4 and 6 there is a rapid decline in the number of progenitors in the mitotic cycle. The mechanism for this rapid decline in progenitor proliferation is mediated in part by TGFβ2. Addition of TGFβ2 to mitotic progenitors in vitro inhibits their proliferation, whereas inhibition of TGFβ signaling extends the period of proliferation in the postnatal retina. These effects are likely mediated through Smad 2/3 activation of the cell cycle inhibitor p27^{kip} (60).

Many of the BMP ligands and their receptors are expressed in a developmentally spatially regulated pattern in the retina (62). In the chick retina, the predominantly expressed BMP is BMP-4; its expression is very strong in the dorsal retina (63). At later embryonic stages, BMP-2 and BMP-7 are expressed at the dorsal–ventral boundary. BMP-5 is also expressed throughout the retina with the strongest expression in the center. The BMP receptors are expressed at these same stages although their highest levels are in the ventral retina. *Bmprla* and *Bmprlb* are strongly expressed in the ventral retina. *Bmprla* is also weakly expressed in the dorsal retina. BmprII is uniformly expressed in the E8 retina (63). In mice, similar expression patterns are found for BMPs and their receptors (64). *Bmprlb* is most highly expressed in the ventral retina, whereas *Bmprla* and BmprII are expressed more evenly throughout the dorsal–ventral axis. BMP-4 is expressed primarily by the cells in the ciliary epithelium and not in the retina proper, although a lower level of expression is present in the amacrine and ganglion cells (64).

The highly asymmetric expression patterns of BMPs and their receptors have led several groups to analyze their role in the regulation of topographic map formation, although this is beyond the scope of this short review. However, recent studies have shown that BMPs are also important regulators of retinal progenitor proliferation and apoptosis. Although mice deficient in the two BMP receptor I genes (*Bmprla* and *Bmprlb*) show early embryonic lethality, conditional deletion of both *Bmprla*$^{-/-}$, *Bmprlb*$^{-/-}$, using a Six3-promoter driving Cre-recombinase, results in mice with very small eyes (64). The retinas in these mice are very thin due to excess apoptosis and reduced cell proliferation. The increase in apoptosis occurs early, about E11.5; at this age, there is normally apoptosis to make room for the optic disk. The loss in proliferation occurs later at E11.5, and correlates with a reduction in *Chx10* and *cyclinD1* expression. *Chx10* regulates retinal progenitor cell number, and *CyclinD1* regulates the G1-S transition in the cell cycle. Additionally, in the *Bmprla*$^{-/-}$; *Bmprlb*$^{-/-}$; *Six3-Cre* mice, *Math5* is not expressed at E11.5. E11.5 is when retinal neurogenesis commences and *Math5* expression indicates the start of neurogenesis. The *Bmprla*$^{-/-}$; *Bmprlb*$^{-/-}$; *Six3-Cre* mice also show reduced FGF15 expression, suggesting that some of the effects of BMPs may be mediated through receptor tyrosine kinase signaling.

12.5 LIF AND CNTF

Ciliary neurotrophic factor (CNTF) was first identified for its role in the regulation of ciliary ganglion neuron development (see Reference 65 for review). This protein is part of a larger family of molecules, known as cytokines, that are primarily known for their functions in hemopoesis. CNTF signals through a tripartite receptor complex that contains an alpha subunit that confers specificity, the gp130, and the beta subunit of the LIF (leukemia inhibitory factor) receptor. CNTF binds to the alpha subunit and induces heterodimerization of the beta receptor subunits; this leads to tyrosine phosphorylation of intracellular proteins called STATs (signal transducers and activators of transcription) by receptor-associated JAK kinases. The activated STATs dimerize and translocate to the nucleus, where they bind to specific DNA sequences and act as transcriptional activators.

In the CNS, CNTF has been most thoroughly studied as a neuronal survival factor, and it can promote rod photoreceptor survival in several animal models of retinal disease [e.g., (66)]. In the developing nervous system, however, CNTF functions to promote gliogenesis (67–69). Treatment of multipotent neural stem cells with CNTF promotes their differentiation into astrocytes, in part through the direct binding and activation of the promoter of the glial-specific protein GFAP by STAT3.

In the developing retina, CNTF (or the related cytokine LIF) has been shown to have several functions, including the suppression of rhodopsin expression in newly generated photoreceptor precursors, and the promotion of Müller glial development (17,70–72). Treatment of explant cultures of retina with CNTF stimulates both STAT3 and ERK phosphorylation, although the latter is in progenitors, with the former being mostly in newly developing postmitotic rods. CNTF stimulates Müller glial development from the progenitors and suppresses opsin expression in Crx+ newly generated rods, and it appears that the two responses are distinct. The suppression of photoreceptor maturation by CNTF can be blocked by STAT3 inhibition, whereas either STAT inhibition or ERK inhibition is sufficient to block the effect of CNTF on Müller glial development (17,73).

12.6 NOTCH AND DSL LIGANDS

The Notch signal transduction pathway is a critical regulator of cell fate in most developing tissues. This signaling pathway is composed of the Notch receptor, a single-pass type I transmembrane protein expressed by the signal-receiving cells, and its family of ligands (Delta/Serrate/Lag, DSL), which are also single-pass type I transmembrane proteins expressed in adjacent signal-sending cells. The Notch pathway was first described for its role in a process called lateral inhibition in *Drosophila* development. In this process, a small group of cells (equivalence group) express both the receptor and ligand, and through a presumably stochastic process, some cells eventually express more ligand than receptor, resulting in lower and higher levels of Notch activity, respectively. This initially unstable pattern becomes reinforced through reiterative Notch signaling, and results in the selection of the ligand expressing cells for further differentiation events, whereas high Notch activity within adjacent cells inhibits their differentiation. Notch signaling can be divided into canonical and noncanonical signal transduction pathways (74). Here, we discuss the core elements of the molecular signal transduction cascade for canonical Notch signal transduction, and its function during vertebrate retinal development.

12.6.1 CORE ELEMENTS OF NOTCH SIGNAL TRANSDUCTION

Notch is a large (approximately 300 kDa) protein containing 36 tandem epidermal growth factor-like (EGF) repeats and 3 Lin-Notch repeats (LNR, unique motifs to Notch genes) in its N-terminal extracellular domain, a transmembrane domain, and a C-terminal intracellular domain, responsible for conducting the Notch signal. Notch undergoes posttranslational modification to generate a functional receptor during its biosynthesis and transport to the cell surface: First, a furin-like convertase cleaves Notch in the extracellular domain at the LNR repeats (S1 cleavage) during its transport to the cell surface; next, the cleaved extracellular domain reassociates with the remaining portion of the molecule (75,76) to make the final heterodimeric receptor.

The heterodimeric Notch receptor is activated when it binds to its cognate DSL-ligand (Delta in this case) in an adjacent cell (figure 12.1). Delta binds the Notch extracellular domain, and the complex is internalized into the Delta-expressing cells through dynamin-mediated transendocytosis (77–79). Transendocytosis of the Notch extracellular domain initiates a process termed RIPping [Regulated Intramembrane Proteolysis; (80)], in which the removal of the large N-terminal extracellular domain by Delta induces a conformational change in the remaining portion of the protein, allowing access of members of the ADAM/TACE family of metalloproteases to cleave it at the S2 site (81–83). The remaining transmembrane domain is then cleaved at the S3 cleavage site located near the cytoplasmic membrane surface by the presenilin–γ-secretase complex (composed of at least four proteins: presenilin, nicastrin, pen-2, and Aph-1) (84,85). This last step releases the Notch internal cytoplasmic domain (NICD) from the membrane and allows it to move to the nucleus. The NICD contains several protein interaction motifs, including the protein interaction RAM and ANK domains, nuclear localization sequences (NLS), and a PEST motif for rapid protein degradation. NICD forms a complex with the DNA-binding protein CSL (human CBF1–RBPjk, *Drosophila* Su(H), *C. elegans* Lag-1) via its RAM and ANK domains. NICD binding to CSL recruits another protein called Mastermind (Mam), which in turn recruits p300 and associated transcriptional initiation factors (86). The NICD–Mam–CSL complex can thus activate transcription of genes with a CSL consensus site in their promoter. Some of the key transcriptional targets are members of the Hes (Hairy/Enhancers of Split)-related family of transcriptional repressors (87).

Notch activation thus leads to the induction of Hes transcription by the NICD–Mam–CSL complex, and Hes proteins in turn inhibit cellular differentiation by two routes. First, Hes proteins repress transcription of specific target genes by binding to DNA sequences found in their promoter regions, called N-boxes, and recruiting transcriptional repressor complexes (88). Second, Hes proteins physically interact with target gene protein products, and directly inhibit their function (89). The most well-characterized Hes targets are members of the proneural bHLH family

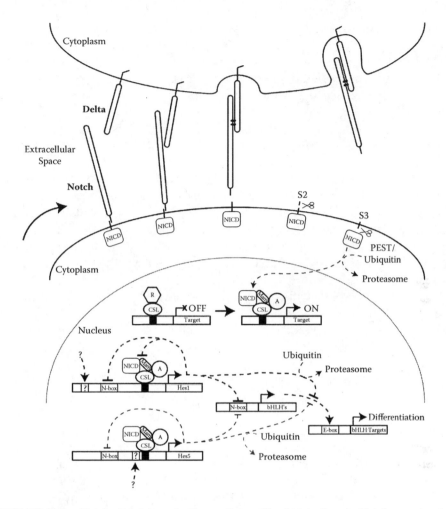

FIGURE 12.1   Notch signal transduction pathway. The heterodimeric Notch receptor is transported to the cell surface of the signal-receiving cell. Delta, from an adjacent signal-sending cell, binds the Notch extracellular domain. Trans-endocytosis of Notch by Delta induces two subsequent cleavage events termed S1 and S2 in the remaining N-terminus, and S3 in a region near the cytoplasmic surface of the transmembrane domain (scissors), that releases the Notch internal cytoplasmic domain (NICD). NICD translocates to the nucleus, forms a complex with Mastermind (Mam), and the DNA-binding protein CSL, switching CSL from a transcriptional repressor to an activator complex. NICD/Mam/CSL induce expression of a set of target effector genes including Hes1 and Hes5. Hes effectors inhibit cellular differentiation by binding to N-box motifs in the promoters of their respective set of target genes, such as proneural bHLH transcription factors (Mash1, Ngn2), and recruiting transcriptional repressor complexes to inhibit their expression. Hes proteins can also bind directly to proneural bHLH proteins, blocking their ability to bind E-box motifs in promoters of their respective set of target genes and initiate cellular differentiation programs. Negative feedback mechanisms include rapid degradation of NICD and Hes proteins by ubiquitin-proteasome pathways, Hes autorepression, and blocking NICD/Mam/CSL function. Hes genes may have different functional efficiencies, and also other regulatory inputs as indicated.

of transcription factors, which function to promote neural differentiation. Proneural bHLH transcription factors, such as Ngn2 and Mash1, bind to E-box sequences found in the promoter regions of their respective set of target genes and activate genetic programs for neuronal differentiation. Thus, by repressing both transcription and function of proneural genes, Notch signaling via Hes genes inhibits progenitor cell differentiation.

The Notch signal transduction pathway contains built-in mechanisms for its downregulation. Free NICD is targeted for rapid degradation by its PEST motif and ubiquitin-proteosome pathways (90,91). Hes genes are also subjected to rapid ubiquitin-proteosome-mediated turnover. Hes1 has a reported half-life of approximately 20–25 min for both its mRNA and protein product (92). Hes1 can feedback to inhibit Notch signaling by physically binding the NICD/Mam/CSL complex and blocking its function (93). Hes1 also contains an N-box in its own promoter, thereby repressing its own transcription (92,94). Thus, the Notch signal is inherently transient and must be continually reactivated to maintain the progenitors in an undifferentiated state.

This basic outline of the Notch signaling pathway is somewhat simplified. Mammals have four Notch paralogs (Notch 1–4), five DSL ligands (Delta-like 1,3,4; Jagged1,2), approximately seven Hes genes (Hes1–7) and three Hey genes (1–3) that exhibit unique and partially overlapping expression patterns during development. Certain target genes, such as Hes1 and Hes5, exhibit greater activation in response to Notch signaling than others (95). The efficacy of target activation is critically dependent on the particular arrangement of CSL binding sites in promoters of target genes (96–99), and each Notch paralog exhibits different preference for binding sites across known target genes (99,100). Additionally, certain Hes/Hey genes may function as better effectors than others. For example, although Hes5 functions similarly compared to Hes1, Hes5 seems to be not as efficient at attenuating bHLH function or inhibiting its own transcription (101,102). Moreover, although Hes6 is a Hes gene family member, it is not a direct target of the NICD–Mam–CSL complex, but instead is activated by the derepression of proneural bHLH transcription factors, and actually functions to repress other Notch effectors, such as Hes1 and Hes5 (103–105). Finally, the efficiency of Hes/Hey effectors may also depend on the available repertoire of cofactors, such as corepressor complexes, and it is also likely that different DSL ligands can activate Notch1–4 family members differently.

12.6.2 Notch Signaling during Vertebrate Retinal Development

It has been known for some time that intercellular inhibitory interactions regulate retinal progenitor cell differentiation (106–108). Notch signaling is thought to mediate most, if not all, cellular inhibitory interactions necessary for patterning neural cells (109,110). The Notch pathway has been extensively studied during vertebrate retinal development, and an emerging consensus of Notch function is that: (a) Notch activity is necessary in progenitor cells to maintain their undifferentiated state throughout the period of retinogenesis, and (b) Notch signaling regulates the differentiation of retinal cell types (111–123).

Many components of the Notch signaling pathway are expressed at some point during eye development. In the mouse, Notch1 is expressed in the neuroblast layer of the developing neural retina, Notch2 is expressed in the pigmented epithelium and optic stalk, whereas Notch3 is expressed throughout the epithelium in both the central retina and peripheral structures where the ciliary body and iris would be located (124–126). Many Notch ligands are expressed in the mouse retina; Jagged1, Jagged2 (Jagged genes are mouse homologs of Serrate), Delta-like1 (DLL1), DLL3, and DLL4 are all expressed in unique and partially overlapping patterns as well (124,125,127,128). Many Hes/Hey genes are also expressed in the mouse retina including Hes1, Hes5, Hes6, and Hey1–3 (104,113,117,118). Additionally, many Notch pathway components, including Notch4, also function during vascular development of the retina (129). Thus, the retina offers a rich source for studying Notch signal transduction during neural development.

Studies of Notch signaling in the mammalian retina often encounter several difficulties including embryonic lethality and functional redundancy between Notch pathway family members. For example, deletion of Notch1 and DLL1 results in early embryonic lethality around the onset of retinal development (130,131). Therefore, genetic studies of Notch signaling during retinal development in mice have primarily relied on manipulations of downstream Notch effectors. However, two studies have recently reported the effects of conditionally knocking out Notch1 during retinal development, bypassing early lethality. These mice have smaller eyes, reduced progenitor cell proliferation, and increased differentiation of cone photoreceptors early (122,123) and rod photoreceptors later (123). Functional compensation between Notch1 and Notch3 has been demonstrated in the cerebral cortex (132), and Notch1 probably compensates for Notch3 deletion to some degree in the retina (126). However, Notch3 likely plays a unique role in retinal development, because patients with mutations in Notch3 (CADASIL disease) exhibit some retinal dysfunction (133). Deletion of Hes1 results in neonatal lethality, but Hes1 mutant animals also exhibit small eyes, premature neuronal differentiation, and fewer Müller glia (113). Deletion of Hes5 is not lethal and leads to 35% fewer Müller glia (117). However, deleting both Hes1 and Hes5 seems to prevent optic cup formation (134).

Gain-of-function studies have also been utilized to study the effects of various Notch pathway components. Retroviral infection (and/or transient DNA transfection) containing constitutively active NICD1, Hes1, Hes5, and Hey2 have shown that Notch signaling prevents neural differentiation and promotes Müller glia formation (116–118,120) However, there are some subtle differences in the resulting phenotype of infected/transfected cells. For example, overexpression of Hes5 promotes Müller glia differentiation at the expense of other later-born cell types (117), whereas overexpression of Hes1 generally inhibits neuronal and glial differentiation, yielding a population of undifferentiated "Müller glia-like" cells in the postnatal retina (120). Altogether, studies of Notch pathway components in the retina are consistent with those of Notch function elsewhere in the nervous system: loss of Notch function promotes neuronal differentiation, whereas gain of Notch function prevents neuronal differentiation.

12.7 FUNCTIONAL ASSAYS FOR STUDYING SIGNAL TRANSDUCTION IN THE RETINA

12.7.1 TISSUE CULTURE OF THE RETINA

Retinal cells can be maintained in primary tissue culture as dissociated cells or explants. The methods work equally well for embryonic retina and neonatal retina, and can be applied to a variety of different species, including chick, rat, mouse, and human. The dissociated cell cultures provide a convenient way to test the effects of soluble factors on retinal progenitor proliferation, as well as differentiation to different lineages. Explant cultures have the advantage that the normal lamination of the retina is preserved and develops reasonably well in vitro. In addition, some of the more highly differentiated properties of retinal cells, such as the dorsal-ventral gradient in cone opsin expression, develop appropriately in explant cultures.

PROTOCOL 1: DISSOCIATED CELL CULTURE OF PRIMARY RETINAL CELLS

1. The embryos are harvested or the pup is sacrificed in accordance with approved protocols for the institution.
2. The eyes are removed and placed in a sterile Petri dish containing cold (4°C) Hanks's Balanced Salt Solution supplemented with 3% D-glucose and 0.01 M HEPES, pH 7.4 (HBSS+).
3. The retinas are dissected from the extraocular tissue in this solution. The pigmented epithelium, lens, and scleral tissue are removed, and the retina is then transferred to a new sterile Petri dish containing (calcium/magnesium free) CMF-HBSS using forceps or a Pasteur pipette.
4. The retinas are then transferred to a 15 mL centrifuge tube containing 5 mL 0.25% trypsin, in HBSS-CMF (0.5 mL stock at 2.5%; plus 4.5 mL HBSS-CMF). The centrifuge tube containing the retinas is then put on a rocking platform in a 37°C incubator and gently rocked for 5–15 min.
5. As a guideline, embryonic chick retinas typically require only 5 min, whereas postnatal mouse retina can require up to 10 min for thorough dissociation. In all cases, however, care must be taken not to overtreat the retinas with trypsin, because this will result in low yields of viable progenitor cells.
6. The trypsin treatment is completed when the retinas are broken up into small pieces, but not yet into single cells. Add 0.5 mL fetal bovine serum (FBS; Invitrogen) to the tube to inactivate the trypsin.
7. Centrifuge the tube at 1500 rpm (approximately $750 \times g$) for 10 min to pellet the cells. Carefully remove the supernatant (leaving a small amount of solution at the bottom, so that the pellet is not disturbed).
8. Resuspend the pellet cells in 2 mL of medium (see step 10) with gentle trituration using a (fire-polished) Pasteur pipette.
9. Determine the number of cells using a hemocytometer; trypan blue can be used to estimate the percentage viability. Plate cells between 50,000 (low density) and 500,000 (high density) cells per well (for a 24-well plate) onto

poly-D-lysine–Matrigel (BD Biosciences)-coated coverslips (see the following text). Cultures are maintained at 37°C and 5% CO_2.

10. The culture medium contains DMEM/F12 (without glutamate or aspartate), 25 µg/mL insulin, 100 µg/mL transferrin, 60 µM putrescine, 30 nM selenium, 20 nM progesterone, 100 U/mL penicillin, 100 µg/mL streptomycin, 0.05 M HEPES, and 1% FBS. The final medium is effective for approximately 1 week when stored at 4°C.

11. Maintain sterile stock solutions so there is no need to filter the final medium. However, if contamination is suspected, the medium can be filtered once without loss of potency.

12. Once the retinal cells are established in vitro, one half of the medium is replaced every other day. The progenitor cells of chick or rodent retina can be maintained in this medium for up to 1 week, and they retain their ability to generate neurons, as demonstrated by double labeling for BrdU/thymidine and neuronal-specific markers (see the following text).

13. A serum concentration of 1% is recommended; however, this can be reduced to 0.1% with little reduction in proliferation. The medium can also be supplemented with growth factors to stimulate the proliferation of the progenitor cells. The same medium can be used for serum-free cultures, but without serum the progenitor cells are more likely to differentiate.

Substrate for adherent culture of retinal cells

1. The retinal progenitor/stem cells proliferate in adherent cultures on glass coverslips coated with poly-D-lysine and Matrigel.

2. To prepare the coverslips, use high molecular weight poly-D-lysine (30–70,000 MW). The poly-D-lysine is dissolved in sterile water at a concentration of 0.5 mg/mL, and 1 mL aliquots are stored in 15 mL conical tubes at −20°C.

3. Before use, a tube is thawed, and 9 mL of sterile water is added for a final concentration of 50 µg/mL.

4. The coverslips used most frequently are 12 mm circular. They are placed into a small vial for sterilization with an autoclave.

5. Twenty-five to thirty coverslips are placed in a large Petri dish for coating. The poly-D-lysine solution is added to the dish, and care is taken to ensure all coverslips are submerged in the solution, and the coverslips are incubated at 37°C for 15–30 min.

6. The poly-D-lysine solution is removed, and the coverslips are washed in sterile water thrice for a minimum 5 min each. Care is taken to wash the coverslips very well, mixing the coverslips with flamed forceps, because poly-D-lysine in solution can be toxic to cells.

7. The coverslips can be dried and stored for up to 2 weeks in the Petri dish at 4°C, or used immediately.

8. When ready to use, put one coverslip in each well of a 24-well plate using flamed forceps. Then proceed to coat them with Matrigel.

9. Matrigel is supplied by the manufacturer as a frozen solution. Thaw the bottle slowly on ice (for several hours) to prevent gel formation. Small (200 µL) aliquots are made using precooled tubes (15 mL) on ice and a prechilled pipette.

If the Matrigel warms during the aliquoting, it will gel and not be effective for the cell cultures. The aliquots are stored at −20°C for up to 6 months.

10. To coat the coverslips, remove one aliquot of Matrigel from the −20°C freezer and place on ice for 15–30 min to thaw (200 μL is used for one 24-well plate).

11. Add 10 mL of ice-cold HBSS+ to the 15 mL tube containing 200 μL of thawed Matrigel. Mix gently.

12. Immediately, put 0.5 mL of the dilute Matrigel solution into each well of a 24-well plate in which you have already placed poly-D-lysine-coated cover-slips, and place the plate in the incubator for 30 min at 37°C.

13. Remove the plate from the incubator and, under the sterile hood, remove nearly all of the liquid from the wells. A small amount of the dilute Matrigel solution is left in the well just to cover the coverslips.

14. Let the plate dry in the hood uncovered for 15–30 min. The Matrigel will dry into a thin coating. Plate cells onto the Matrigel, or store the plate at 4°C for no more than 2 days before plating the cells.

PROTOCOL 2: EXPLANT CULTURE OF RETINA

1. For the explant cultures, the retina is cultured at the gas–liquid interface on a Millicell filter insert in either a 6-well or 24-well plate as follows.

2. Once the animal is sacrificed and the eyes removed, the retina is dissected away from the RPE, lens, and choroid while suspended in HBSS.

3. Next use iridectomy scissors to make four cuts around the circumference of the retina, so that it can be opened up like a flower and flattened.

4. Use a wide-bore transfer pipette to move the retinas into a dish containing explant media. We typically cut the tip off a sterile plastic pipette so that the retinas can be drawn into the pipette, without causing damage.

5. Place a sterile Millipore filter (Millicell-CM 0.4 mM filter inserts for 24-well are 12 mm, Catalog #PICM01280; for 6-well they are 30 mm; Catalog #PICM03050) onto a small Petri dish lid and use a pipette (as previously described) to transfer retinas one at a time in a drop of media onto the filter.

6. Looking through the microscope, use forceps to make sure that the gan-glion cell side is facing up. Then spread the liquid around the membrane a bit, and flatten the retina as much as possible with forceps.

7. Once it is relatively unfolded, use a pipette to remove the rest of the media. This usually helps to flatten the retina further. Use forceps to uncurl edges. If the filter is too dry, this will rip the retina. If it is too wet, the retina will not stay unfolded.

8. We usually put 2–4 retinas onto each filter, if we are using the larger filter inserts in 6-well plates. However, if you want to keep track of individual retinas, you can put one retina each onto a single 24-well-size filter.

9. Next, place the filter insert, with the retinas on top, into a well of a 6-well plate containing 1 mL of prewarmed explant media (300 μL/well for the 24-well plate). The media will just come up to the bottom of the membrane.

Media should not be added to the top of the explant, but rather they should sit at the gas–liquid interface.

10. One half of the medium should be changed every day. Make sure no bubbles accumulate under the filter.

11. The retinas can be maintained for at least 14 days in this culture system, and they develop relatively normal lamination and differentiated cell types. However, the retinas will not grow to the same degree as in vivo, and so there may be quantitative differences and changes in the normal ratios of cell types from the normal retina.

12. For histological processing of explant cultures, we remove the medium from the well and add 4% paraformaldehyde in PBS for 2–4 h.

13. Next, the fixative is removed and replaced with PBS. This usually is sufficient to dislodge the retinas from the filter insert, and they can then be moved to new dishes for immunofluorescent labeling.

14. If the retina sticks to the membrane, you can fix the retinas on the membrane. If the retinas are to be sectioned on a cryostat, you can cut the membrane out of the plastic insert and freeze the membrane with the retina. If the retina is to be processed as a whole-mount, use a scalpel blade to gently scrape the retina off the membrane.

15. We use standard methods for antibody labeling of cryostat sections; however, the labeling of retinal whole mounts is worth detailing.

PROTOCOL 3: IMMUNOHISTOCHEMISTRY OF FLATMOUNT RETINAS

1. Put each retina in its own well of a 24-well plate. Otherwise, they may stick together, which impairs antibody penetration.

2. Dilute the primary antibody in PBS/0.3% Triton (PBST) with 5% serum. Add 200 μL per well. Incubate with gentle rocking overnight (or up to 2 days) at 4°C.

3. Wash 3 × 30 min in PBS

4. Dilute the secondary antibody in PBS/0.3% Triton (PBST) with 5% serum. Add 200 μL per well.

5. Incubate overnight in cold room.

6. Wash 3 × 30 min in PBS

7. To coverslip the whole mounts, we use Secure-Seal Imaging Chamber SS1-13 (Grace Bio-Labs; Bend, Oregon). These act as spacers, so that the explants are not crushed.

Explant media (15 mL)

8 mL sterile H_2O

3 mL 5× DMEM/F12 without L-aspartic acid and L-glutamic acid (U.S. Biological custom media D9807–05)

1.5 mL hormone supplement (see the following text)

1.5 mL FBS

300 μL 30% D-glucose

225 μL 7.5% $NaHCO_3$

150 μL N-2 supplement (Invitrogen)
150 μL penicillin/ streptomycin (Invitrogen)
75 μL 1 M HEPES
60 μL L-glutamine

Hormone supplement (200 mL)

40 mL DMEM/F12 without L-aspartic acid and L-glutamic acid
4 mL 30% glucose
3 mL 7.5% NaHCO$_3$
1 mL 1M HEPES
20 μL 10× putrescine stock (10× = 96.6 μg/mL; Sigma)
20 μL 2 mM progesterone stock (made in ethanol)
60 μL 1 mM selenium stock
200 mg holo-transferrin (Sigma)
20 mL insulin solution (dissolve 50 mg of insulin in 2 mL of 0.1 N HCl, then
    bring up to 20 mL with sterile, distilled water; Sigma)

12.7.2 METHODS TO STUDY THE NOTCH PATHWAY IN RETINAL CULTURES

We have recently developed a simple approach to study the temporal effects of
manipulations in Notch signaling, by using a γ-secretase inhibitor that blocks the
presenilin–γ-secretase complex from releasing the NICD. All mammalian Notch
paralogs (Notch 1–4) require presenilin–γ-secretase-mediated S3 cleavage for
canonical signal transduction (135). One γ-secretase inhibitor in particular (*N*-[*N*-
(3,5-difluorophenacetyl)-l-alanyl]-*S*-phenylglycine *t*-butyl ester, or DAPT) has been
shown to phenocopy Notch pathway mutants in both zebra fish and *Drosophila*
(136,137). DAPT treatment causes a rapid decrease in Hes1 and Hes5 gene expression in retinal progenitor cells and downregulates Hes1 and Hes5 reporter activity
(99,121). We have used DAPT in the chick and mouse retina to synchronize the differentiation of retinal progenitor cells.

    We have also developed a method for monitoring the activity of the Notch
pathway in living cells using the mouse Hes1 and Hes5 promoters, driving desta-
bilized eGFP (Nelson et al. 2006). Hes1 and Hes5 cis-regulatory elements contain
the NICD/Mam/CSL binding sites conferring responsiveness to Notch signaling
activity, as well as other potential regulatory inputs (discussed in Ong et al. 2006).
We have developed several methods based on electroporation to transfect DNA into
retina explants from different model systems (chicken or mouse), and into dissociated retinal cells.

PROTOCOL 4: PHARMACOLOGICAL INHIBITION
OF NOTCH SIGNAL TRANSDUCTION

1. Retinas from animals of the desired embryonic age are collected in HBSS+
   (see Reference 121 and protocol 1) and cultured either as dissociated cells
   (protocol 1) or explants (protocol 2).

2. Add DAPT or DMSO vehicle to the wells of the dissociated cells or the explants (*N*-[*N*-(3,5-difluorophenacetyl)-l-alanyl]-*S*-phenylglycine *t*-butyl ester; EMD Biosciences #565770; 10 mM stock in DMSO) to desired final concentration. In a dose response assay, we found that concentrations between 10 μM and 50 μM inhibit Notch signaling effectively, whereas 100 μM DAPT precipitates in the medium.

3. We assay for the inactivation of the Notch signal using QPCR for Hes1 and Hes5 gene expression changes. We have found that Hes5 is downregulated over 10-fold in as little as 3 h.

4. The effects of Notch inactivation can be assayed using QPCR for candidate genes that have already been identified as part of the neuronal differentiation pathway. Alternatively, we have also used cDNA microarrays to identify additional components of the pathway (Nelson et al., 2007).

PROTOCOL 5: MONITORING NOTCH ACTIVITY WITH QPCR

1. Transfer retinas into 500 μL of TriZOL (Invitrogen) in a 1.5 mL RNase-free tube.

2. Homogenize tissue thoroughly with a Pellet Pestle Motor and fresh, RNase-free Pellet Pestles (Kimble–Kontes): samples can be frozen (−80°C) at this point indefinitely.

3. Add 200 μL RNase-free chloroform, vortex, spin to separate layers, and transfer the top, aqueous layer to a fresh tube.

4. We typically reextract with another round of chloroform, and even have found that including an additional phenol:chloroform:isoamyl alcohol and chloroform extraction can be helpful in cleaning up lysates.

5. Add an equal volume of 100% isopropanol (RNase-free), mix, and spin at maximum speed for 10 min: samples can again be stored at −80°C at this point.

6. Wash pellet with 70% ethanol (RNase-free), decant, and air-dry (do not overdry).

7. Resuspend in 40 μL of RNase-free H_2O.

8. Digest genomic DNA by adding RiboLock RNase-inhibitor (1–2 μL; Fermentas), 5 μL 10× buffer, and 4–5 μL of RQ1 RNase-free DNase (Promega).

9. Incubate at 37°C for 30–45 min.

10. Remove genomic DNA by using the RNeasy-cleanup procedure, part of the RNeasy mini kit (Qiagen), according to manufacturer's instructions.

11. Elute in 20 μL of RNase-free H_2O (an additional reelution with the flow-through can result in more total RNA).

12. For cDNA synthesis, we use a standard oligo-dT primed cDNA synthesis reaction with SuperScript II Reverse Transcriptase (RT; Invitrogen). Standard positive RT reaction mix:

    10 μL total RNA from above
    1 μL oligo-dT primer (0.5 μg/μL)
    1 μL dNTPs (10 mM)

    Denature at 65°C, 5 min, place on ice, and then add the following:

    4 μL 5× SSII First Strand Buffer (Invitrogen)

2 µL DTT (100 mM)

1 µL RiboLock RNase-inhibitor

1 µL SuperScript II RT

Incubate for 50–75 min at 42°C, heat-kill RT at 70°C for 15 min (always include a no-RT control). We use 10–15 µL of the total RNA from step 11 for a positive reaction, and 5 µL for a negative reaction, scaled accordingly. Dilute reactions 1:3 or 1:4 with H_2O to prepare them for normalization via QPCR (store at −20 to −80°C).

13. The reaction mix for QPCR reaction is as follows:

1 µL cDNA

1 µL forward primer 5 min (20 mM)

1 µL reverse primer 3 min (20 mM)

7 µL H_2O

10 µL SYBR Green PCR Master Mix (Applera Corp.)

_____

20 µL total volume

14. The amount of template varies depending on the initial concentration of RNA from the samples. To compare samples, we normalize by assaying levels of control genes, such as GAPDH and/or β-actin. Sample concentrations are then adjusted according to the ratio of the cycle numbers that the control transcripts exhibit log scale increases in amplification measured by fluorescence readout. A threshold is set at the level that the fluorescence increase has reached a maximal slope. The cycle number difference in transcript levels measured between the experimental and control cDNA samples is used to calculate the fold difference. This fold difference is used in conjunction with the original sample volumes to dilute the more concentrated sample to that of the less concentrated sample. After sample concentrations are adjusted, they should be tested again to determine how well they were normalized.

15. Each positive control sample is run in triplicate (one negative RT reaction is run per sample, and an appropriate master mix scheme should be utilized whenever possible). For prenormalized samples, GAPDH threshold cycle levels fell in a relatively broad range across all samples (figure 12.3a). Following the aforementioned normalization procedure, samples are adjusted to arise at approximately 14 cycles, and are rechecked to verify new levels. The resulting normalized GAPDH levels fell into a much narrower range (figure 12.3b). Each retinal pair is now ready for further QPCR analysis with additional primer pairs for candidate genes.

16. QPCR primer sets should be designed to amplify 100–200 bp amplicons, and should always be checked for specificity. We have found that validated QPCR primer datasets available on the Web can be quite useful and reliable, such as Primer Bank (http://pga.mgh.harvard.edu/primerbank/). Additionally, GAPDH and/or β-actin should always be included in each run to allow for more precise normalization of sample concentrations and accurate values in test primer sets. Finally, although this simplified protocol is effective at assaying fold changes between conditions, if the actual transcript level

is to be measured, then a series of standard dilutions, addition of known amounts of template, and even multiple primer sets should be used.

17. We use the following primer sets for chick and mouse GAPDH, Hes1, and Hes5:

Chicken

    GAPDH (GenBank NM_204305)

        Forward   CATCCAAGGAGTGAGCCAAG

        Reverse   TGGAGGAAGAAATTGGAGGA

    Hes1   (c-hairy-1: GenBank AF032966)

        Forward   GAAGTCCTCCAAACCCATCA

        Reverse   AGGTGCTTCACCGTCATCTC

    Hes5   (Hes5–1, Fior and Henrique, 2005; Nelson et al., 2006: GenBank AY916777)

        Forward   CCAGAGACACCAACCCAACT

        Reverse   CAGAGCTTCTTTGAGGCACC

Mouse:

    GAPDH (GenBank NM_008084)

        Forward   GGCATTGCTCTCAATGACAA

        Reverse   CTTGCTCAGTGTCCTTGCTG

    Hes1   (GenBank NM_008235)

        Forward   CCAGCCAGTGTCAACACGA

        Reverse   AATGCCGGGAGCTATCTTTCT

    Hes5   (GenBank NM_010419)

        Forward   AGTCCCAAGGAGAAAAACCGA

        Reverse   GCTGTGTTTCAGGTAGCTGAC

PROTOCOL 6: MONITORING NOTCH SIGNALING ACTIVITY IN RETINAL PROGENITOR CELLS WITH FLUORESCENT REPORTER CONSTRUCTS

Electroporation: Chick explant cultures

1. Retinal explants are prepared from embryonic chicken, as described previously.

2. Transfection solution is prepared by mixing 10 µL of Hes5 or Hes1 cDNA [5 µg/µL] with approximately 3–4 µL of loading buffer (sterile filtered 50% glycerol/1× PBS, with Fast green for visualization): this solution increases the density of the DNA, and is enough for approximately 7–10 explants (1.5–2 µL/explant).

3. For chick explants (figure 12.2A), we found that DNA can be targeted into specific regions (central or peripheral) using standard electrodes.

4. After dissection, the retinas are submerged in 1 mL of HBSS in a well of a 24-well plate.

5. The DNA (1.5–2 µL) suspended in loading buffer is pipetted onto the retina, such that it flows over the region to be transfected.

6. Electrodes are immediately placed around the explant, and positioned to transfect the desired region.

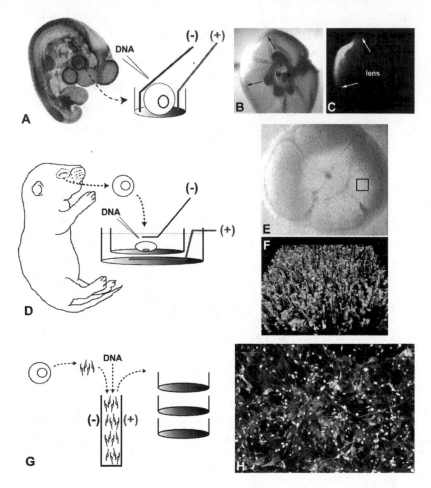

FIGURE 12.2 Strategies for transfecting retinal cells. Chick retinal explants (A), mouse retinal explants (B), and dissociated retinal cells can be transfected with GFP reporter constructs by electroporation, cultured, and used to monitor different components of active Notch signaling (Nelson, B.R. et al., Notch activity is downregulated just prior to retinal ganglion cell differentiation. *Dev Neurosci*, 2006. 28(1-2): 128-41). See text for details.

7. Electroporation conditions are as follows: 3 pulses, 25 V, 50 ms pulse length, and BTX square wave electroporater. We have found that both standard electrodes (as depicted) and TweezerTrodes (7 mm, BTX) work well.

Electroporation: Mouse explant cultures

1. Retinal explants are prepared from embryonic or newborn mouse, as described earlier.
2. For mouse explants (figure 12.2B), we have found that owing to their comparably flatter morphology, a vertical electrode system works better. We have constructed a special chamber by drilling a hole into the side of a 6-well plate, placing one electrode through this hole and shaping it such that

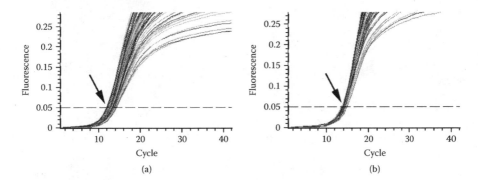

(a)   (b)

FIGURE 12.3 Quantitative polymerase chain reaction (QPCR). Control transcripts, such as GAPDH and/or β-actin, can be used to normalize relative sample concentrations by QPCR. Prenormalized samples show a broad range of C(t) values (a, arrow), but following normalization procedures, samples now fall into a much narrower range (b, arrow), facilitating quantification of target gene level differences across many samples.

it lies flat along the bottom: this will be the positive electrode, and should be exposed along the entire bottom.

3. A Millicell culture-plate insert (0.4 μm, 30 mm, Millipore) is placed into the well, and both the inner and outer chambers are filled with HBSS+ solution.
4. The mouse explant is submerged and positioned over the positive electrode such that the central, mitotic surface is up.
5. Transfection solution is pipetted over the explant (1.5–2 μL/explant), the negative electrode is immediately lowered into the well over the explant and pulsed: 3–5 pulses, 35 V, 50 μs, pulse length, and BTX square wave electroporator. We have found that 3–5 mouse explants can be transfected simultaneously by this method.
6. Transfected explants are then cultured as described in protocol 2.

Electroporation of dissociated cells

1. For dissociated cell transfection (figure 12.2C), prepare dissociated cells as described in protocol 1.
2. Transfer cells into a disposable 2 mm cuvette (approximately 400 μL total volume, BioRad), add DNA (10–20 μg), and mix thoroughly
3. We typically test a range of conditions, because chick and mouse cells have slightly different characteristics, and the specific embryonic or postnatal age of the cells is also a factor; a general starting point is 400–550 V, three pulses, 50 μs pulse length, and BTX square wave electroporator.
4. Following the electroporation, we allow cells to rest momentarily, and then divide into 3–4 wells of a 24-well plate containing coated coverslips, and culture, as described in protocol 1.

PROTOCOL 7: PARAFFIN IN SITU HYBRIDIZATION (ISH)

1. RNA Probe (Riboprobe) Synthesis
2. Tissue preparation

3. In situ hybridization
4. Antibody staining and color reaction
5. Optional post-ISH immunolabeling
6. Solutions and reagents

RiboProbe synthesis

1. We recommend 1.5–2 kb antisense riboprobes with a minimum length of 500 bp for this protocol.
2. Standard 20 µL Riboprobe Synthesis Reaction: 7 µL RNase-free H_2O, 4 µL 5× buffer, 2 µL 100 mMDTT, 3 µL 10× NTP w/DIG, 1 µL RNasin, 1 µL RNA Polymerase (T3, T7, or SP6), 2 µL linearized template DNA (phenol-extracted, ethanol-precipitated). Incubate at 37°C overnight for T3 or T7 reactions, or at 42°C overnight for SP6 reactions.
3. A DNase treatment is performed by adding 1 µL of RNase-free DNase to the synthesis reaction and incubating for 30 min at 37°C.
4. Riboprobe is then precipitated in 2.5 volumes (50 µL) chilled 100% ethanol, washed in 100 µL of 70% ethanol, and resuspended in 40 µL 0.5× TE with 0.5% SDS. Riboprobe suspension is incubated at room temperature for 15 to 30 min and gently vortexed to ensure resuspension. Riboprobe is stored at −80°C

Tissue preparation

1. Tissues are dissected in chilled HBSS+ or PBS and placed in Modified Carnoy`s fix 12–16 h at 4°C, with agitation.
2. Tissues are dehydrated in ascending solutions of RNase-free ethanol, at 4°C with agitation, 1 h each at 70, 80, 90, and 100%. Tissues are washed in fresh 100% ethanol overnight at 4°C. Tissue is stable in 100% ethanol at 4°C for up to 1 week.
3. To process tissue for paraffin embedding, it is washed twice in xylene at room temperature, with agitation. Tissue is then incubated in 3 × 1 h washes of paraffin at 65°C. Temperature should not exceed 73°C. Tissue is then oriented in paraffin and allowed to cool to room temperature. Paraffin tissue blocks can be stored at room temperature or 4°C for several years.
4. Tissue in blocks of paraffin is sectioned at 6–8 µm. Sections are floated briefly in a 50°C water bath and collected on SuperfrostPlus (Fisher) slides. Slides are allowed to dry 12 h or more at room temperature. Slides can be stored at room temperature.

In situ hybridization

1. RNase-free glassware, coverslips, and slide boxes are prepared by baking at 170°C for a minimum of 6 h.
2. Paraffin tissue slides are prepared by baking at 65–70°C for 12–24 h, to adhere tissue. Then, tissue slides are dewaxed by washing twice in

xylene and twice in 100% ethanol, for 5 min each. Slides are then air-dried with a fan at room temperature for 10 min.

3. A prehybridization step is performed by applying 120–150 µL of pre-warmed hybridization buffer onto each slide, and incubating under an RNase-free coverslip for 1 h at 68–70°C in a chamber humidified with 50% formamide and 2× SSC. After prehybridization, coverslips are removed and residual hybridization buffer is removed from slides with a lint-free tissue, avoiding contact with the tissue sections.

4. To prepare probe solution for hybridization, riboprobe is diluted into prewarmed hybridization buffer just before hybridization. Probe concentration should range from 1:100 to 1:1000, and must be optimized for each probe and tissue. Hybridization is performed by applying 120–150 µL of probe solution onto each slide and incubating under an RNase-free coverslip for 6–24 h at 68–72°C in a chamber humidified with 50% formamide and 2× SSC.

5. Slides are removed from the hybridization oven and placed in a slide rack in wash buffer at 68–72°C. Coverslips are carefully removed from the submerged slides using forceps. Slides are washed 3 × 30 min submerged in wash buffer at 68–72°C

6. Slides are washed twice in PBSTw at room temperature for 10 min each.

Antibody staining and color reaction

1. Blocking solution containing 2% Roche Blocking Reagent and 20% goat serum in PBSTw is applied to slides for 1 h in a humid chamber at room temperature.

2. Slides are incubated in an anti-DIG alkaline phosphatase-conjugated antibody (Roche Molecular Chemicals) diluted 1:2000 in blocking solution overnight at 4°C.

3. Slides are washed 4 × 30 min at room temperature with agitation in PBSTw containing 0.1% skim milk.

4. Slides are washed 2 × 10 min in NTMT to equilibrate to alkaline pH (9.0–9.5).

5. Slides are incubated in 20% NBT/BCIP liquid substrate system (Sigma Cat #B1911) diluted in NTMT for 2 to 5 h in the dark at room temperature. Substrate solution is changed every 1.5 h.

6. Slides are incubated in a bath of 2% NBT/BCIP liquid substrate system diluted in NTMT for 12 to 24 h in the dark at room temperature.

7. If further color development is required, slides are incubated in 5–50% NBT/BCIP liquid substrate system (Sigma Cat #B1911) diluted in NTMT for 2–48 h in the dark at room temperature.

8. Color reaction is stopped with PBS, and tissue is postfixed with 4% PFA for 20–30 min.

9. Post ISH immunolabeling can be performed, as described in the following text, or slides can be mounted in Fluoromount and stored at 4°C.

Post ISH immunolabeling

We find that several antibodies work well after this ISH procedure, and can be very useful in determining which cell types express a given transcript. These include progenitor cell markers, such as anti-PH3 antibodies, to mark mitotic cells (M-phase), and anti-BrdU antibodies to label progenitors during DNA replication (S-phase cells), if a pulse of BrdU was administered before sacrifice. For BrdU staining, we have found that DNase treatment can replace the standard acid treatment, and this preserves other antigens. Simply incubate slides overnight at room temp with a mixture of primary antibodies and DNaseI (1:100, stock 100 U/µL in 10 mM Tris, pH 7.5, 5 mM MgCl, 1 mM DTT, Sigma DNaseI from pancreas extract), followed by standard washes and secondary labels. Differentiation markers include antineurofilament, Tuj1, Islet1, and Calretinin. Other antibodies that have worked for us afterward include Pax6, p27, and Prox1. Epitope tags that also work well are Myc (9e10 antibody), and GFP (from University of Alberta). Counterstaining with propidium iodide or DAPI can also be carried out to label all cells in the section.

Solutions

1. PBSTw: 1 × PBS, 0.1%Tween20
2. Modified Carnoy's Fix:
       60% ethanol, 11.1% formaldehyde (30% of 37% stock), 10% glacial acetic acid
3. 10× salt solution:
       2 M NaCl (23.38 g), 0.089 M Tris HCl (2.808 g), 0.011 M Tris base (0.265 g), 0.05 M $NaH_2PO_4H_2O$ (1.38 g), 0.05 M $NaHPO_4$ (1.42 g), 0.05 M EDTA (20 mL of 0.5 M stock), and RNase-free H_2O up to 200 mL
4. 10× Yeast tRNA: 10 mg/mL in RNase-free H_2O. Aliquot and store at −80°C.
5. Hybridization buffer:
       1× salt solution, 50% deionized formamide, 10% dextran sulfate, 1× Denhardt's, 3% Roche Blocking Reagent, 0.05% Tween20, RNase-free H_2O to 45 mL, aliquot and store at −20°C. Add 1/10 volume of 10× yeast tRNA (see the previous text) right before use.
6. Wash solution:
       50% formamide, 1× SSC, 0.1% Tween20
7. NTMT:
       0.1 M NaCl (20 mL of 5 M stock), 0.1 M Tris, pH 9.5 (50 mL of 2 M stock), 0.5 M $MgCl_2$ (5 mL of 1 M stock), 0.1% Tween20 (1 mL), dH_2O up to 1 L

REFERENCES

1. Sanes, D.H., Reh, T.A., and Harris, W.A., *Development of the nervous system.* Elsevier Academic Press: Burlington, chap. 1, 2006.
2. Reh, T.A. and T. Tully, Regulation of tyrosine hydroxylase-containing amacrine cell number in larval frog retina, *Dev Biol*, 1986, 114(2): 463–9.

3. Reh, T.A., Cell-specific regulation of neuronal production in the larval frog retina, *J Neurosci*, 1987, 7(10): 3317–24.

4. Watanabe, T. and M.C. Raff, Retinal astrocytes are immigrants from the optic nerve, *Nature*, 1988, 332(6167): 834–7.

5. Reh, T.A., Cellular interactions determine neuronal phenotypes in rodent retinal cultures, *J Neurobiol*, 1992, 23(8): 1067–83.

6. Anchan, R.M. et al., EGF and TGF-alpha stimulate retinal neuroepithelial cell proliferation in vitro, *Neuron*, 1991, 6(6): 923–36.

7. Lillien, L. and C. Cepko, Control of proliferation in the retina: temporal changes in responsiveness to FGF and TGF alpha, *Development*, 1992, 115(1): 253–66.

8. Kelley, M.W., J.K. Turner, and T.A. Reh, Retinoic acid promotes differentiation of photoreceptors in vitro, *Development*, 1994, 120(8): 2091–102.

9. Levine, E.M., S. Fuhrmann, and T.A. Reh, Soluble factors and the development of rod photoreceptors, *Cell Mol Life Sci*, 2000, 57(2): 224–34.

10. Anchan, R.M. and T.A. Reh, Transforming growth factor-beta-3 is mitogenic for rat retinal progenitor cells in vitro, *J Neurobiol*, 1995, 28(2): 133–45.

11. Cohen, S., The stimulation of epidermal proliferation by a specific protein (EGF), *Dev Biol*, 1965, 12(3): 394–407.

12. Citri, A. and Y. Yarden, EGF-ERBB signalling: towards the systems level, *Nat Rev Mol Cell Biol*, 2006, 7(7): 505–16.

13. Wong, R.W. and L. Guillaud, The role of epidermal growth factor and its receptors in mammalian CNS, *Cytokine Growth Factor Rev*, 2004, 15(2–3): 147–56.

14. Taylor, M. and T.A. Reh, Induction of differentiation of rat retinal, germinal, neuroepithelial cells by dbcAMP, *J Neurobiol*, 1990, 21(3): 470–81.

15. Reh, T.A. and E.M. Levine, Multipotential stem cells and progenitors in the vertebrate retina, *J Neurobiol*, 1998, 36(2): 206–20.

16. Close, J.L. et al., Epidermal growth factor receptor expression regulates proliferation in the postnatal rat retina, *Glia*, 2006, 54(2): 94–104.

17. Rhee, K.D. et al., Cytokine-induced activation of signal transducer and activator of transcription in photoreceptor precursors regulates rod differentiation in the developing mouse retina, *J Neurosci*, 2004, 24(44): 9779–88.

18. Lillien, L. and D. Wancio, Changes in epidermal growth factor receptor expression and competence to generate glia regulate timing and choice of differentiation in the retina, *Mol Cell Neurosci*, 1998, 10(5–6): 296–308.

19. Itoh, N. and D.M. Ornitz, Evolution of the Fgf and Fgfr gene families, *Trends Genet*, 2004, 20(11): 563–9.

20. Martinez, S., The isthmic organizer and brain regionalization, *Int J Dev Biol*, 2001, 45(1): 367–71.

21. Pittack, C., G.B. Grunwald, and T.A. Reh, Fibroblast growth factors are necessary for neural retina but not pigmented epithelium differentiation in chick embryos, *Development*, 1997, 124(4): 805–16.

22. Crossley, P.H. and G.R. Martin, The mouse Fgf8 gene encodes a family of polypeptides and is expressed in regions that direct outgrowth and patterning in the developing embryo, *Development*, 1995, 121(2): 439–51.

23. McCabe, K.L., E.C. Gunther, and T.A. Reh, The development of the pattern of retinal ganglion cells in the chick retina: mechanisms that control differentiation, *Development*, 1999, 126(24): 5713–24.

24. Herzog, W. et al., Fgf3 signaling from the ventral diencephalon is required for early specification and subsequent survival of the zebrafish adenohypophysis, *Development*, 2004, 131(15): 3681–92.

25. Reifers, F. et al., Overlapping and distinct functions provided by fgf17, a new zebrafish member of the Fgf8/17/18 subgroup of Fgfs, *Mech Dev*, 2000, 99(1–2): 39–49.

26. Vogel-Hopker, A. et al., Multiple functions of fibroblast growth factor-8 (FGF-8) in chick eye development, *Mech Dev*, 2000, 94(1–2): 25–36.
27. Walshe, J. and I. Mason, Unique and combinatorial functions of Fgf3 and Fgf8 during zebrafish forebrain development, *Development*, 2003, 130(18): 4337–49.
28. Martinez-Morales, J.R. et al., Differentiation of the vertebrate retina is coordinated by an FGF signaling center, *Dev Cell*, 2005, 8(4): 565–74.
29. McCabe, K.L., C. McGuire, and T.A. Reh, Pea3 expression is regulated by FGF signaling in developing retina, *Dev Dyn*, 2006, 235(2): 327–35.
30. Fischer, A.J., B.D. Dierks, and T.A. Reh, Exogenous growth factors induce the production of ganglion cells at the retinal margin, *Development*, 2002, 129(9): 2283–91.
31. Brown, S.P., S. He, and R.H. Masland, Receptive field microstructure and dendritic geometry of retinal ganglion cells, *Neuron*, 2000, 27(2): 371–83.
32. Yancopoulos, G.D. et al., Vascular-specific growth factors and blood vessel formation, *Nature*, 2000, 407(6801): 242–8.
33. Yang, K. and C.L. Cepko, Flk-1, a receptor for vascular endothelial growth factor (VEGF), is expressed by retinal progenitor cells, *J Neurosci*, 1996, 16(19): 6089–99.
34. Yourey, P.A. et al., Vascular endothelial cell growth factors promote the in vitro development of rat photoreceptor cells, *J Neurosci*, 2000, 20(18): 6781–8.
35. Hashimoto, T. et al., VEGF activates divergent intracellular signaling components to regulate retinal progenitor cell proliferation and neuronal differentiation, *Development*, 2006, 133(11): 2201–10.
36. Haigh, J.J. et al., Cortical and retinal defects caused by dosage-dependent reductions in VEGF-A paracrine signaling, *Dev Biol*, 2003, 262(2): 225–41.
37. Das, I. et al., Trk C signaling is required for retinal progenitor cell proliferation, *J Neurosci*, 2000, 20(8): 2887–95.
38. Harada, C. et al., Effect of p75NTR on the regulation of naturally occurring cell death and retinal ganglion cell number in the mouse eye, *Dev Biol*, 2006, 290(1): 57–65.
39. Ingham, P.W. and M. Placzek, Orchestrating ontogenesis: variations on a theme by Sonic hedgehog, *Nat Rev Genet*, 2006, 7(11): 841–50.
40. Macdonald, R. et al., Midline signalling is required for Pax gene regulation and patterning of the eyes, *Development*, 1995, 121(10): 3267–78.
41. Chiang, C. et al., Cyclopia and defective axial patterning in mice lacking Sonic hedgehog gene function, *Nature*, 1996, 383(6599): 407–13.
42. Levine, E.M. et al., Sonic hedgehog promotes rod photoreceptor differentiation in mammalian retinal cells in vitro, *J Neurosci*, 1997, 17(16): 6277–88.
43. Jensen, A.M. and V.A. Wallace, Expression of Sonic hedgehog and its putative role as a precursor cell mitogen in the developing mouse retina, *Development*, 1997, 124(2): 363–71.
44. Wang, Y.P. et al., Development of normal retinal organization depends on Sonic hedgehog signaling from ganglion cells, *Nat Neurosci*, 2002, 5(9): 831–2.
45. Wang, Y. et al., Retinal ganglion cell-derived Sonic hedgehog locally controls proliferation and the timing of RGC development in the embryonic mouse retina, *Development*, 2005, 132(22): 5103–13.
46. Zhang, X.M. and X.J. Yang, Regulation of retinal ganglion cell production by Sonic hedgehog, *Development*, 2001, 128(6): 943–57.
47. Stenkamp, D.L. et al., Function for Hedgehog genes in zebrafish retinal development, *Dev Biol*, 2000, 220(2): 238–52.
48. Stenkamp, D.L. et al., Embryonic retinal gene expression in sonic-you mutant zebrafish, *Dev Dyn*, 2002, 225(3): 344–50.
49. Spence, J.R. et al., The hedgehog pathway is a modulator of retina regeneration, *Development*, 2004, 131(18): 4607–21.

50. Moshiri, A. and T.A. Reh, Persistent progenitors at the retinal margin of ptc+/- mice, *J Neurosci*, 2004, 24(1): 229–37.

51. Moshiri, A., C.R. McGuire, and T.A. Reh, Sonic hedgehog regulates proliferation of the retinal ciliary marginal zone in posthatch chicks, *Dev Dyn*, 2005, 233(1): 66–75.

52. Perron, M. et al., A novel function for Hedgehog signalling in retinal pigment epithelium differentiation, *Development*, 2003, 130(8): 1565–77.

53. Furimsky, M. and V.A. Wallace, Complementary Gli activity mediates early patterning of the mouse visual system, *Dev Dyn*, 2006, 235(3): 594–605.

54. Massague, J. and R.R. Gomis, The logic of TGFbeta signaling, *FEBS Lett*, 2006, 580(12): 2811–20.

55. Penton, A., S.B. Selleck, and F.M. Hoffmann, Regulation of cell cycle synchronization by decapentaplegic during Drosophila eye development, *Science*, 1997, 275(5297): 203–6.

56. Horsfield, J. et al., Decapentaplegic is required for arrest in G1 phase during Drosophila eye development, *Development*, 1998, 125(24): 5069–78.

57. Fischer, A.J. et al., BMP4 and CNTF are neuroprotective and suppress damage-induced proliferation of Muller glia in the retina, *Mol Cell Neurosci*, 2004, 27(4): 531–42.

58. Davis, A.A., M.M. Matzuk, and T.A. Reh, Activin A promotes progenitor differentiation into photoreceptors in rodent retina, *Mol Cell Neurosci*, 2000, 15(1): 11–21.

59. Roberts, V.J. et al., Hybridization histochemical and immunohistochemical localization of inhibin/activin subunits and messenger ribonucleic acids in the rat brain, *J Comp Neurol*, 1996, 364(3): 473–93.

60. Close, J.L., B. Gumuscu, and T.A. Reh, Retinal neurons regulate proliferation of postnatal progenitors and Muller glia in the rat retina via TGF beta signaling, *Development*, 2005, 132(13): 3015–26.

61. Obata, H. et al., Expression of transforming growth factor-beta superfamily receptors in rat eyes, *Acta Ophthalmol Scand*, 1999, 77(2): 151–6.

62. Belecky-Adams, T. and R. Adler, Developmental expression patterns of bone morphogenetic proteins, receptors, and binding proteins in the chick retina, *J Comp Neurol*, 2001, 430(4): 562–72.

63. Zhang, X.M. and X.J. Yang, Temporal and spatial effects of Sonic hedgehog signaling in chick eye morphogenesis, *Dev Biol*, 2001, 233(2): 271–90.

64. Murali, D. et al., Distinct developmental programs require different levels of Bmp signaling during mouse retinal development, *Development*, 2005, 132(5): 913–23.

65. Stahl, N. and G.D. Yancopoulos, The tripartite CNTF receptor complex: activation and signaling involves components shared with other cytokines, *J Neurobiol*, 1994, 25(11): 1454–66.

66. LaVail, M.M. et al., Protection of mouse photoreceptors by survival factors in retinal degenerations, *Invest Ophthalmol Vis Sci*, 1998, 39(3): 592–602.

67. Bonni, A. et al., Regulation of gliogenesis in the central nervous system by the JAK-STAT signaling pathway, *Science*, 1997, 278(5337): 477–83.

68. Rajan, P. and R.D. McKay, Multiple routes to astrocytic differentiation in the CNS, *J Neurosci*, 1998, 18(10): 3620–9.

69. Nakashima, K. et al., Astrocyte differentiation mediated by LIF in cooperation with BMP2, *FEBS Lett*, 1999, 457(1): 43–6.

70. Neophytou, C. et al., Muller-cell-derived leukaemia inhibitory factor arrests rod photoreceptor differentiation at a postmitotic pre-rod stage of development, *Development*, 1997, 124(12): 2345–54.

71. Kirsch, M. et al., Ciliary neurotrophic factor blocks rod photoreceptor differentiation from postmitotic precursor cells in vitro, *Cell Tissue Res*, 1998, 291(2): 207–16.

72. Graham, D.R., P.A. Overbeek, and J.D. Ash, Leukemia inhibitory factor blocks expression of Crx and Nrl transcription factors to inhibit photoreceptor differentiation, *Invest Ophthalmol Vis Sci*, 2005, 46(7): 2601–10.

73. Goureau, O., K.D. Rhee, and X.J. Yang, Ciliary neurotrophic factor promotes Muller glia differentiation from the postnatal retinal progenitor pool, *Dev Neurosci*, 2004, 26(5–6): 359–70.

74. Arias, A.M., New alleles of Notch draw a blueprint for multifunctionality, *Trends Genet*, 2002, 18(4): 168–70.

75. Blaumueller, C.M. et al., Intracellular cleavage of Notch leads to a heterodimeric receptor on the plasma membrane, *Cell*, 1997, 90(2): 281–91.

76. Logeat, F. et al., The Notch1 receptor is cleaved constitutively by a furin-like convertase, *Proc Natl Acad Sci U.S.A.*, 1998, 95(14): 8108–12.

77. Seugnet, L., P. Simpson, and M. Haenlin, Requirement for dynamin during Notch signaling in Drosophila neurogenesis, *Dev Biol*, 1997, 192(2): 585–98.

78. Parks, A.L. et al., Ligand endocytosis drives receptor dissociation and activation in the Notch pathway, *Development*, 2000, 127(7): 1373–85.

79. Chitnis, A., Why is delta endocytosis required for effective activation of notch? *Dev Dyn*, 2006, 235(4): 886–94.

80. Brown, M.S. et al., Regulated intramembrane proteolysis: a control mechanism conserved from bacteria to humans, *Cell*, 2000, 100(4): 391–8.

81. Brou, C. et al., A novel proteolytic cleavage involved in Notch signaling: the role of the disintegrin-metalloprotease TACE, *Mol Cell*, 2000, 5(2): 207–16.

82. Mumm, J.S. et al., A ligand-induced extracellular cleavage regulates gamma-secretase-like proteolytic activation of Notch1, *Mol Cell*, 2000, 5(2): 197–206.

83. Vooijs, M. et al., Ectodomain shedding and intramembrane cleavage of mammalian Notch proteins is not regulated through oligomerization, *J Biol Chem*, 2004, 279(49): 50864–73.

84. Schroeter, E.H., J.A. Kisslinger, and R. Kopan, Notch-1 signalling requires ligand-induced proteolytic release of intracellular domain, *Nature*, 1998, 393(6683): 382–6.

85. Chyung, J.H., D.M. Raper, and D.J. Selkoe, Gamma-secretase exists on the plasma membrane as an intact complex that accepts substrates and effects intramembrane cleavage, *J Biol Chem*, 2005, 280(6): 4383–92.

86. Wallberg, A.E. et al., p300 and PCAF act cooperatively to mediate transcriptional activation from chromatin templates by notch intracellular domains in vitro, *Mol Cell Biol*, 2002, 22(22): 7812–9.

87. Tun, T. et al., Recognition sequence of a highly conserved DNA binding protein RBP-J kappa, *Nucl Acid Res*, 1994, 22(6): 965–71.

88. Chen, H. et al., Conservation of the Drosophila lateral inhibition pathway in human lung cancer: a hairy-related protein (HES-1) directly represses achaete-scute homolog-1 expression, *Proc Natl Acad Sci U.S.A.*, 1997, 94(10): 5355–60.

89. Sasai, Y. et al., Two mammalian helix-loop-helix factors structurally related to Drosophila hairy and Enhancer of split, *Genes Dev*, 1992, 6(12B): 2620–34.

90. Oberg, C. et al., The Notch intracellular domain is ubiquitinated and negatively regulated by the mammalian Sel-10 homolog, *J Biol Chem*, 2001, 276(38): 35847–53.

91. McGill, M.A. and C.J. McGlade, Mammalian numb proteins promote Notch1 receptor ubiquitination and degradation of the Notch1 intracellular domain, *J Biol Chem*, 2003, 278(25): 23196–203.

92. Hirata, H. et al., Oscillatory expression of the bHLH factor Hes1 regulated by a negative feedback loop, *Science*, 2002, 298(5594): 840–3.

93. King, I.N. et al., Hrt and Hes negatively regulate Notch signaling through interactions with RBP-Jkappa, *Biochem Biophys Res Commun*, 2006, 345(1): 446–52.

94. Takebayashi, K. et al., Structure, chromosomal locus, and promoter analysis of the gene encoding the mouse helix-loop-helix factor HES-1: Negative autoregulation through the multiple N box elements, *J Biol Chem*, 1994, 269(7): 5150–6.

95. Nishimura, M. et al., Structure, chromosomal locus, and promoter of mouse Hes2 gene, a homologue of Drosophila hairy and Enhancer of split, *Genomics*, 1998, 49(1): 69–75.

96. Cave, J.W. et al., A DNA transcription code for cell-specific gene activation by notch signaling, *Curr Biol*, 2005, 15(2): 94–104.

97. Castro, B. et al., Lateral inhibition in proneural clusters: cis-regulatory logic and default repression by Suppressor of Hairless, *Development*, 2005, 132(15): 3333–44.

98. Lamar, E. and C. Kintner, The Notch targets Esr1 and Esr10 are differentially regulated in Xenopus neural precursors, *Development*, 2005, 132(16): 3619–30.

99. Ong, C.T. et al., Target selectivity of vertebrate notch proteins: Collaboration between discrete domains and CSL-binding site architecture determines activation probability, *J Biol Chem*, 2006, 281(8): 5106–19.

100. Shimizu, K. et al., Functional diversity among Notch1, Notch2, and Notch3 receptors. *Biochem Biophys Res Commun*, 2002, 291(4): 775–9.

101. Akazawa, C. et al., Molecular characterization of a rat negative regulator with a basic helix-loop-helix structure predominantly expressed in the developing nervous system, *J Biol Chem*, 1992, 267(30): 21879–85.

102. Takebayashi, K. et al., Structure and promoter analysis of the gene encoding the mouse helix-loop-helix factor HES-5. Identification of the neural precursor cell-specific promoter element, *J Biol Chem*, 1995, 270(3): 1342–9.

103. Bae, S. et al., The bHLH gene Hes6, an inhibitor of Hes1, promotes neuronal differentiation, *Development*, 2000, 127(13): 2933–43.

104. Koyano-Nakagawa, N, et al., Hes6 acts in a positive feedback loop with the neurogenins to promote neuronal differentiation, *Development*, 2000, 127(19): 4203–16.

105. Fior, R. and D. Henrique, A novel hes5/hes6 circuitry of negative regulation controls Notch activity during neurogenesis, *Dev Biol*, 2005, 281(2): 318–33.

106. Reh, T.A. and I.J. Kljavin, Age of differentiation determines rat retinal germinal cell phenotype: induction of differentiation by dissociation, *J Neurosci*, 1989, 9(12): 4179–89.

107. Adler, R. and M. Hatlee, Plasticity and differentiation of embryonic retinal cells after terminal mitosis, *Science*, 1989, 243(4889): 391–3.

108. Altshuler, D. and C. Cepko, A temporally regulated, diffusible activity is required for rod photoreceptor development in vitro, *Development*, 1992, 114(4): 947–57.

109. Lewis, J., Neurogenic genes and vertebrate neurogenesis, *Curr Opin Neurobiol*, 1996, 6(1): 3–10.

110. Lowell, S. et al., Stimulation of human epidermal differentiation by delta-notch signalling at the boundaries of stem-cell clusters, *Curr Biol*, 2000, 10(9): 491–500.

111. Dorsky, R.I., D.H. Rapaport, and W.A. Harris, Xotch inhibits cell differentiation in the Xenopus retina, *Neuron*, 1995, 14(3): 487–96.

112. Austin, C.P. et al., Vertebrate retinal ganglion cells are selected from competent progenitors by the action of Notch, *Development*, 1995, 121(11): 3637–50.

113. Tomita, K. et al., Mash1 promotes neuronal differentiation in the retina, *Genes Cells*, 1996, 1(8): 765–74.

114. Henrique, D. et al., Maintenance of neuroepithelial progenitor cells by Delta-Notch signalling in the embryonic chick retina, *Curr Biol*, 1997, 7(9): 661–70.

115. Dorsky, R.I. et al., Regulation of neuronal diversity in the Xenopus retina by Delta signalling, *Nature*, 1997, 385(6611): 67–70.

116. Furukawa, T. et al., rax, Hes1, and notch1 promote the formation of Muller glia by postnatal retinal progenitor cells, *Neuron*, 2000, 26(2): 383–94.

117. Hojo, M. et al., Glial cell fate specification modulated by the bHLH gene Hes5 in mouse retina, *Development*, 2000, 127(12): 2515–22.
118. Satow, T. et al., The basic helix-loop-helix gene hesr2 promotes gliogenesis in mouse retina, *J Neurosci*, 2001, 21(4): 1265–73.
119. Silva, A.O., C.E. Ercole, and S.C. McLoon, Regulation of ganglion cell production by Notch signaling during retinal development, *J Neurobiol*, 2003, 54(3): 511–24.
120. Takatsuka, K. et al., Roles of the bHLH gene Hes1 in retinal morphogenesis, *Brain Res*, 2004, 1004(1–2): 148–55.
121. Nelson, B.R. et al., Notch activity is downregulated just prior to retinal ganglion cell differentiation, *Dev Neurosci*, 2006, 28(1–2): 128–41.
122. Yaron, O. et al., Notch1 functions to suppress cone-photoreceptor fate specification in the developing mouse retina, *Development*, 2006, 133(7): 1367–78.
123. Jadhav, A.P., H.A. Mason, and C.L. Cepko, Notch 1 inhibits photoreceptor production in the developing mammalian retina, *Development*, 2006, 133(5): 913–23.
124. Lindsell, C.E. et al., Expression patterns of Jagged, Delta1, Notch1, Notch2, and Notch3 genes identify ligand-receptor pairs that may function in neural development, *Mol Cell Neurosci*, 1996, 8(1): 14–27.
125. Bao, Z.Z. and C.L. Cepko, The expression and function of Notch pathway genes in the developing rat eye, *J Neurosci*, 1997, 17(4): 1425–34.
126. Kitamoto, T. et al., Functional redundancy of the Notch gene family during mouse embryogenesis: analysis of Notch gene expression in Notch3-deficient mice, *Biochem Biophys Res Commun*, 2005, 331(4): 1154–62.
127. Benedito, R. and A. Duarte, Expression of Dll4 during mouse embryogenesis suggests multiple developmental roles, *Gene Expr Patterns*, 2005, 5(6): 750–5.
128. Nelson, B.R. and T.A. Reh, unpublished data, 2006.
129. Hofmann, J.J. and M. Luisa Iruela-Arispe, Notch expression patterns in the retina: An eye on receptor-ligand distribution during angiogenesis, *Gene Expr Patterns*, 2006, 7: 461–470.
130. Conlon, R.A., A.G. Reaume, and J. Rossant, Notch1 is required for the coordinate segmentation of somites, *Development*, 1995, 121(5): 1533–45.
131. de la Pompa, J.L. et al., Conservation of the Notch signalling pathway in mammalian neurogenesis, *Development*, 1997, 124(6): 1139–48.
132. Mason, H.A. et al., Notch signaling coordinates the patterning of striatal compartments, *Development*, 2005, 132(19): 4247–58.
133. Parisi, V. et al., Early visual function impairment in CADASIL, *Neurology*, 2003, 60(12): 2008–10.
134. Hatakeyama, J. et al., Hes genes regulate size, shape and histogenesis of the nervous system by control of the timing of neural stem cell differentiation, *Development*, 2004, 131(22): 5539–50.
135. Saxena, M.T. et al., Murine notch homologs (N1-4) undergo presenilin-dependent proteolysis, *J Biol Chem*, 2001, 276(43): 40268–73.
136. Geling, A. et al., A gamma-secretase inhibitor blocks Notch signaling in vivo and causes a severe neurogenic phenotype in zebrafish, *EMBO Rep*, 2002, 3(7): 688–94.
137. Micchelli, C.A. et al., Gamma-secretase/presenilin inhibitors for Alzheimer's disease phenocopy Notch mutations in Drosophila, *Faseb J*, 2003, 17(1): 79–81.
138. Medina, M. and C.G. Dotti, RIPped out by presenilin-dependent gamma-secretase, *Cell Signal*, 2003, 15(9): 829–41.

13 Probing the Hedgehog Pathway in Retinal Development

Brian McNeill, Dana Wall,
Shawn Beug, and Valerie A. Wallace

CONTENTS

13.1   OVERVIEW

A major area in eye development research is the identification of genetic pathways that regulate eye morphogenesis. The Hedgehog (Hh) family of signaling molecules is among the cell-extrinsic factors that govern retinal development. The Hh pathway exhibits a remarkable degree of interspecies conservation and is crucial for eye development in species as diverse as fly, fish, and rodents. Extraretinal Hh secreted from the ventral midline plays a pivotal role in regulating eye morphogenesis. Neuron-derived Hh signals endogenous to the retina function as short-range signals that exert local effects on retinal progenitor cells, as well as extraretinal effects on axon guidance and optic nerve development. In this chapter we overview the function of the Hh signaling pathway in eye development.

13.2   Hh SIGNALING IN EYE DEVELOPMENT

13.2.1   HH PATHWAY

13.2.1.1   Hh: An Important Molecule for Metazoan Development

Hh was initially identified through a *Drosophila* mutagenesis screen as a segment polarity gene that when mutated resulted in an alteration in the denticle patterning in larvae, yielding an overall nonsegmented "hairy" phenotype that resembled a hedgehog (1). Subsequent discovery and characterization of the mouse orthologs *Sonic*, *Desert*, and *Indian* Hedgehog (*Shh*, *Dhh*, and *Ihh*, respectively) showed high conservation to *hh*, and these orthologs also have important roles in vertebrate embryonic development.

Hh proteins encode secreted intercellular proteins that regulate diverse developmental events through morphogenic and spatiotemporal regulation of downstream patterning genes. They mediate their effects on patterning through short- and long-range signaling. In the case of the latter, Hh proteins can be distributed in a gradient and function as classical morphogens. In the spinal cord, for example, Hh induces different cell fates along the dorsal ventral axis through concentration-dependent regulation of transcription factor expression (2–7).

13.2.1.2   Hh Production and Extracellular Transport

Hh proteins undergo extensive posttranslational modifications. All of the patterning activity of the Hh proteins resides in the NH_2-terminal half of the protein (Hh-N), which is generated by proteolysis catalyzed by amino acid residues in the carboxy domain of Hh (Hh-C). Subsequently, a cholesterol moiety is covalently added at the carboxy terminal of Hh-N (8). Addition of a palmitate group at the NH_2 terminal of Hh-N is catalyzed by an acyltransferase to generate a dual lipid modified form of the protein, Hh-Np (9–12). The cholesterol modification is essential for membrane tethering, as well as long-range patterning mediated by Shh, whereas palmitoylation results in the production of a more potent form of Hh and promotes long-range transport of the protein in tissues (11,13–17). For simplicity, we will use Hh when referring to lipid-modified Hh-Np.

A number of other gene products play an important role in regulating movement of Hh proteins in tissues. Secretion of Hh proteins requires the activity of a multipass transmembrane protein Dispatched (Disp). Disp contains a sterol-sensing domain that is known to be involved in cholesterol transport and is necessary for secretion of cholesterol-modified Hh (18). The requirement for Disp for Hh patterning is highly conserved, as Hh patterning is severely disrupted in *Disp* mutant flies and mice (8,19–21). Perturbation of heparin sulfate proteoglycan (HSPG) biosynthesis and function in both *Drosophila* and mouse has revealed a requirement for these acidic glycoproteins in both short- and long-range Hh activity (22–29). Hh proteins can also bind other plasma membrane-associated proteins, resulting in either positive or negative feedback of Hh signaling. The transmembrane proteins iHog and Boi (Cdo and Boc in mammals) (30–33) increase the binding affinity of Hh to its receptor Patched (Ptc), resulting in amplification of the Hh signal. In contrast, Megalin (gp330), a member of the low-density lipoprotein receptor family, Gas1, and Hh-interacting protein (Hip) mediate endocytosis of Hh proteins, targeting them for degradation or transcytosis, thus reducing the long-range movement of Hh proteins (34–36).

13.2.1.3   Hh Signal Transduction

Activation of Hh signaling requires several factors that are involved in both negative and positive feedback mechanisms (figure 13.1; reviewed in references 37,38). In the absence of Hh protein, the Hh receptor Ptc, a 12-transmembrane protein, antagonizes the activity of Smoothened (Smo), a 7-transmembrane protein that is the obligate signal transduction component of the Hh pathway. Smo controls the activity of the cubitous interruptus (ci in *Drosophila*) Zinc finger domain transcription factors that regulate target gene expression downstream of Smo. In mammals, Hh target gene induction is controlled by 3 ci homologues, Gli1, Gli2, and Gli3, where Gli2 and Gli3 possess both activator and repressor activities while Gli1 acts solely as a transcriptional activator and serves to amplify the Hh signal. When the pathway is inactive, ci/Gli inactivated by phosphorylation and, in the case of ci and Gli3, by cleavage to shorter transcriptional repressor forms of the proteins. Hh binding to Ptc derepresses Smo activity, which is associated with Smo phosphorylation and stabilization at the plasma membrane. Smo activation promotes Gli-mediated Hh target gene activation by repressing phosphorylation and cleavage and promoting nuclear translocation of transcriptional activator forms of Gli, in part by inducing Gli dissociation from *Suppressor of Fused (Su(Fu))*. Ultimately, Gli induces expression of Hh target genes such as *Ptc* and *Gli1*, with *Ptc* involved in a negative feedback

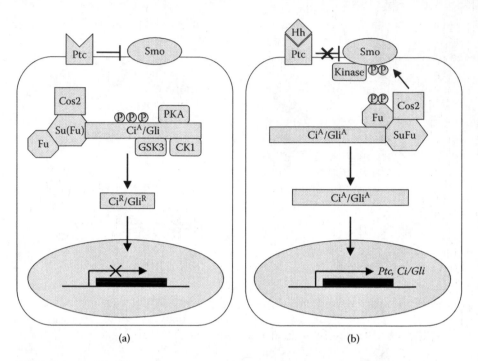

FIGURE 13.1 The Hh signal transduction pathway. (a) In the absence of the Hh ligand, Ptc represses the activity of Smo. As a consequence, ci/Gli is associated with Cos2, Fu, and SuFu, which phosphorylates Ci/Gli through recruitment of kinases CK1, GSk3, and PKA. Phosphorylated ci/Gli is targeted for cleavage into its repressor form and is translocated into the nucleus, where it functions as a transcriptional repressor. (b) Hh binding to Ptc relieves the repressive effect of Ptc on Smo, allowing Smo to be phosphorylated by the action of GRK in vertebrates and PKA and CK1 in *Drosophila*. Phosphorylated Smo associates with Cos2 (*Drosophila*), ultimately resulting in blockade of ci/Gli phosphorylation with the full-length ci/Gli isoform shuttled into the nucleus to activate Hh target genes. Moreover, inhibition of SuFu by the activation of Fused is required for maximal Gli transactivation. Hh, hedgehog; Ptc, patched; Smo, smoothened; Ci, cubitus interruptus; PKA, protein kinase A; Cos2, Costal-2; CK1, Casein kinase 1; GSK3, glycogen synthase kinase 3; GRK, G-protein-coupled receptor kinase 2; Fu, fused; SuFu, suppressor of fused; Gli, Glioma-associated oncogene.

signal. Thus, Hh-responding cells can be identified by the presence of *Ptc* and *Gli* transcripts. Although the Hh orthologs exhibit different spatiotemporal expression patterns, they all bind to the same receptors and activate the same signal transduction pathway.

13.2.2 EARLY EYE PATTERNING AND REQUIREMENT FOR HH SIGNALING

13.2.2.1 Eye Morphogenesis in Mammals

The first morphological appearance of eye development in the mouse begins at embryonic day 9 (E9) as bilateral evaginations, termed optic vesicles, form in the neural tube at the level of the ventral diencephalon (figure 13.2). The optic vesicles grow out laterally and contact the surface ectoderm, inducing the formation of the

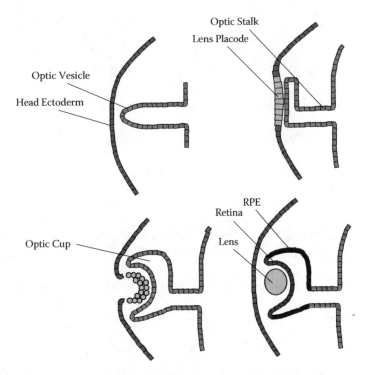

FIGURE 13.2 (See accompanying color CD.) Eye morphogenesis. The optic vesicle is formed by the evagination of the forebrain. It grows laterally to make contact with the head ectoderm, and induces the formation of the lens placode. Through invagination, the lens placode forms the lens, while the optic vesicle forms the optic cup. As the optic cup differentiates, the inner layer forms the retina, and the outer layer forms the retinal pigmented epithelium.

lens. At this stage, the vesicles invaginate, resulting in the formation of a proximal optic stalk and a distal bilayered cup; the outer layer of the cup differentiates as the retinal pigment epithelium (RPE), and the inner layer differentiates as the neural retina. The developing eyecup remains attached to the rest of the brain via the optic stalk, which will become the optic nerve once it is invaded by blood vessels and the axons of retinal ganglion cells (RGC).

13.2.2.2 Shh Expression from the Midline and the Establishment of Bilateral Optic Vesicles

Shh signaling from the axial mesoderm (prechordal plate and notochord) is required for the induction of midline structures along the rostral–caudal axis of the central nervous system. Misexpression studies in *Xenopus* and zebrafish revealed that midline-derived *Shh* is necessary for normal forebrain patterning (39–42). Similarly, the ventral midline fails to invaginate in mice with defects in the Shh pathway, leading to the loss of midline structures and the formation of a contiguous forebrain, a condition known as holoprosencephaly (HPE) (19,43–46). HPE arises from the failure of the embryonic telencephalon to completely separate into individual cerebral hemispheres. *Shh* mutants frequently display the most severe form of HPE, cyclopia, in which the ventral diencephalon precursor cells completely fail to fully bifurcate into a bilateral eye field, generating a single optic vesicle. Coincident with the critical role of Gli3 in vertebrate development, telencephalic midline formation occurs in the absence of *Gli1* and *Gli2*, and the majority of the dorsal telencephalon is ventralized in the *Gli3* mouse mutant (47–50). Thus, the main role of Shh in the ventral forebrain may be to antagonize Gli3 function to enable the establishment of the bilateral eye fields (50).

Shh expression from the ventral midline is dependent on upstream Nodal signaling. In the *cyclops* (*cyc*) mutant zebrafish, the ventral forebrain fails to develop, giving rise to cyclopia (39,51–54). *Shh* appears to be a direct target of the nodal-related gene *cyc*. In *cyc* mutants, Shh expression in the forebrain is absent (42), and *Shh* expression in the ventral brain and floorplate can be rescued with microinjection of *cyc* mRNA. Similarly, electroporation of *cyc* downstream genes into the chicken neural tube demonstrated that the Shh promoter is responsive to *cyc* signaling. Therefore, *Nodal* signaling is necessary for the initiation of *Shh* in the prechordal plate.

13.2.2.3   Shh Signaling and Proximal-Distal Patterning of the Optic Vesicle

In addition to its role in the establishment of the bilateral eye field, midline-derived Shh signals are also important for patterning of the optic vesicle along the proximodistal axis. Perturbation of signaling from axial tissues to neuroepithelial cells results in conversion of the optic stalk to the neural retina (53,54). The boundary between the presumptive optic stalk and neural retina is established through reciprocal transcriptional repression between homeodomain transcription factors, including Pax6, Pax2, Vax1, and Vax2 (55–57). Pax6 has vital roles in retinal cell fate specification, and its expression is largely excluded from the optic stalk (58,59). The optic stalk specific genes *Pax2* and *Vax2* are sensitive to Shh dosage and are restricted to the posterior-ventral half of the optic vesicle.

Multiple experiments have revealed a requirement for Shh in optic cup morphogenesis. In *cyc* zebrafish, the optic stalk marker Pax2 is absent, resulting in a loss of the proximal cell population (39). Similarly, loss of the *Shh* pathway in mice and chick leads to the absence of *Pax2* and the optic stalk with an expansion of the optic cup (60,61). On the other hand, overexpression of *Shh* expands the *Pax2* and *Vax1* expression domains with a concomitant reduction of *Pax6* expression in the optic cup, resulting in expansion of the ventral optic stalk tissue at the expense of the retina (39,54,55,61,62). Moreover, optic stalk fates are induced adjacent to regions of highest Shh activity. Specifically, the proximodistal fate is specified by transcriptional repression between *Pax6* and *Vax2*, which is regulated by Shh-dependent subcellular localization of Vax2 (61–65). In the absence of Shh, Vax2 is translocated to the cytoplasm, relieving Vax2-mediated repression of the critical *Pax6* enhancer (65). In the presence of Shh, Vax2 is retained in the nucleus and represses *Pax6* promoter activity (65). Thus, Vax2-positive cells close to the midline source of Shh adopt an astroglial cell fate. Lastly, maintenance of dorsoventral components of the optic vesicle is regulated by the antagonistic interaction between ventrally derived Shh and dorsally derived BMP4 (61,66–68). Taken together, these results indicate that specification of optic stalk and retinal tissue is regulated through interaction between the BMP and Shh pathways coupled with temporal and spatial regulation of Shh-dependent Vax2 subcellular localization that demarks the Pax6 territory whereby cells exposed to Shh specify the optic stalk and the rest of the optic vesicle develops into the neural retina.

13.2.3 RETINAL DEVELOPMENT

The neural retina is derived from the inner epithelial layer of the optic cup and consists of six types of neurons and one glial cell type that are arranged in three distinct nuclear layers: the outer nuclear layer (ONL), inner nuclear layer (INL), and ganglion cell layer (GCL) (figure 13.3). The different cell types are generated in a phylogenetically conserved but overlapping order with retinal ganglion cells (RGCs) and horizontal cells exiting the cell cycle first, followed in overlapping phases by cone-photoreceptors, amacrine cells, rod-photoreceptors, bipolar cells and, finally, Müller glial cells [figure 13.4b; (69,70)]. It is important to emphasize that there may be a considerable lag between the day that a cell type is born (undergoes terminal cell cycle exit) and the onset of expression of end-stage differentiation markers characteristic of that cell type. Whereas RGCs differentiate immediately following cell cycle exit (71), the expression of rhodopsin in rod photoreceptors is delayed by 5.5–6.5 days following cell cycle exit (72). Cell lineage tracing revealed that these cell types are all derived from a population of multipotent retinal progenitor cells (RPCs) residing in the inner layer of the optic cup (73–76).

Cellular differentiation in the retina is also spatially regulated. In the chick and rodent retina, neuronal differentiation is initiated in the central retina in the vicinity of the optic disk and spreads out in a wavelike fashion toward the periphery (77). In frogs and fish, neuronal differentiation in initiated in a patch of cells located in the ventral-temporal region of the eye and spreads around the retina in a pattern reminiscent of the opening of a fan (78).

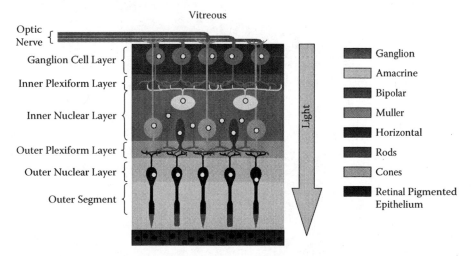

FIGURE 13.3   (See accompanying color CD.) Histology of the mature mammalian retina composed of seven retinal cell types and organized into three cellular layers.

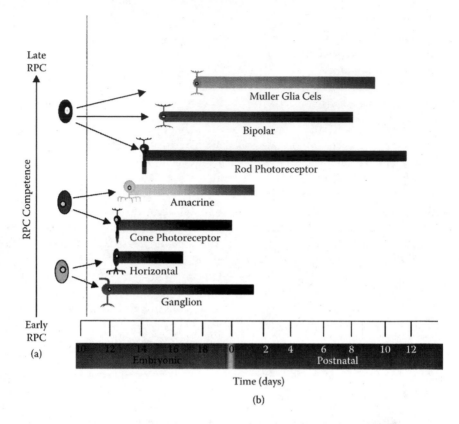

FIGURE 13.4 (See accompanying color CD.) Retinal development. (a) The competence model of retinal cell-fate determination, which illustrates the progression of the RPC through different competence states in which they can produce a subset of retinal cell types. (b) The generally conserved order of birth of the seven cell types that comprise the mature retina.

The process of retina development involves a tight coordination of proliferation and differentiation to ensure the timely generation of specific classes of cells at distinct positions and in appropriate numbers (79). The competence model of retinal development has been proposed by Cepko and colleagues to describe how the temporal changes in cell type specification are achieved by a multipotential progenitor population (reviewed in Reference 80). This model dictates that RPCs progress irreversibly through a series of competence states characterized by their ability to produce a limited number of cell types at each stage of retinogenesis (figure 13.4a). In this model, RPCs cannot be made to progress precociously to a later competence state or to revert to an earlier competence state. With the exception of RGCs (81), the model is consistent with data from heterochronic cell mixing experiments that reveal stage-specific restriction of RPC potential (82–84).

RPC progression through these competence states is regulated by cell-intrinsic and cell-extrinsic mechanisms. RPCs exhibit cell-intrinsic changes during development in their proliferative capacity, gene expression, and competence to generate specific cell types (82,83,85–87). Indeed, the generation of late-born cell types in the rodent retina appears largely to be controlled cell intrinsically (88). The cell-intrinsic regulation of cell diversification is regulated, in part, by homeobox and basic helix-loop-helix (bHLH) transcription factors, which have been shown to be involved in the specification of many neuronal subtypes throughout the CNS, including the retina (89). Key transcription factors involved in retinal cell specification include Math5, Math3, Mash1, Hes1, Hes5, Pax6, Chx10, and Prox1, although the list of transcription factors that influence cell fate or even the development of specific subclasses of retinal cell types is growing (90,91). These intrinsic factors generally do not induce cell-specific differentiation alone, but rather work in combination with other factors to ensure proper cell fate (92,93). How extrinsic cues impinge on these transcription factors to influence cell fate is, however, largely unknown.

There is also a large body of work demonstrating the importance of cell-extrinsic mechanisms on the regulation of RPC diversification. The Delta-Notch signaling pathway has been shown to play a role in regulating the timing of cell cycle exit and neuronal and glial cell fate in the vertebrate retina (94–99). Several studies also indicate a role for feedback inhibition of retinal cell type generation by differentiated neurons of the same class (83,100–104). With the exception of RGC development, which is regulated in part by RGC-derived Shh and GDF11 signaling (66,105,106), the signaling molecules that regulate these feedback loops are largely unknown. Finally, cell-extrinsic signals have also been shown to regulate the timing of differentiation of committed, postmitotic progenitors. CNTF, LIF, and taurine, as well as other unidentified factors, regulate rod photoreceptor differentiation (107–111).

13.2.4 Hh in Retinal Development

13.2.4.1 Sonic Hedgehog in Retinal Ganglion Cell Development

During vertebrate retinal development, *Shh* expression is coupled to RGC differentiation (for a complete cross-species comparison of ocular Hh expression, see reference 112). In mouse and chick, *Shh* is initially expressed in RGCs, the first-born neurons of the retina, and sweeps across the retina signaling to undifferentiated RPCs. Work by Waid and McLoon (104) in chick retinal development has implicated an unidentified diffusible factor secreted from differentiated RGCs that prevents undifferentiated cells from adopting the ganglion cell fate. Recent evidence has emerged likening Shh as a candidate for this unidentified diffusible factor. Ablation of *Shh* signaling in the mouse retina results in an overproduction of RGCs, inferring a role for Hh in suppressing the RGC fate (105). Furthermore, modulation of Hh pathway activation in mouse and chick retinal explants has shown an inverse correlation between RGC development and the level of Hh signaling (66,105). Therefore, in mammalian development, Shh secreted from RGCs contributes to negative feedback inhibition for further RGC genesis (66,105).

During zebrafish retinal development, Shh acts as an inductive cue to help propagate *Shh* expression across the retina (113). However, the role of Shh signaling in regulating RGC development in fish is more controversial, with some studies indicating that

midline-derived rather than RGC-derived Shh is required for the timely propagation of
RGC development across the retina (113–115).

13.2.4.2 RGC-Derived Shh Acts as a Mitogen for RPCs

In addition to the role of Shh in regulating RGC differentiation, Shh signaling from
RGCs also acts to locally stimulate RPC proliferation. In vitro treatment of perinatal retinal cell cultures from mouse with biologically active Shh protein results in
increased BrdU incorporation, indicative of an increase in RPC proliferation (116).
Furthermore, in $Ptc^{+/-}$ mice, which exhibit a cell-intrinsic increase in Hh pathway
activation, RPC proliferation is extended and neuronal differentiation is delayed
(117,118). Conversely, in vivo ablation of Shh signaling in the retina results in a
decreased RPC population with precocious cell cycle exit and premature neuronal
differentiation (105). Shh-mediated control of cell proliferation has been observed
in other systems of eye development as well. Disruption of Hh signaling in *Xenopus*
and chick results in microphthalmia, a condition characterized by a smaller than normal eye (112,119). However, contradictory evidence has emerged in characterizing
Shh as a mitogen in chick RPCs. Treatment of the chick retinal ciliary margin zone,
an area containing residual retinal stem cells, with Shh results in increased proliferation (120,121). However, another group failed to observe increased cell cycling in
chick retinal explant cultures treated with biologically active Shh at various stages of
development (66). Further studies will be helpful in reconciling the role of Shh as a
mitogen for chick RPCs.

The proliferative effects of Hh pathway activation in the retina are dependent
on the stage of embryogenesis. Shh has limited mitogenic capability in murine RPC
cultures at early embryonic ages (i.e., E14) versus a substantial mitogenic effect at
later developmental and early postnatal stages (116). This suggests that other developmental cues may play a more pronounced role in regulating proliferation at early
time points, and/or the RPCs derived from early time points are not as competent
to respond to Shh signaling. As development proceeds, the RPCs progress in competence to respond to Shh signaling and, thus, Shh signaling exerts a major role on
RPC proliferation in the perinatal retina.

13.2.4.3 Differing Functions of Hh in Murine and Fish Retinal Development

In zebrafish, Shh signaling is required for cell cycle progression, and also for cell
cycle exit. Mutants for *Shh* (*sonic you*) in the fish retina are characterized by a reduction in proliferation at 34 h post fertilization (115), inferring a mitogenic role for Hh
in fish. Other studies have suggested an active Hh pathway promotes cell cycle exit,
in part, by activating p57^{Kip2} (122). A recent report on fish and frog retinal development has attempted to reconcile these differences (123). Locker et al. have demonstrated that Shh shortens the length of G1 and G2 phases of the cell cycle and causes
cells to exit the cell cycle more quickly (123). Therefore, Hh may function to stimulate quiescent RPCs to fast-cycling amplifying progenitors that, in turn, exit the cell
cycle earlier [reviewed in reference 124].

The divergent role of Shh signaling in regulating RGC development and RPC
proliferation in fish and mammals could be explained by considering the differing

lengths of retinal histogenesis between species. The zebrafish retina develops in 3 days, whereas the mouse retina requires ~3 weeks to reach maturity. Therefore, as development proceeds at a faster rate in fish, cell cycle exit would take precedence over maintaining cells in the cell cycle. Retinal development in the mouse requires a longer period; therefore, it is necessary to keep cells proliferating throughout retinal development to ensure all cell types are produced in sufficient number to establish the appropriate size of the adult retina.

13.2.4.4   Mechanism of Hh-Induced Activity in the Retina

The mechanism of Shh-induced proliferation of RPCs is likely mediated by several effectors. The Gli series of transcription factors are known to have roles in modulating targets of the Hh pathway. In fish retinal development, Gli3 has been implicated as the major regulator of Shh-induced neurogenesis (125). Morpholino-mediated knockdown of *Gli3* in the fish eye leads to inhibition of *ath5* expression and a reduction in RGC production (125). All three *Gli* genes are expressed in the mouse retina (126), but their relative roles in mediating the Hh signal in the retina are not known.

Activation of the Hh pathway in mouse retinal explants results in induction of cell cycle components such as *CyclinD1*, providing a mechanism for increased cell cycling downstream of Hh signaling (105). *CyclinD1* is a reported direct target of the Gli transcription factors in other tissues (127), further implicating it as a responsive Hh target gene. Shh treatment also induces *Hes1* expression (105), a well-characterized target of the Notch signaling pathway. In chick, ectopic expression of *Hes1* is sufficient to induce RPC proliferation (128) and, thus, *Hes1* is a potential Hh target providing another mechanism for Hh-induced cellular proliferation. Mouse mutants for *Hes1* and conditional mutants for Shh in the retina are phenotypically similar and characterized by RGC overproduction and a reduction in RPCs (129). This functional redundancy for Hes1 and Shh suggests that Hh effects may be mediated, in part, through Hes1 during retinal development.

There is also evidence for tissue specificity of cell cycle targets of the Hh pathway. *N-Myc* has been characterized as a target of the pathway in particular Hh responsive tissues, for example, in the development of the cerebellum and hair follicle, in which Hh-mediated proliferation results in upregulation of Myc mRNA and protein (130,131). Furthermore, targeted suppression of *N-Myc* compromised a proliferative phenotype induced by an activated Hh pathway in cerebellar granule neuronal precursors (132). In contrast, *N-Myc* expression is not modulated by Shh in the retina (105). Therefore, the mechanism of Hh-regulated cellular proliferation appears to be tissue specific, as different targets of the pathway are implicated in different tissues.

13.2.5   The Role of Hh Orthologs in Eye Development

Ihh and *Dhh* are the other two mammalian Hh homologues. The expression of *Ihh* has been localized to a single layer of cells just outside the eye, adjacent to the RPE (133). While expression of *Dhh* has been detected in the neural retina and RPE by RT-PCR (134–137), the localization of *Dhh* transcripts in the rodent eye by in situ hybridization (ISH) has not been reported. The function of either gene in mammalian visual system development remains unclear; however, it has been suggested that, at least in mice, Ihh signaling may only be involved in periocular development and is not required for Hh target gene expression in the retina, optic disk, and nerve (138).

RPE-associated Hh gene expression appears to be conserved, as *shh* and *tiggy-winkle hedgehog* (*twhh*) are expressed in the RPE in zebrafish (139) and *X-shh, banded hh* (*X-bhh*), and *cephalic hh* (*X-chh*) are expressed in the central region of the developing RPE in *Xenopus* (140). In zebra fish, RPE-derived Hh signaling is required for photoreceptor differentiation (139) and, in the frog, Hh signaling is required for RPE development (140). Finally, Tsonis et al. (141) have shown that expression of Hh pathway components is induced in the developing and regenerating lens in the newt, indicating a role for activation of this pathway in the lens.

13.2.6 Extraretinal Effects of Shh

In addition to acting as a short-range signal to regulate RPC behavior, RGC-derived Shh plays a role in other aspects of visual system development, namely, axon guidance and optic nerve development.

13.2.6.1 Shh Controls RGC Axon Guidance

Coordination of RGC axon guidance into the optic nerve is tightly regulated during eye morphogenesis. During retinal development, RGCs migrate toward their final position on the vitreal side of the neural retina. The axons of RGC then extend into the nerve fiber layer on the vitreal surface of the retina and grow toward the optic disk, where they exit the eye and project through the optic nerve (142,143). Precise RGC axon pathfinding within the optic cup is guided by both attractive and repulsive guidance cues secreted by the neural retina and the optic disk.

A role for Shh in axon pathfinding was first described in zebrafish where mutants that exhibited defects in retinotectal projections were shown to have mutations in the *Shh* gene (144–146). In mice, guidance of RGC axons to the optic disk requires Netrin1 expression in astrocyte precursor cells located at the disk. Shh signaling in the retina promotes normal RGC axon guidance, in part, by promoting the development of the optic disk astrocytes (147). Moreover, Shh mediates concentration-dependent effects on RGC axon pathfinding such that at high concentrations Shh functions as a chemorepellent and at low concentrations it functions as a chemoattractant (148,149). Thus, this concentration-dependent effect of Shh on axon outgrowth might be one mechanism that prevents the invasion of RGC axons into the neural retina, where the concentration of Shh is higher and guides them along the nerve fiber layer, where the level of Shh protein is lower (149).

Shh also has a role in the formation of the retinotopic map of RGC projections to the optic tectum. Astroglial cells located at the chiasm regulate ipsilateral and contralateral axonal projections by differentially guiding RGC axons at the optic chiasm between the preoptic area and ventral hypothalamus (150,151). Perturbation of Shh signaling at the optic chiasm results in axon projection errors. For example, ectopic expression of Shh throughout the midline prevents axons from reaching the optic chiasm and instead misdirects a proportion of axons into the ipsilateral optic tract (148,152). Shh functions indirectly in this context, by exerting concentration-dependent effects on the expression of Netrin and Slit molecules (153–156).

Thus, astroglial cells exposed to higher concentrations of *Shh* express chemorepellent *Slit2/3*, whereas cells that are exposed to low levels express chemoattractant *Slit1* (155). However, it is unknown whether this concentration-dependent effect is a temporal phenomenon such that early Shh exposure leads to expression of chemoattractant genes and later exposure of cumulative Shh leads to expression of chemorepulsive cues. These results indicate that *Shh* expression at the optic chiasm guides RGC axon pathfinding by affecting cell fate specification of the astroglial population close to the midline source of Shh.

13.2.6.2 Shh Regulates Optic Nerve Gliogenesis

RGC axon-derived signaling regulates glial cell development in the optic nerve. The optic nerve contains two glial cell types: astrocytes derived from neuroepithelial cells that line the optic stalk and oligodendrocytes that migrate into the optic nerve from the optic chiasm. Signals emanating from RGC axons are required for astrocyte proliferation and oligodendrocyte development (157,158). Hh signaling has been shown to be required for astrocyte development in the mouse and chick optic nerve (61,133,147,159). The source of the Hh signal in this context is Shh from RGC axons, as axotomy-induced RGC axon degeneration inhibits Hh signaling and astrocyte proliferation in the neonatal optic nerve (133), and inactivation of *Shh* in RGCs during embryogenesis results in a failure of astrocyte development in the nerve (147). Hh signaling in the nerve promotes astrocyte development, in part, by maintaining expression of *Pax2* and suppressing neuronal differentiation (61,159). Similar to astrocytes, oligodendrocyte development and survival depends on RGC-derived signals (158,160,161). Removal of RGC axons by enucleation or blockade of *Shh* signaling in the optic chiasm in chick abrogates oligodendrocyte specification at the chiasm, resulting in a failure of oligodendrocyte precursor cell invasion into the optic nerve (162). Thus, axon-derived Shh signaling plays a key role in optic nerve development by regulating the development of both glial cell populations.

13.2.6.3 Shh May Regulate Neuogenesis in the Adult Brain

RGC axon-derived Hh signaling may also modulate neuogenesis in the adult brain. Shh is anterogradely transported along the length of RGC axons to the superior colliculus, where these axons synapse (163). Although the functional significance of axonally transported Shh on target cells in the superior colliculus is not known, there is evidence for a proliferative effect of Shh signaling on progenitors located in the subventricular zone of the forebrain, hippocampal stem cell niches, and in the production of new olfactory neurons (164–167). In the case of the hippocampus, it is likely that the source of Shh that drives stem cell proliferation is from the terminals of axons that project to this region (164,166). Similarly, in *Drosophila*, hh axonal transport by retinal neurons and its secretion at the growth cones regulates neurogenesis in the brain (168–170). Thus, it is tempting to speculate that Shh traveling through RGC axons to their synaptic targets in the brain might perform a similar function in the developing vertebrate visual system.

13.3 FUNCTIONAL ASSAYS FOR HEDGEHOG PATHWAY ACTIVATION IN THE RETINA

13.3.1 MANIPULATION OF RETINAL TISSUE

Assessing the factors involved in the development of the retina is aided by the ability to physically isolate and culture the retina in the presence of various reagents. The retina can be cultured in vitro at a range of developmental stages, allowing the study of extrinsic factors on both embryonic and postnatal retinal development. Retinas can be cultured as explants, reaggregates (pellets), or dissociated monolayers, depending on the experimental question. The pellet cultures are homogeneous in size at the outset of the culture, thereby allowing for a more accurate quantification of the effects of various manipulations on cell number. Pellet cultures can also be established by mixing different cell populations, for example, embryonic and postnatal retinal cells, and are therefore a useful system for investigating the effects of stage-specific environmental cues on the developmental competence of retinal progenitors. The experimental steps involved in isolating and culturing the retina as an explant, cell pellet, or dissociated monolayer are as follows.

PROTOCOL 1: RETINAL EXPLANT CULTURE

1. Eyes are dissected in CO_2-independent medium, and the neural retina is separated from the lens and RPE.
2. The retina is flattened on a 13 mm polycarbonate filter (previously sterilized with ethanol) by creating four equidistant incisions into the tissue and then transferred to a 24-well plate with 500 μL of serum-free culture medium (SFM) (1:1 Dulbecco's Modified Eagle's Medium/F12 containing insulin (10 μg/mL), N-acetyl-L-cysteine (60 μg/mL) and gentamycin (25 μg/mL), apotransferrin (100 μg/mL), bovine serum albumin (100 μg/mL, fraction V Sigma A4161), progesterone (60 ng/mL), putrescine (16 μg/mL), and sodium selenate (40 ng/mL)).
3. Additional drugs/reagents may be added to the culture medium to test the effect on retinal growth and cell type development.
4. The explant is maintained in a humidified atmosphere at 37°C and 8% CO_2.
5. The culture medium may be refreshed after 3 days.
6. The retinal explants may be fixed in 4% paraformaldehyde in PBS buffer for histological processing or harvested for molecular analysis.

PROTOCOL 2: RETINAL PELLET CULTURE

1. The retina is dissected by removing the lens and RPE and incubated in 0.125% trypsin (Boehringer Mannheim Cat. Number 109819) in Ca^{2+}/Mg^{2+}-free PBS for 10 min.
2. The digestion is stopped by the addition of 10% fetal calf serum (FCS)/MEM with 0.04% DNAse, and the suspension is triturated with a Pasteur pipette to achieve a single-cell suspension.

3. The cells are centrifuged at 1000 × g for 5 min, and the pellet is resuspended in 1 mL of 20% FCS/MEM containing 0.02% insulin.
4. The total cell number is counted, and the cell density adjusted to 2.75 × 10^5 cells per 300 μL medium and transferred to an Eppendorf tube. The cells are centrifuged at 800 × g for 8 min, and the pellet is allowed to incubate for 1 h at room temperature before it is transferred to polycarbonate filters and cultured in medium, as before, at 37°C and 8% CO_2.
5. The pellets can be cultured for up to 14 days and analyzed by histological or molecular biological approaches.

PROTOCOL 3: DISSOCIATED RETINA CELL MONOLAYER CULTURE

1. Following isolation of the neural retina (as before), the retina is dissociated by incubating in 0.125% trypsin in PBS without Ca^{2+}/Mg^{2+} for 10 min and triturating into a single cell suspension.
2. Cell density is adjusted to 50,000 cells/75 μL of medium and cells are plated in 24-well tissue culture plates. The cells are allowed to adhere as a small volume, and then the well is flooded with 0.5 mL of medium.
3. The cells are cultured at 37°C and 8% CO_2 and harvested as desired.

13.3.2 METHODS TO MANIPULATE THE HH PATHWAY IN PRIMARY RETINAL CELLS

There are numerous reagents available to activate or inhibit the Hh signaling pathway. Chemical activators of the pathway include recombinant Shh-N protein and numerous Shh agonists (171) that are easily added to in vitro retinal culture medium to stimulate the pathway. Cyclopamine is an example of a widely used chemical inhibitor of the pathway that acts to inhibit the activity of Smo (45,172). The use of expression plasmids encoding components of the Hh signaling pathway is also helpful in dissecting the role of the various members of the pathway. As primary retinal cell cultures are difficult to transfect using lipofectamine and Ca^{2+}phosphate-based methods, other methods of transfection such as retroviral infection (173) and electroporation (174) have proven a useful means of introducing transgenes into retinal cells for in vitro analysis. Transgenic mouse models are also valuable resources for delineating the role of the Hh pathway in retinal development in vivo.

13.3.3 RETROVIRAL-MEDIATED TRANSFECTION OF RETINAL CELLS

Retroviral vectors have been used successfully to deliver transgenes to mitotic cells in the rodent retina [reviewed in reference (173)]. These vectors use the same machinery for infection and single copy stable integration of viral DNA in the host chromosome as their wild-type counterparts and therefore represent a cell-heritable model of transgene expression. Replication-incompetent retroviral vectors can also be generated by removal of the *gag, pol,* and *env* genes from the viral coding sequences. These replication-incompetent retroviral vectors are particularly useful for lineage-tracing studies, as only the descendants of the infected progenitor cells will inherit the retrovirus DNA, allowing for infected clones to be identified on the basis of reporter gene expression from the retroviral vector. Although the majority of the retroviral vectors only infect mitotic cells, it should be noted that recent generations of HIV-based retroviral vectors that can infect postmitotic cells have been developed (175) and are very useful for manipulating gene expression in differentiated cells of the neural retina.

PROTOCOL 4: PRODUCTION OF RETROVIRUS AND RETROVIRAL INFECTION OF RETINAL CELLS

1. A cDNA of interest is cloned into a plasmid-based retroviral construct that contains the viral LTRs and a marker gene, such as beta-galactosidase or GFP, but that lacks the *gag, pol,* and *env* genes.
2. The resulting retroviral vector is then transfected using calcium phosphate precipitation into a packaging cell line that supplies the *gag, pol,* and *env* genes, and the transfected cells secrete infectious virus into the cell culture medium. Phoenix-Eco cells are a widely used packaging cell line for generating ecotropic retroviruses that are capable of infecting mouse and rat cells, and complete protocols for virus generation are available (http://www.stanford.edu/group/nolan/protocols/pro_helper_free.html).
3. The retrovirus-containing medium is concentrated by high-speed centrifugation (21,000 rpm for 2 h at 4°C, Beckman JA25.50 rotor), aliquoted, and stored at −80°C.
4. The concentrated retrovirus stocks can be titered by infecting NIH 3T3 cells and counting the number of cells that express the reporter gene. Typical titers range from 10^7–10^8 cfu/mL.
5. A high degree of infection of retinal progenitors in explants is achieved by two applications, 1 h apart, of 6 µL of the retrovirus solution to the top of the explant that has been flattened onto a polycarbonate filter.
6. For experiments that require clonal analysis of infected cells in vitro, the virus stock is diluted to 10^6 CFU/mL, and explants are infected only once with virus.
7. For experiments that require clonal analysis of retrovirally infected cells in vivo, 1 µL of a 10^6 CFU/mL solution of retrovirus is injected subretinally (between the RPE and the NR) into the eye of postnatal day one mouse pups.
8. The clonal relationship of infected cells is easily resolved because the descendants of single-infected progenitors are organized in columns.

PROTOCOL 5: ELECTROPORATION

1. DNA plasmids can be prepared for electroporation using a Qiagen maxi plasmid kit.
2. The neural retina is dissected in CO_2-independent medium by removing the RPE while leaving the lens intact.
3. The retina is placed into a 2 mm gap cuvette (BTX Harvard Appartus) with 100–150 µL of plasmid DNA resuspended in TE buffer with a minimum concentration of 500 µg/mL.
4. The retina is pulsed five times at 30 V with a path length of 50 ms and interval of 950 ms using an ECM 830 Electro Square Porator (BTX Harvard Appartus).

5. Following electroporation, the lens is detached from the retina, and the retina is transferred to polycarbonate filters and cultured as before.
6. Expression of the transgene can be detected as early as 12 h following electroporation. Figure 13.5 shows an example of the electroporation of a constitutively active Smoothened plasmid with cotransfection of a poly-ubiquitin promoter-driven GFP construct to localize transfected cells. Successful electroporation is verified by the expression of GFP, and the expression of Hh pathway targets, *Gli1* and *CyclinD1*, illustrates positive pathway activation.

13.3.4 IN VIVO TRANSGENIC MOUSE MODELS

Knockout mouse models for Shh results in embryonic lethality, making it difficult to assess the in vivo role of Shh signaling throughout the period of retinal development. To address this problem, our laboratory has generated an in vivo mouse model of cre-Loxp-mediated conditional inactivation of *Shh* in the retina (figure 13.6). The conditional *Shh* allele contains LoxP sites that flank exon 2 of the *Shh* gene (176). Exon 2 encodes most of the N-terminal active Shh fragment and, thus, cre-mediated excision of the floxed exon results in a null allele. The Cre recombinase activity is driven by a variant of the Pax6 promoter that is only expressed in the periphery of

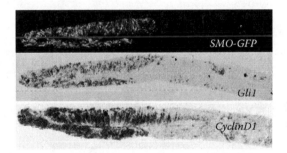

FIGURE 13.5   (See accompanying color CD.) Coelectroporation of active smo and GFP plasmids in retinal explants is visualized by positive GFP expression, and Hh pathway activation is confirmed with expression of Hh target genes, *Gli1* and *CyclinD1*, in the transfected cells.

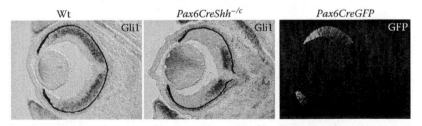

FIGURE 13.6   (See accompanying color CD.) In situ hybridization for *Gli1* confirms Hh pathway inactivation in the peripheral retina in the *Pax6Cre;Shh⁻/ᶜ* mutant. Shh is a combination of a null allele and a floxed allele; therefore, Shh inactivation only requires cre recombinase activity at one allele. GFP fluorescence for the *Pax6Cre-GFP* transgene illustrates where the cre transgene is expressed.

the retina (177). Therefore, mice harboring the *Pax6-Cre* transgene and a floxed *Shh* gene will have exon 2 of *Shh* excised exclusively in the peripheral region of the retina (105). This results in specific blockade of the Shh pathway in the retina, as determined by a loss of *Gli1* expression in the peripheral retina. This mouse model provides a useful means of deciphering the effects of blockade of Shh signaling on many aspects of retinal development, including proliferation, cell-type specification, and target gene expression. Other useful mouse models employed to elucidate the role of the Hh pathway in the retina include mice that are heterozygous for knockout of the *Patched* gene, which results in an overactive Hh pathway, as well as mutants for the Gli mediators of the pathway. The study of Gli mutants is useful for delineating the role of each of the Gli proteins in Hh-mediated retinal development.

13.3.5 METHODS TO ANALYZE THE EFFECT OF THE HH PATHWAY

Following manipulation of the Hh pathway in the retina, there are many ways to study the consequences of pathway activation or inhibition. Elucidating the function of Hh in the retina requires analyzing cellular proliferation, cell-type specification, and target gene expression in response to altering pathway activity. Cultured retinal explants or retinal tissue derived from transgenic mouse models can be analyzed by fixing the tissue in 4% paraformaldehyde PBS and cryosectioning. Histological techniques such as immunohistochemistry, in situ hybridization, and Brdu labeling can then be used to study the effects of Hh in situ on retinal development.

PROTOCOL 6: IN SITU HYBRIDIZATION

This protocol is divided into five sections:

1. Preparation of antisense DIG-labeled RNA probes
2. Tissue fixation and sectioning
3. Hybridization and antibody staining
   a. Hybridization
   b. Post-hybridization
   c. Immunohistochemistry
4. Color reaction
5. Solutions and reagents

Preparation of antisense DIG-labeled RNA probes

1. 10 µg of DNA is linearized with the appropriate restriction enzyme.
2. Linearized plasmid is purified by phenol/chloroform extraction followed by ethanol precipitation.
3. The DNA pellet is resuspended in 30 µL RNAse-free water.
4. The in vitro synthesis of the antisense RNA is performed using the following transcription mix: 5 µL of 5× Transcription Buffer, 1 µL of RNAse OUT RNAse inhibitor (40 units/µL, Invitrogen), 2.5 µL of 10× DIG-UTP RNA Labeling Mix (Roche), 1 µL of T3 or T7 RNA polymerase (20 units/µL, Invitrogen), 1 µg of linearized DNA, and sterile water to a final volume of 25 µL.

5. The transcription reaction is allowed to proceed for 1 h at 37°C.
6. An additional 1 μL of T3 or T7 RNA polymerase is added, and the incubation is continued for an additional hour.
7. Synthesized DIG-labeled RNA is precipitated with 10% 3 M sodium acetate pH 5.2 in ethanol on dry ice.
8. The RNA pellet is washed in 1 volume of 70% ethanol and resuspended in 100 μL of DEPC-treated water.
9. DIG-labeled riboprobes are stored at −70°C.

Tissue fixation and sectioning

1. Tissues are fixed in 4% PFA/0.1 M phosphate buffer, pH 7.4 overnight at 4°C.
2. After fixation, tissues are washed 3× in PBS, and cryoprotected in 30% sucrose/1 × PBS (Dulbecco's) overnight at 4°C.
3. Fixed tissues are equilibrated in a 50:50 mixture of 30% sucrose:OCT (TissueTek 4583) for at least one hour.
4. Tissues are embedded in sucrose:OCT, frozen in liquid nitrogen, and stored at −70°C until cryosectioned using superfrost plus slides.
5. Slides are stored with desiccant at −20°C.

Hybridization and antibody staining

Hybridization
    1. Sections are warmed to room temperature.
    2. Probe is diluted (1:500 to 1:2000) in hybridization solution.
    3. The probe is denatured at 70°C for 10 min, vortexed, and 100–150 μL was placed onto each slide and coverslipped.
    4. Slides are placed in a humidified chamber with 50% Formamide/1× salt and incubated at 65°C overnight.
Posthybridization
    1. Slides are washed in prewarmed wash buffer 3× 30 min at 65°C with rocking.
    2. Slides are washed 2× 30 min in 1× MABT at room temperature.
Immunohistochemistry
    1. Slides are treated with blocking solution at RT for at least 1 h in a PBS humidified chamber.
    2. Slides are incubated in 250–500 μL anti-DIG antibody (BM cat.#1093274) diluted 1:1500 in blocking solution overnight at 4°C.
    3. Slides are washed 4× 20 min in 1× MABT at room temperature with rocking.
    4. Slides are equilibrated in staining buffer (without NBT and BCIP) 2× 10 min at room temperature.
Color reaction
    1. Slides are incubated in staining buffer with NBT (4.5 μL/mL) and BCIP (3.5 μL/mL) in the dark at room temperature.
    2. The staining reaction is terminated by washing several times in PBS.
    3. Slides are mounted in (1:1) glycerol/PBS, and stored at 4°C.

Solutions and reagents
1. 4% PFA in 0.1 M phosphate buffer, pH 7.4
2. 30% sucrose/PBS
3. rRNA (concentration 10 mg/ml)
4. 10× salt
   NaCl (114 g), Tris-HCl, pH 7.5 (14.04 g), Tris base (1.34 g), NaH$_2$PO$_4$•2H$_2$O (7.8 g), Na$_2$HPO$_4$ (7.1 g), 0.5 M EDTA (100 mL), dH$_2$O up to 1 L, autoclave and store at room temperature
5. 100× Denhardt's
   2% w/v BSA, 2% w/v Ficoll™ (400), 2% w/v polyvinyl pyrolidone; aliquot and store at −20°C
6. Hybridization buffer
   1× salt (see earlier text), 50% deionized formamide, 10% dextran sulfate, rRNA (1 mg/mL) (see 3), 1× Denhardt's (see 5), H$_2$O; aliquot and store at −20°C
7. Wash buffer
   1× SSC, 50% formamide, 0.1% Tween20, dH$_2$O; should be made fresh
8. 5× MABT, pH 7.5
   0.5 M maleic acid, NaOH (to adjust pH), 0.75 M NaCl, 0.5% Tween20, dH$_2$O; store at 4°C.
9. Blocking reagent (Roche)
   Blocking reagent (cat. #1096176) is prepared as 10% stocks in 1× MABT; aliquoted and stored at −20°C.
10. Blocking solution
    20% sheep serum, 2% blocking reagent (see 9), 1× MABT, dH$_2$O; store at −20°C.
11. Staining buffer (alkaline-phosphatase) (see table 13.1).

TABLE 13.1
Alkaline-Phosphatase Staining Buffer

Reagent	Prestain	Stain[a]
100 mM NaCl$_2$	6 mL of 5 M	2 mL of 5 M
50 mM MgCl$_2$	15 mL of 1 M	Omit
100 mM Tris, pH 9–9.5	30 ml of 1 M	10 mL of 1 M
0.1% Tween20	300 μL	Omit
dH$_2$O	250 mL	83 mL

[a] To prepare 70 mL of staining buffer, add 10% polyvinyl alcohol (PVA) to 66.5 mL of stain. Dissolve PVA by heating to 85°C with constant stirring. Allow solution to cool to 55°C before adding MgCl$_2$ (3.5 mL) and Tween (70 μL). Once at room temperature, add NBT (4.5 μL/mL of staining buffer, Roche, # 1383213) and BCIP (3.5 μL/mL of staining buffer, Roche # 1383221), and add buffer to slides. Incubate in the dark, and do not agitate solution once staining reaction has started.

PROTOCOL 7: BrdU LABELING AND IMMUNOSTAINING

1. Cells in S phase of the cell cycle are BrdU labeled in vivo by administering two intraperitoneal injections at a concentration of 0.1 g/kg body weight BrdU dissolved in CO_2-independent medium. Retinal explants are pulse-labeled 4–16 h with 10 µM BrdU at 37°C.
2. Labeled tissues are fixed and sectioned according to Protocol 6: Tissue fixation and sectioning.
3. Slides are fixed in 70% ethanol for 5 min.
4. Sections are rehydrated with 1× PBS for 5 min.
5. Slides are incubated in 2N HCl for 20 min at 37°C and neutralized in 0.1 M Tris (pH 8.8) + 0.1% Tween20 for 10 min.
6. Sections are blocked with 10% FCS/PBS for 1 h.
7. Samples are incubated in anti-BrdU antibody (BD Bioscience, # 347580) (1:100) in 10% FCS/PBS for 1 h.
8. Slides are washed 3× 10 min in 1× PBS.
9. Samples are incubated in secondary antibodies for 1 h.
10. Slides are washed 3× 10 min in 1× PBS and coverslipped.

REFERENCES

1. Nusslein-Volhard, C. and Wieschaus, E., Mutations affecting segment number and polarity in Drosophila, *Nature* 287(5785), 795 801, 1980.
2. Briscoe, J., Chen, Y., Jessell, T.M., and Struhl, G., A hedgehog-insensitive form of patched provides evidence for direct long-range morphogen activity of sonic hedgehog in the neural tube, *Mol Cell* 7(6), 1279–91, 2001.
3. Ericson, J., Briscoe, J., Rashbass, P., van Heyningen, V., and Jessell, T.M., Graded sonic hedgehog signaling and the specification of cell fate in the ventral neural tube, *Cold Spring Harb Symp Quant Biol* 62, 451–66, 1997.
4. Gritli-Linde, A., Lewis, P., McMahon, A.P., and Linde, A., The whereabouts of a morphogen: direct evidence for short- and graded long-range activity of hedgehog signaling peptides, *Dev Biol* 236(2), 364–86, 2001.
5. Hynes, M., Ye, W., Wang, K., Stone, D., Murone, M., Sauvage, F., and Rosenthal, A., The seven-transmembrane receptor smoothened cell-autonomously induces multiple ventral cell types, *Nat Neurosci* 3(1), 41–6, 2000.
6. Lewis, P.M., Dunn, M.P., McMahon, J.A., Logan, M., Martin, J.F., St-Jacques, B., and McMahon, A.P., Cholesterol modification of sonic hedgehog is required for long-range signaling activity and effective modulation of signaling by Ptc1, *Cell* 105(5), 599–612, 2001.
7. Persson, M., Stamataki, D., te Welscher, P., Andersson, E., Bose, J., Ruther, U., Ericson, J., and Briscoe, J., Dorsal-ventral patterning of the spinal cord requires Gli3 transcriptional repressor activity, *Genes Dev* 16(22), 2865–78, 2002.
8. Porter, J.A., Young, K.E., and Beachy, P.A. Choloesterol modification of hedgehog signaling proteins in animal development, *Science* 274(5285), 255–9, 1996.
9. Chamoun, Z., Mann, R.K., Nellen, D., von Kessler, D.P., Bellotto, M., Beachy, P.A., and Basler, K., Skinny hedgehog, an acyltransferase required for palmitoylation and activity of the hedgehog signal, *Science* 293(5537), 2080–4, 2001.
10. Lee, J.D. and Treisman, J.E., Sightless has homology to transmembrane acyltransferases and is required to generate active Hedgehog protein, *Curr Biol* 11(14), 1147–52, 2001.

11. Chen, M.H., Li, Y.J., Kawakami, T., Xu, S.M., and Chuang, P.T., Palmitoylation is required for the production of a soluble multimeric Hedgehog protein complex and long-range signaling in vertebrates, *Genes Dev* 18(6), 641–59, 2004.

12. Micchelli, C.A., The, I., Selva, E., Mogila, V., and Perrimon, N., Rasp, a putative transmembrane acyltransferase, is required for Hedgehog signaling, *Development* 129(4), 843–51, 2002.

13. Gallet, A., Rodriguez, R., Ruel, L., and Therond, P.P., Cholesterol modification of hedgehog is required for trafficking and movement, revealing an asymmetric cellular response to hedgehog, *Dev Cell* 4(2), 191–204, 2003.

14. Gallet, A., Ruel, L., Staccini-Lavenant, L., and Therond, P.P., Cholesterol modification is necessary for controlled planar long-range activity of Hedgehog in Drosophila epithelia, *Development* 133(3), 407–18, 2006.

15. Callejo, A., Torroja, C., Quijada, L., and Guerrero, I., Hedgehog lipid modifications are required for Hedgehog stabilization in the extracellular matrix, *Development* 133(3), 471–83, 2006.

16. Dawber, R.J., Hebbes, S., Herpers, B., Docquier, F., and van den Heuvel, M., Differential range and activity of various forms of the Hedgehog protein, *BMC Dev Biol* 5, 21, 2005.

17. Goetz, J.A., Singh, S., Suber, L.M., Kull, F.J., and Robbins, D.J., A highly conserved amino-terminal region of sonic hedgehog is required for the formation of its freely diffusible multimeric form, *J Biol Chem* 281(7), 4087–93, 2006.

18. Burke, R., Nellen, D., Bellotto, M., Hafen, E., Senti, K.A., Dickson, B.J., and Basler, K., Dispatched, a novel sterol-sensing domain protein dedicated to the release of cholesterol-modified hedgehog from signaling cells, *Cell* 99(7), 803–15, 1999.

19. Caspary, T., Garcia-Garcia, M.J., Huangfu, D., Eggenschwiler, J.T., Wyler, M.R., Rakeman, A.S., Alcorn, H.L., and Anderson, K.V., Mouse Dispatched homolog1 is required for long-range, but not juxtacrine, Hh signaling, *Curr Biol* 12(18), 1628–32, 2002.

20. Kawakami, T., Kawcak, T., Li, Y.J., Zhang, W., Hu, Y., and Chuang, P.T., Mouse dispatched mutants fail to distribute hedgehog proteins and are defective in hedgehog signaling, *Development* 129(24), 5753–65, 2002.

21. Ma, Y., Erkner, A., Gong, R., Yao, S., Taipale, J., Basler, K., and Beachy, P.A., Hedgehog-mediated patterning of the mammalian embryo requires transporter-like function of dispatched, *Cell* 111(1), 63–75, 2002.

22. The, I., Bellaiche, Y., and Perrimon, N., Hedgehog movement is regulated through tout velu-dependent synthesis of a heparan sulfate proteoglycan, *Mol Cell* 4(4), 633–9, 1999.

23. Bornemann, D.J., Duncan, J.E., Staatz, W., Selleck, S., and Warrior, R., Abrogation of heparan sulfate synthesis in Drosophila disrupts the Wingless, Hedgehog and Decapentaplegic signaling pathways, *Development* 131(9), 1927–38, 2004.

24. Han, C., Belenkaya, T.Y., Wang, B., and Lin, X., Drosophila glypicans control the cell-to-cell movement of Hedgehog by a dynamin-independent process, *Development* 131(3), 601–11, 2004.

25. Takei, Y., Ozawa, Y., Sato, M., Watanabe, A., and Tabata, T., Three Drosophila EXT genes shape morphogen gradients through synthesis of heparan sulfate proteoglycans, *Development* 131(1), 73–82, 2004.

26. Grobe, K., Inatani, M., Pallerla, S.R., Castagnola, J., Yamaguchi, Y., and Esko, J.D., Cerebral hypoplasia and craniofacial defects in mice lacking heparan sulfate Ndst1 gene function, *Development* 132(16), 3777–86, 2005.

27. Danesin, C., Agius, E., Escalas, N., Ai, X., Emerson, C., Cochard, P., and Soula, C., Ventral neural progenitors switch toward an oligodendroglial fate in response to increased Sonic hedgehog (Shh) activity: involvement of Sulfatase 1 in modulating Shh signaling in the ventral spinal cord, *J Neurosci* 26(19), 5037–48, 2006.

28. Koziel, L., Kunath, M., Kelly, O.G., and Vortkamp, A., Ext1-dependent heparan sulfate regulates the range of Ihh signaling during endochondral ossification, *Dev Cell* 6(6), 801–13, 2004.
29. Han, C., Belenkaya, T.Y., Khodoun, M., Tauchi, M., Lin, X., and Lin, X., Distinct and collaborative roles of Drosophila EXT family proteins in morphogen signalling and gradient formation, *Development* 131(7), 1563–75, 2004.
30. Tenzen, T., Allen, B.L., Cole, F., Kang, J.S., Krauss, R.S., and McMahon, A.P., The cell surface membrane proteins Cdo and Boc are components and targets of the Hedgehog signaling pathway and feedback network in mice, *Dev Cell* 10(5), 647–56, 2006.
31. Yao, S., Lum, L., and Beachy, P., The ihog cell-surface proteins bind Hedgehog and mediate pathway activation, *Cell* 125(2), 343–57, 2006.
32. Zhang, W., Kang, J.S., Cole, F., Yi, M.J., and Krauss, R.S., Cdo functions at multiple points in the Sonic Hedgehog pathway, and Cdo-deficient mice accurately model human holoprosencephaly, *Dev Cell* 10(5), 657–65, 2006.
33. Okada, A., Charron, F., Morin, S., Shin, D.S., Wong, K., Fabre, P.J., Tessier-Lavigne, M., and McConnell, S.K., Boc is a receptor for sonic hedgehog in the guidance of commissural axons, *Nature* 444(7117), 369–73, 2006.
34. Chuang, P.T. and McMahon, A.P., Vertebrate Hedgehog signalling modulated by induction of a Hedgehog-binding protein, *Nature* 397(6720), 617–21, 1999.
35. McCarthy, R.A., Barth, J.L., Chintalapudi, M.R., Knaak, C., and Argraves, W.S., Megalin functions as an endocytic sonic hedgehog receptor, *J Biol Chem* 277(28), 25660–7, 2002.
36. Lee, C.S., Buttitta, L., and Fan, C.M., Evidence that the WNT-inducible growth arrest-specific gene 1 encodes an antagonist of sonic hedgehog signaling in the somite, *Proc Natl Acad Sci U.S.A.* 98(20), 11347–52, 2001.
37. Jia, J. and Jiang, J., Decoding the Hedgehog signal in animal development, *Cell Mol Life Sci* 63(11), 1249–65, 2006.
38. Ingham, P.W. and Placzek, M., Orchestrating ontogenesis: variations on a theme by sonic hedgehog, *Nat Rev Genet* 7(11), 841–50, 2006.
39. Ekker, S.C., Ungar, A.R., Greenstein, P., von Kessler, D.P., Porter, J.A., Moon, R.T., and Beachy, P.A., Patterning activities of vertebrate hedgehog proteins in the developing eye and brain, *Curr Biol* 5(8), 944–55, 1995.
40. Hu, D. and Helms, J.A., The role of sonic hedgehog in normal and abnormal craniofacial morphogenesis, *Development* 126(21), 4873–84, 1999.
41. Nasevicius, A. and Ekker, S.C., Effective targeted gene 'knockdown' in zebrafish, *Nat Genet* 26(2), 216–20, 2000.
42. Krauss, S., Concordet, J.P., and Ingham, P.W., A functionally conserved homolog of the Drosophila segment polarity gene hh is expressed in tissues with polarizing activity in zebrafish embryos, *Cell* 75(7), 1431–44, 1993.
43. Chiang, C., Litingtung, Y., Lee, E., Young, K.E., Corden, J.L., Westphal, H., and Beachy, P.A., Cyclopia and defective axial patterning in mice lacking Sonic hedgehog gene function, *Nature* 383(6599), 407–13, 1996.
44. Wijgerde, M., McMahon, J.A., Rule, M., and McMahon, A.P., A direct requirement for Hedgehog signaling for normal specification of all ventral progenitor domains in the presumptive mammalian spinal cord, *Genes Dev* 16(22), 2849–64, 2002.
45. Incardona, J.P., Gaffield, W., Kapur, R.P., and Roelink, H., The teratogenic Veratrum alkaloid cyclopamine inhibits sonic hedgehog signal transduction, *Development* 125(18), 3553–62, 1998.
46. Fuccillo, M., Rallu, M., McMahon, A.P., and Fishell, G., Temporal requirement for hedgehog signaling in ventral telencephalic patterning, *Development* 131(20), 5031–40, 2004.

47. Park, H.L., Bai, C., Platt, K.A., Matise, M.P., Beeghly, A., Hui, C.C., Nakashima, M., and Joyner, A.L., Mouse Gli1 mutants are viable but have defects in SHH signaling in combination with a Gli2 mutation, *Development* 127(8), 1593–605, 2000.

48. Rallu, M., Machold, R., Gaiano, N., Corbin, J.G., McMahon, A.P., and Fishell, G., Dorsoventral patterning is established in the telencephalon of mutants lacking both Gli3 and Hedgehog signaling, *Development* 129(21), 4963–74, 2002.

49. Ruiz i Altaba, A., Gli proteins encode context-dependent positive and negative functions: implications for development and disease, *Development* 126(14), 3205–16, 1999.

50. Furimsky, M. and Wallace, V.A., Complementary Gli activity mediates early patterning of the mouse visual system, *Dev Dyn* 235(3), 594–605, 2006.

51. Hatta, K., Kimmel, C.B., Ho, R.K., and Walker, C., The cyclops mutation blocks specification of the floor plate of the zebrafish central nervous system, *Nature* 350(6316), 339–41, 1991.

52. Strahle, U. and Blader, P., Early neurogenesis in the zebrafish embryo, *Faseb J* 8(10), 692–8, 1994.

53. Barth, K.A. and Wilson, S.W., Expression of zebrafish nk2.2 is influenced by sonic hedgehog/vertebrate hedgehog-1 and demarcates a zone of neuronal differentiation in the embryonic forebrain, *Development* 121(6), 1755–68, 1995.

54. Macdonald, R., Barth, K.A., Xu, Q., Holder, N., Mikkola, I., and Wilson, S.W., Midline signalling is required for Pax gene regulation and patterning of the eyes, *Development* 121(10), 3267–78, 1995.

55. Hallonet, M., Hollemann, T., Pieler, T., and Gruss, P., Vax1, a novel homeobox-containing gene, directs development of the basal forebrain and visual system, *Genes Dev* 13(23), 3106–14, 1999.

56. Nornes, H.O., Dressler, G.R., Knapik, E.W., Deutsch, U., and Gruss, P., Spatially and temporally restricted expression of Pax2 during murine neurogenesis, *Development* 109(4), 797–809, 1990.

57. Bertuzzi, S., Hindges, R., Mui, S.H., O'Leary, D.D., and Lemke, G., The homeodomain protein vax1 is required for axon guidance and major tract formation in the developing forebrain, *Genes Dev* 13(23), 3092–105, 1999.

58. Walther, C. and Gruss, P., Pax-6, a murine paired box gene, is expressed in the developing CNS, *Development* 113(4), 1435–49, 1991.

59. Grindley, J.C., Davidson, D.R., and Hill, R.E., The role of Pax-6 in eye and nasal development, *Development* 121(5), 1433–42, 1995.

60. Huh, S., Hatini, V., Marcus, R.C., Li, S.C., and Lai, E., Dorsal-ventral patterning defects in the eye of BF-1-deficient mice associated with a restricted loss of shh expression, *Dev Biol* 211(1), 53–63, 1999.

61. Zhang, X.M. and Yang, X.J., Temporal and spatial effects of Sonic hedgehog signaling in chick eye morphogenesis, *Dev Biol* 233(2), 271–90, 2001.

62. Take-uchi, M., Clarke, J.D., and Wilson, S.W., Hedgehog signalling maintains the optic stalk-retinal interface through the regulation of Vax gene activity, *Development* 130(5), 955–68, 2003.

63. Sasagawa, S., Takabatake, T., Takabatake, Y., Muramatsu, T., and Takeshima, K., Axes establishment during eye morphogenesis in Xenopus by coordinate and antagonistic actions of BMP4, Shh, and RA, *Genesis* 33(2), 86–96, 2002.

64. Mui, S.H., Kim, J.W., Lemke, G., and Bertuzzi, S., Vax genes ventralize the embryonic eye, *Genes Dev* 19(10), 1249–59, 2005.

65. Kim, J.W. and Lemke, G., Hedgehog-regulated localization of Vax2 controls eye development, *Genes Dev* 20(20), 2833–47, 2006.

66. Zhang, X.M. and Yang, X.J., Regulation of retinal ganglion cell production by Sonic hedgehog, *Development* 128(6), 943–57, 2001.

67. Golden, J.A., Bracilovic, A., McFadden, K.A., Beesley, J.S., Rubenstein, J.L., and Grinspan, J.B., Ectopic bone morphogenetic proteins 5 and 4 in the chicken forebrain lead to cyclopia and holoprosencephaly, *Proc Natl Acad Sci U.S.A.* 96(5), 2439–44, 1999.
68. McMahon, J.A., Takada, S., Zimmerman, L.B., Fan, C.M., Harland, R.M., and McMahon, A.P., Noggin-mediated antagonism of BMP signaling is required for growth and patterning of the neural tube and somite, *Genes Dev* 12(10), 1438–52, 1998.
69. Young, R.W., Cell differentiation in the retina of the mouse, *Anat Rec* 212(2), 199–205, 1985.
70. Kholodenko, I.V., Buzdin, A.A., Kholodenko, R.V., Baibikova, J.A., Sorokin, V.F., Yarygin, V.N., and Sverdlov, E.D., Mouse retinal progenitor cell (RPC) cocultivation with retinal pigment epithelial cell culture affects features of RPC differentiation, *Biochemistry (Mosc)* 71(7), 767–74, 2006.
71. Waid, D.K. and McLoon, S.C., Immediate differentiation of ganglion cells following mitosis in the developing retina, *Neuron* 14(1), 117–24, 1995.
72. Morrow, E.M., Belliveau, M.J., and Cepko, C.L., Two phases of rod photoreceptor differentiation during rat retinal development, *J Neurosci* 18(10), 3738–48, 1998.
73. Holt, C.E., Bertsch, T.W., Ellis, H.M., and Harris, W.A., Cellular determination in the Xenopus retina is independent of lineage and birth date, *Neuron* 1, 15–26, 1988.
74. Turner, D. and Cepko, C., A common progenitor for neurons and glia persists in rat retina late in development, *Nature* 328, 131–6, 1987.
75. Wetts, R. and Fraser, S., Multipotent precursors can give rise to almost all major cell types of the frog retina, *Science* 239, 1142–45, 1988.
76. Holt, C.E., Bertsch, T.W., Ellis, H.M., and Harris, W.A., Cellular determination in the Xenopus retina is independent of lineage and birth date, *Neuron* 1(1), 15–26, 1988.
77. McCabe, K.L., Gunther, E.C., and Reh, T.A., The development of the pattern of retinal ganglion cells in the chick retina: mechanisms that control differentiation, *Development* 126(24), 5713–24, 1999.
78. Hu, M. and Easter, S.S., Retinal neurogenesis: the formation of the initial central patch of postmitotic cells, *Dev Biol* 207(2), 309–21, 1999.
79. Dyer, M.A. and Cepko, C.L., Regulating proliferation during retinal development, *Nat Rev Neurosci* 2(5), 333–42, 2001.
80. Livesey, F.J. and Cepko, C.L., Vertebrate neural cell-fate determination: lessons from the retina, *Nat Rev Neurosci* 2(2), 109–18, 2001.
81. James, J., Das, A.V., Bhattacharya, S., Chacko, D.M., Zhao, X., and Ahmad, I., In vitro generation of early-born neurons from late retinal progenitors, *J Neurosci* 23(23), 8193–203, 2003.
82. Watanabe, T. and Raff, M.C., Rod photoreceptor development in vitro: intrinsic properties of proliferating neuroepithelial cells change as development proceeds in the rat retina, *Neuron* 4(3), 461–7, 1990.
83. Belliveau, M.J. and Cepko, C.L., Extrinsic and intrinsic factors control the genesis of amacrine and cone cells in the rat retina, *Development* 126(3), 555–66, 1999.
84. Belliveau, M.J., Young, T.L., and Cepko, C.L., Late retinal progenitor cells show intrinsic limitations in the production of cell types and the kinetics of opsin synthesis, *J Neurosci* 20(6), 2247–54, 2000.
85. Alexiades, M.R. and Cepko, C., Quantitative analysis of proliferation and cell cycle length during development of the rat retina, *Dev Dyn* 205(3), 293–307, 1996.
86. Alexiades, M.R. and Cepko, C.L., Subsets of retinal progenitors display temporally regulated and distinct biases in the fates of their progeny, *Development* 124(6), 1119–31, 1997.
87. Brown, N.L., Kanekar, S., Vetter, M.L., Tucker, P.K., Gemza, D.L., and Glaser, T., Math5 encodes a murine basic helix-loop-helix transcription factor expressed during early stages of retinal neurogenesis, *Development* 125(23), 4821–33, 1998.

88. Cayouette, M., Barres, B.A., and Raff, M., Importance of intrinsic mechanisms in cell fate decisions in the developing rat retina, *Neuron* 40(5), 897–904, 2003.

89. Hatakeyama, J., Tomita, K., Inoue, T., and Kageyama, R., Roles of homeobox and bHLH genes in specification of a retinal cell type, *Development* 128(8), 1313–22, 2001.

90. Yang, Z., Ding, K., Pan, L., Deng, M., and Gan, L., Math5 determines the competence state of retinal ganglion cell progenitors, *Dev Biol* 264(1), 240–54, 2003.

91. Matter-Sadzinski, L., Puzianowska-Kuznicka, M., Hernandez, J., Ballivet, M., and Matter, J.M., A bHLH transcriptional network regulating the specification of retinal ganglion cells, *Development* 132(17), 3907–21, 2005.

92. Hatakeyama, J. and Kageyama, R., Retinal cell fate determination and bHLH factors, *Semin Cell Dev Biol* 15(1), 83–9, 2004.

93. Mu, X., Fu, X., Sun, H., Liang, S., Maeda, H., Frishman, L.J., and Klein, W.H., Ganglion cells are required for normal progenitor- cell proliferation but not cell-fate determination or patterning in the developing mouse retina, *Curr Biol* 15(6), 525–30, 2005.

94. Dorsky, R.I., Rapaport, D.H., and Harris, W.A., Xotch inhibits cell differentiation in the Xenopus retina, *Neuron* 14(3), 487–96, 1995.

95. Henrique, D., Hirsinger, E., Adam, J., Le Roux, I., Pourquie, O., Ish-Horowicz, D., and Lewis, J., Maintenance of neuroepithelial progenitor cells by Delta-Notch signalling in the embryonic chick retina, *Curr Biol* 7(9), 661–70, 1997.

96. Dorsky, R.I., Chang, W.S., Rapaport, D.H., and Harris, W.A., Regulation of neuronal diversity in the Xenopus retina by Delta signalling, *Nature* 385(6611), 67–70, 1997.

97. Furukawa, T., Mukherjee, S., Bao, Z.Z., Morrow, E.M., and Cepko, C.L., Rax, Hes1, and notch1 promote the formation of Muller glia by postnatal retinal progenitor cells, *Neuron* 26(2), 383–94, 2000.

98. Yaron, O., Farhy, C., Marquardt, T., Applebury, M., and Ashery-Padan, R., Notch1 functions to suppress cone-photoreceptor fate specification in the developing mouse retina, *Development* 133(7), 1367–78, 2006.

99. Jadhav, A.P., Cho, S.H., and Cepko, C.L., Notch activity permits retinal cells to progress through multiple progenitor states and acquire a stem cell property, *Proc Natl Acad Sci U.S.A.* 103(50), 18998–9003, 2006.

100. Belliveau, M.J. and Cepko, C.L., Extrinsic and intrinsic factors control the genesis of amacrine and cone cells in the rat retina, *Development* 126(3), 555–66, 1999.

101. Reh, T., Cell-specific regulation of neuronal production in the larval frog retina, *J Neurosci* 7, 3317–24, 1987.

102. Waid, D.K. and McLoon, S.C., Ganglion cells influence the fate of dividing retinal cells in culture, *Development* 125(6), 1059–66, 1998.

103. Reh, T.A., Cell-specific regulation of neuronal production in the larval frog retina, *J Neurosci* 7(10), 3317–24, 1987.

104. Waid, D.K. and McLoon, S.C., Ganglion cells influence the fate of dividing retinal cells in culture, *Development* 125(6), 1059–66, 1998.

105. Wang, Y., Dakubo, G.D., Thurig, S., Mazerolle, C.J., and Wallace, V.A., Retinal ganglion cell-derived sonic hedgehog locally controls proliferation and the timing of RGC development in the embryonic mouse retina, *Development* 132(22), 5103–13, 2005.

106. Kim, J., Wu, H.H., Lander, A.D., Lyons, K.M., Matzuk, M.M., and Calof, A.L., GDF11 controls the timing of progenitor cell competence in developing retina, *Science* 308(5730), 1927–30, 2005.

107. Neophytou, C., Vernallis, A.B., Smith, A., and Raff, M.C., Muller-cell-derived leukaemia inhibitory factor arrests rod photoreceptor differentiation at a postmitotic pre-rod stage of development, *Development* 124(12), 2345–54, 1997.

108. Altshuler, D. and Cepko, C., A temporally regulated, diffusible activity is required for rod photoreceptor development in vitro, *Development* 114(4), 947–57, 1992.

109. Altshuler, D., Lo Turco, J.J., Rush, J., and Cepko, C., Taurine promotes the differentiation of a vertebrate retinal cell type in vitro, *Development* 119(4), 1317–28, 1993.

110. Kirsch, M., Schulz-Key, S., Wiese, A., Fuhrmann, S., and Hofmann, H., Ciliary neurotrophic factor blocks rod photoreceptor differentiation from postmitotic precursor cells in vitro, *Cell Tissue Res* 291(2), 207–16, 1998.

111. Schulz-Key, S., Hofmann, H.D., Beisenherz-Huss, C., Barbisch, C., and Kirsch, M., Ciliary neurotrophic factor as a transient negative regulator of rod development in rat retina, *Invest Ophthalmol Vis Sci* 43(9), 3099–108, 2002.

112. Amato, M.A., Boy, S., and Perron, M., Hedgehog signaling in vertebrate eye development: a growing puzzle, *Cell Mol Life Sci* 61(7–8), 899–910, 2004.

113. Neumann, C.J. and Nuesslein-Volhard, C., Patterning of the zebrafish retina by a wave of sonic hedgehog activity, *Science* 289(5487), 2137–9, 2000.

114. Kay, J.N., Link, B.A., and Baier, H., Staggered cell-intrinsic timing of ath5 expression underlies the wave of ganglion cell neurogenesis in the zebrafish retina, *Development* 132(11), 2573–85, 2005.

115. Stenkamp, D.L., Frey, R.A., Mallory, D.E., and Shupe, E.E., Embryonic retinal gene expression in sonic-you mutant zebrafish, *Dev Dyn* 225(3), 344–50, 2002.

116. Jensen, A.M. and Wallace, V.A., Expression of Sonic hedgehog and its putative role as a precursor cell mitogen in the developing mouse retina, *Development* 124(2), 363–71, 1997.

117. Black, G.C., Mazerolle, C.J., Wang, Y., Campsall, K.D., Petrin, D., Leonard, B.C., Damji, K.F., Evans, D.G., McLeod, D., and Wallace, V.A., Abnormalities of the vitreoretinal interface caused by dysregulated Hedgehog signaling during retinal development, *Hum Mol Genet* 12(24), 3269–76, 2003.

118. Moshiri, A. and Reh, T.A., Persistent progenitors at the retinal margin of ptc+/- mice, *J Neurosci* 24(1), 229–37, 2004.

119. Yang, X.J., Roles of cell-extrinsic growth factors in vertebrate eye pattern formation and retinogenesis, *Semin Cell Dev Biol* 15(1), 91–103, 2004.

120. Moshiri, A., McGuire, C.R., and Reh, T.A., Sonic hedgehog regulates proliferation of the retinal ciliary marginal zone in posthatch chicks, *Dev Dyn* 233(1), 66–75, 2005.

121. Spence, J.R., Madhavan, M., Ewing, J.D., Jones, D.K., Lehman, B.M., and Del Rio-Tsonis, K., The hedgehog pathway is a modulator of retina regeneration, *Development* 131(18), 4607–21, 2004.

122. Shkumatava, A. and Neumann, C.J., Shh directs cell-cycle exit by activating p57Kip2 in the zebrafish retina, *EMBO Rep* 6(6), 563–9, 2005.

123. Locker, M., Agathocleous, M., Amato, M.A., Parain, K., Harris, W.A., and Perron, M., Hedgehog signaling and the retina: insights into the mechanisms controlling the proliferative properties of neural precursors, *Genes Dev* 20(21), 3036–48, 2006.

124. Agathocleous, M., Locker, M., Harris, W.A., and Perron, M., A general role of Hedgehog in the regulation of proliferation, *Cell Cycle* 6(2), 156–9, 2007.

125. Tyurina, O.V., Guner, B., Popova, E., Feng, J., Schier, A.F., Kohtz, J.D., and Karlstrom, R.O., Zebrafish Gli3 functions as both an activator and a repressor in Hedgehog signaling, *Dev Biol* 277(2), 537–56, 2005.

126. Hui, C.C., Slusarski, D., Platt, K.A., Holmgren, R., and Joyner, A.L., Expression of three mouse homologs of the Drosophila segment polarity gene cubitus interruptus, Gli, Gli-2, and Gli-3, in ectoderm- and mesoderm-derived tissues suggests multiple roles during postimplantation development, *Dev Biol* 162(2), 402–13, 1994.

127. Hu, M.C., Mo, R., Bhella, S., Wilson, C.W., Chuang, P.T., Hui, C.C., and Rosenblum, N.D., GLI3-dependent transcriptional repression of Gli1, Gli2 and kidney patterning genes disrupts renal morphogenesis, *Development* 133(3), 569–78, 2006.

128. Hashimoto, T., Zhang, X.M., Chen, B.Y., and Yang, X.J., VEGF activates divergent intracellular signaling components to regulate retinal progenitor cell proliferation and neuronal differentiation, *Development* 133(11), 2201–10, 2006.

129. Takatsuka, K., Hatakeyama, J., Bessho, Y., and Kageyama, R., Roles of the bHLH gene Hes1 in retinal morphogenesis, *Brain Res* 1004(1–2), 148–55, 2004.

130. Kenney, A.M., Widlund, H.R., and Rowitch, D.H., Hedgehog and PI-3 kinase signaling converge on Nmyc1 to promote cell cycle progression in cerebellar neuronal precursors, *Development* 131(1), 217–28, 2004.

131. Mill, P., Mo, R., Hu, M.C., Dagnino, L., Rosenblum, N.D., and Hui, C.C., Shh controls epithelial proliferation via independent pathways that converge on N-Myc, *Dev Cell* 9(2), 293–303, 2005.

132. Hatton, B.A., Knoepfler, P.S., Kenney, A.M., Rowitch, D.H., de Alboran, I.M., Olson, J.M., and Eisenman, R.N., N-myc is an essential downstream effector of Shh signaling during both normal and neoplastic cerebellar growth, *Cancer Res* 66(17), 8655–61, 2006.

133. Wallace, V.A. and Raff, M.C., A role for Sonic hedgehog in axon-to-astrocyte signalling in the rodent optic nerve, *Development* 126(13), 2901–9, 1999.

134. Levine, E.M., Roelink, H., Turner, J., and Reh, T.A., Sonic hedgehog promotes rod photoreceptor differentiation in mammalian retinal cells in vitro, *J Neurosci* 17(16), 6277–88, 1997.

135. Takabatake, T., Ogawa, M., Takahashi, T.C., Mizuno, M., Okamoto, M., and Takeshima, K., Hedgehog and patched gene expression in adult ocular tissues, *FEBS Lett* 410(2–3), 485–9, 1997.

136. Levine, E.M., Roelink, H., Turner, J., and Reh, T.A., Sonic hedgehog promotes rod photoreceptor differentiation in mammalian retinal cells in vitro, *J Neurosci* 17(16), 6277–88, 1997.

137. Takabatake, T., Ogawa, M., Takahashi, T.C., Mizuno, M., Okamoto, M., and Takeshima, K., Hedgehog and patched gene expression in adult ocular tissues, *FEBS Lett* 410(2–3), 485–9, 1997.

138. Dakubo, G.D. and Wallace, V.A., Hedgehogs and retinal ganglion cells: organizers of the mammalian retina, *Neuroreport* 15(3), 479–82, 2004.

139. Stenkamp, D.L., Frey, R.A., Prabhudesai, S.N., and Raymond, P.A., Function for Hedgehog genes in zebrafish retinal development, *Dev Biol* 220(2), 238–52, 2000.

140. Perron, M., Boy, S., Amato, M.A., Viczian, A., Koebernick, K., Pieler, T., and Harris, W.A., A novel function for Hedgehog signalling in retinal pigment epithelium differentiation, *Development* 130(8), 1565–77, 2003.

141. Tsonis, P.A., Vergara, M.N., Spence, J.R., Madhavan, M., Kramer, E.L., Call, M.K., Santiago, W.G., Vallance, J.E., Robbins, D.J., and Del Rio-Tsonis, K., A novel role of the hedgehog pathway in lens regeneration, *Dev Biol* 267(2), 450–61, 2004.

142. Thanos, S. and Mey, J., Development of the visual system of the chick. II. Mechanisms of axonal guidance, *Brain Res Brain Res Rev* 35(3), 205–45, 2001.

143. Oster, S.F., Deiner, M., Birgbauer, E., and Sretavan, D.W., Ganglion cell axon pathfinding in the retina and optic nerve, *Semin Cell Dev Biol* 15(1), 125–36, 2004.

144. Culverwell, J. and Karlstrom, R.O., Making the connection: retinal axon guidance in the zebrafish, *Semin Cell Dev Biol* 13(6), 497–506, 2002.

145. Russell, C., The roles of Hedgehogs and fibroblast growth factors in eye development and retinal cell rescue, *Vision Res* 43(8), 899–912, 2003.

146. Hutson, L.D. and Chien, C.B., Wiring the zebrafish: axon guidance and synaptogenesis, *Curr Opin Neurobiol* 12(1), 87–92, 2002.

147. Dakubo, G.D., Wang, Y.P., Mazerolle, C., Campsall, K., McMahon, A.P., and Wallace, V.A., Retinal ganglion cell-derived sonic hedgehog signaling is required for optic disc and stalk neuroepithelial cell development, *Development* 130(13), 2967–80, 2003.

148. Trousse, F., Marti, E., Gruss, P., Torres, M., and Bovolenta, P., Control of retinal ganglion cell axon growth: a new role for Sonic hedgehog, *Development* 128(20), 3927–36, 2001.

149. Kolpak, A., Zhang, J., and Bao, Z.Z., Sonic hedgehog has a dual effect on the growth of retinal ganglion axons depending on its concentration, *J Neurosci* 25(13), 3432–41, 2005.
150. Marcus, R.C., Blazeski, R., Godement, P., and Mason, C.A., Retinal axon divergence in the optic chiasm: uncrossed axons diverge from crossed axons within a midline glial specialization, *J Neurosci* 15(5 Pt 2), 3716–29, 1995.
151. Wang, L.C., Dani, J., Godement, P., Marcus, R.C., and Mason, C.A., Crossed and uncrossed retinal axons respond differently to cells of the optic chiasm midline in vitro, *Neuron* 15(6), 1349–64, 1995.
152. Macdonald, R., Scholes, J., Strahle, U., Brennan, C., Holder, N., Brand, M., and Wilson, S.W., The Pax protein Noi is required for commissural axon pathway formation in the rostral forebrain, *Development* 124(12), 2397–408, 1997.
153. Plump, A.S., Erskine, L., Sabatier, C., Brose, K., Epstein, C.J., Goodman, C.S., Mason, C.A., and Tessier-Lavigne, M., Slit1 and Slit2 cooperate to prevent premature midline crossing of retinal axons in the mouse visual system, *Neuron* 33(2), 219–32, 2002.
154. Kennedy, T.E., Serafini, T., de la Torre, J.R., and Tessier-Lavigne, M., Netrins are diffusible chemotropic factors for commissural axons in the embryonic spinal cord, *Cell* 78(3), 425–35, 1994.
155. Barresi, M.J., Hutson, L.D., Chien, C.B., and Karlstrom, R.O., Hedgehog regulated Slit expression determines commissure and glial cell position in the zebrafish forebrain, *Development* 132(16), 3643–56, 2005.
156. Tamada, A., Shirasaki, R., and Murakami, F., Floor plate chemoattracts crossed axons and chemorepels uncrossed axons in the vertebrate brain, *Neuron* 14(5), 1083–93, 1995.
157. Burne, J.F. and Raff, M.C., Retinal ganglion cell axons drive the proliferation of astrocytes in the developing rodent optic nerve, *Neuron* 18(2), 223–30, 1997.
158. Barres, B.A. and Raff, M.C., Axonal control of oligodendrocyte development, *J Cell Biol* 147(6), 1123–8, 1999.
159. Soukkarieh, C., Agius, E., Soula, C., and Cochard, P., Pax2 regulates neuronal-glial cell fate choice in the embryonic optic nerve, *Dev Biol*, 303(2), 800–13, 2006.
160. Fernandez, P.A., Tang, D.G., Cheng, L., Prochiantz, A., Mudge, A.W., and Raff, M.C., Evidence that axon-derived neuregulin promotes oligodendrocyte survival in the developing rat optic nerve, *Neuron* 28(1), 81–90, 2000.
161. Wang, S., Sdrulla, A.D., diSibio, G., Bush, G., Nofziger, D., Hicks, C., Weinmaster, G., and Barres, B.A., Notch receptor activation inhibits oligodendrocyte differentiation, *Neuron* 21(1), 63–75, 1998.
162. Gao, L. and Miller, R.H., Specification of optic nerve oligodendrocyte precursors by retinal ganglion cell axons, *J Neurosci* 26(29), 7619–28, 2006.
163. Traiffort, E., Moya, K.L., Faure, H., Hassig, R., and Ruat, M., High expression and anterograde axonal transport of aminoterminal sonic hedgehog in the adult hamster brain, *Eur J Neurosci* 14(5), 839–50, 2001.
164. Palma, V., Lim, D.A., Dahmane, N., Sanchez, P., Brionne, T.C., Herzberg, C.D., Gitton, Y., Carleton, A., Alvarez-Buylla, A., and Ruiz i Altaba, A., Sonic hedgehog controls stem cell behavior in the postnatal and adult brain, *Development* 132(2), 335–44, 2005.
165. Machold, R., Hayashi, S., Rutlin, M., Muzumdar, M.D., Nery, S., Corbin, J.G., Gritli-Linde, A., Dellovade, T., Porter, J.A., Rubin, L.L., Dudek, H., McMahon, A.P., and Fishell, G., Sonic hedgehog is required for progenitor cell maintenance in telencephalic stem cell niches, *Neuron* 39(6), 937–50, 2003.
166. Lai, K., Kaspar, B.K., Gage, F.H., and Schaffer, D.V., Sonic hedgehog regulates adult neural progenitor proliferation in vitro and in vivo, *Nat Neurosci* 6(1), 21–7, 2003.
167. Charytoniuk, D., Traiffort, E., Hantraye, P., Hermel, J.M., Galdes, A., and Ruat, M., Intrastriatal sonic hedgehog injection increases Patched transcript levels in the adult rat subventricular zone, *Eur J Neurosci* 16(12), 2351–7, 2002.

168. Huang, Z. and Kunes, S., Hedgehog, transmitted along retinal axons, triggers neurogenesis in the developing visual centers of the Drosophila brain, *Cell* 86(3), 411–22, 1996.

169. Huang, Z. and Kunes, S., Signals transmitted along retinal axons in Drosophila: Hedgehog signal reception and the cell circuitry of lamina cartridge assembly, *Development* 125(19), 3753–64, 1998.

170. Umetsu, D., Murakami, S., Sato, M., and Tabata, T., The highly ordered assembly of retinal axons and their synaptic partners is regulated by Hedgehog/Single-minded in the Drosophila visual system, *Development* 133(5), 791–800, 2006.

171. Frank-Kamenetsky, M., Zhang, X.M., Bottega, S., Guicherit, O., Wichterle, H., Dudek, H., Bumcrot, D., Wang, F.Y., Jones, S., Shulok, J., Rubin, L.L., and Porter, J.A., Small-molecule modulators of Hedgehog signaling: identification and characterization of Smoothened agonists and antagonists, *J Biol* 1(2), 10, 2002.

172. Cooper, M.K., Porter, J.A., Young, K.E., and Beachy, P.A., Teratogen-mediated inhibition of target tissue response to Shh signaling, *Science* 280(5369), 1603–7, 1998.

173. Cepko, C.L., Ryder, E., Austin, C., Golden, J., Fields-Berry, S., and Lin, J., Lineage analysis using retroviral vectors, *Methods* 14(4), 393–406, 1998.

174. Matsuda, T., and Cepko, C.L., Electroporation and RNA interference in the rodent retina in vivo and in vitro, *Proc Natl Acad USA* 101(1), 16–22, 2004.

175. Naldini, L. and Verma, I.M., Lentiviral vectors, *Adv Virus Res* 55, 599–609, 2000.

176. Lewis, K.E. and Eisen, J.S., Hedgehog signaling is required for primary motoneuron induction in zebrafish, *Development* 128(18), 3485–95, 2001.

177. Marquardt, T., Ashery-Padan, R., Andrejewski, N., Scardigli, R., Guillemot, F., and Gruss, P., Pax6 is required for the multipotent state of retinal progenitor cells, *Cell* 105(1), 43–55, 2001.

14 Thrombospondin-1 Signal Transduction and Retinal Vascular Homeostasis

Nader Sheibani, Yixin Tang, Terri A. DiMaio,
Shuji Kondo, Elizabeth A. Scheef,
and Christine M. Sorenson

CONTENTS

14.1   INTRODUCTION

Thrombospondin-1 (TSP1) is a member of the TSP family of matricellular proteins, currently containing five members (TSP1–5) (1,2). The highly restricted spatial and temporal expression of TSPs indicates important roles for these molecules during development (3). TSP1, or platelet TSP, was the first member of this family identified in platelets α-granules (4). TSP expression during eye development has been previously demonstrated throughout the optic vesicle, lens, retina, Bruch's membrane, and nerve fiber layer. In the adult retina, very little expression is detectable in cell bodies and ganglion cells. It is now known that TSP1 is expressed by a variety of cells in culture, including endothelial cells, astrocytes, pericytes, retinal pigmented epithelial cells, fibroblasts, and smooth muscle cells. In fact, production of TSP1 and TSP2 by astrocytes was recently shown to drive synaptogenesis in retinal neurons (5). TSP1 interacts with at least a dozen cell surface receptors expressed in various combinations on different cell types. It is the interaction of TSP1 with these cell-specific receptors that elicits a cell-type-specific response.

The first biological function described for TSP1 was its antiangiogenic activity in the late 1980s (6,7). In fact, TSP1 was the first endogenous inhibitor of angiogenesis identified whose expression is downregulated during malignant transformation, and associated with the angiogenic switch in many solid tumors (8). There are now many studies that demonstrate downregulation of TSP1 is associated with a more invasive tumor phenotype, and that restoration of TSP1 expression is sufficient to block tumor growth (9). TSP1 is also shown to mediate these effects by directly acting on the endothelium, inducing apoptosis of endothelial cells in vivo and inhibiting endothelial cell migration and proliferation in vitro (10–13). Later, the antiangiogenic activity of TSP1 was mapped to its procollagen and Type I homology domains whose interactions with CD36, a TSP receptor highly expressed on microvascular endothelial cells, mediate the angioinhibitory activity of TSP (13,14). Mice deficient in TSP1 or CD36 exhibit increased vascular density in many tissues, providing additional support for the role of TSP1 and its receptor in angiogenesis (9,15,16). However, the physiological roles TSP1 plays during vascular development and angiogenesis, and how these activities are mediated, require further investigation.

Using the postnatal developing retinal vasculature and oxygen-induced ischemic retinopathy in the wild-type and TSP1-deficient (TSP1$^{-/-}$) mice, we investigated the potential roles of TSP1 in vascular development and neovascularization. To gain further insight into the molecular and cellular mechanisms involved, we also isolated retinal endothelial cells from wild-type and TSP1$^{-/-}$ mice. These studies support an important role for TSP1 in retinal vascular homeostasis.

14.2   POSTNATAL DEVELOPMENT OF RETINAL VASCULATURE

The mouse retinal vasculature develops postnatally (17). During the first week of life, a superficial layer of vessels initiates from near the optic disk and spreads radially to the periphery. During the second and third weeks of life, these vessels sprout deep into the retina forming the deep and intermediate vessels of the retina (figure 14.1). The retinal vasculature continues to undergo pruning and remodeling for the next 3 weeks, and by 6 weeks of life, the mature retinal vasculature is established.

Our laboratory has been evaluating the physiological role TSP1 plays in retinal vascularization. We quantitatively assessed the vascular density and integrity of retinal vasculature by trypsin-digest whole-mount preparation of retinal vasculature (figure 14.2). Using this technique, we observed no significant differences in the retinal vasculature of wild-type and TSP1$^{-/-}$ mice up to 3 weeks of age (18). However, we observed an increase in vascular density of TSP1$^{-/-}$ mice at 6 weeks of age compared to wild-type mice. This was mainly attributed to increased number of endothelial cells in the absence of TSP1. A significant reduction in the rate of apoptosis was observed in retinal vasculature of TSP1$^{-/-}$ mice at 4 weeks of age (18). In the absence of TSP1 the numbers and density of pericytes were not significantly affected. Thus, TSP1 expression is essential for appropriate pruning and remodeling of retinal vasculature.

The regression of hyaloid vasculature, an apoptosis-dependent process, that occurs after birth, was also delayed in the absence of TSP1. The papillary membrane and hyaloid vessels (hyaloid arteries, tunica vasculosa lentis, and vasa hyaloidea propria) are embryonic vessels that provide nourishment to the immature lens, retina, and vitreous (19). However, they regress during the later stages of ocular development. Therefore, TSP1 activity is essential in removal of vasculature in response to excess perfusion.

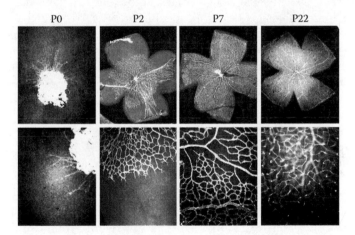

FIGURE 14.1 Postnatal development of mouse retinal vasculature. Retinas from wild-type mice were dissected and stained with a rabbit antimouse collagen IV antibody (Chemicon) showing retinal blood vessels. Lower panels are higher magnifications of upper panels.

FIGURE 14.2 Trypsin digest preparation of wild-type mouse whole-mount retinas. Different magnifications are shown (A: ×25; B: ×40; C: ×100).

14.2.1 Preparation of Retinal Whole-Mount Trypsin Digests

Protocol 1: Preparation of Retinal Trypsin Digests

Retinas from age-matched wild-type and TSP1$^{-/-}$ mice were digested with a solution of trypsin while the retinal vascular network was left intact. The preparations are placed on charged slides, dried, and then stained with periodic acid-schiff (PAS) and hematoxylin.

1. Eyes are enucleated from P21 or P42 mice and fixed in 4% paraformaldehyde for at least 24 h. The eyes are bisected equatorially, and the entire retina is removed under the dissecting microscope.
2. Retinas are washed overnight in distilled water, and then incubated in 3% trypsin (trypsin 1:250, Difco; prepared in 0.1 M Tris, 0.1 M maleic acid, pH 7.8, containing 0.2 M NaF) for approximately 1–1.5 h. This is carefully monitored under a dissecting microscope, and a brush is used to beat the tissue to loosen the digesting tissue.
3. Following completion of digestion, retinal vessels are flattened by four radial cuts and mounted on glass slides for PAS and hematoxylin staining.
4. Nuclear morphology is used to distinguish pericytes from endothelial cells. The nuclei of endothelial cells are oval or elongated and lie within the vessel wall along the axis of the capillary, whereas pericyte nuclei are small, spherical, stain densely, and generally have a protuberant position on the capillary wall.
5. The stained and intact retinal whole mounts are coded and subsequent counting is performed masked.
6. The number of endothelial cells and pericytes are determined by counting respective nuclei under the microscope at a magnification of ×400. A mounting reticle (10 μm × 10 μm) is placed in one of the viewing oculars to facilitate counting. Only retinal capillaries are included in the cell count, which is performed in the midzone of the retina.
7. Count the number of endothelial cells and pericytes in four reticles from the four quadrants of each retina. Generally, retinas from five mice are used. The total number of endothelial cells and pericytes for each retina is determined by adding the numbers from the four reticles. The ratio of endothelial cells to pericytes is then calculated. To evaluate the density of cells in the capillaries, the mean number of endothelial cells or pericytes is recorded in four reticles from the four quadrants of each retina.

To evaluate whether CD36, the known angioinhibitory receptor of TSP1, is involved in these processes, we examined retinal vasculature in CD36$^{-/-}$ mice. Our evaluation of retinal vasculature of CD36$^{-/-}$ mice detected no significant effect on postnatal vascularization of retina at 3 or 6 weeks of age. This suggests the effects of TSP1 on retinal vasculature during pruning and remodeling may be mediated through other receptors that interact with TSP1. Another TSP1 receptor that interacts with the carboxyl-terminal domain of TSP1 is integrin-associated protein, or CD47. There is evidence that some of the antiangiogenic activity of TSP1 may be mediated through other TSP1 receptors, including CD47 and β1-integrin (20–23). We next examined the postnatal development of retinal vasculature in wild-type and CD47$^{-/-}$ mice. We observed a similar phenotype in CD47$^{-/-}$ mice as our TSP1$^{-/-}$ mice. The 6-week-old CD47$^{-/-}$ mice exhibited increased retinal vascular density compared to wild-type mice. Thus, CD47 may be the receptor that is essential for TSP1-mediated vascular remodeling and pruning. This hypothesis was further supported by the work of Freyberg et al. showing that apoptosis of endothelial cells induced by proatherogenic flow conditions is mediated through TSP1 and CD47 (24). Roberts and colleagues also recently showed that CD47 is essential for nitric-oxide-mediated angioinhibitory activity of TSP1 using vascular cells prepared from transgenic mice lacking TSP1, CD36, or CD47 (25).

14.2.2 THROMBOSPONDIN-1 AND OXYGEN-INDUCED ISCHEMIC RETINOPATHY

The mouse oxygen-induced ischemic retinopathy (OIR) is a highly reproducible model of retinal neovascularization (26). In this model the postnatal day 7 (P7) mice and their mother (C57BL/6J, Jackson Labs) are exposed to 75% oxygen for 5 days. The exposure to high oxygen downregulates expression of vascular endothelial growth factor (VEGF), which is essential for survival of existing vessels and formation of additional vasculature. The lack of VEGF results in obliteration of retinal vasculature in the center of the retina and prevent its further vascularization. We observed that mice lacking TSP1 were significantly protected from hyperoxia-mediated vessel obliteration (18). The mice are then returned to room air for 5 days. At this point, the retina becomes ischemic because of excessive loss of blood vessels during exposure to high oxygen, and upregulates VEGF expression to promote growth of new blood vessels. We did not observe a significant difference in the degree of neovascularization in the presence or absence of TSP1 (18). We next determined whether regression of the newly formed vessels, which occurs after P17 during OIR, is affected in the absence of TSP1. We observed a delay in the regression of the newly formed vessels in P21 TSP1$^{-/-}$ mice compared to wild-type mice.

To gain further insight into the mechanisms involved during ischemia-driven retinal neovascularization, we examined the expression of TSP1 and VEGF during room air and OIR in retinas prepared from wild-type and TSP1$^{-/-}$ mice. Northern blot analysis of RNAs prepared from retinas of wild-type P7, P15 room air, or P15 OIR (5 days of hyperoxia and 3 days of normoxia) mice indicated that TSP1 expression was increased approximately twofold from P7 to P15 in room air without a significant effect on TSP1 expression in P15 mice during OIR. Thus, expression of TSP1 in the retina was not significantly affected during OIR. In contrast, VEGF levels were significantly upregulated in wild-type and TSP1$^{-/-}$ mice during OIR (18). These results suggested that a threshold level exists for VEGF and TSP1 such that changes in VEGF levels determine the fate of endothelial cells; i.e., when VEGF increases above this threshold level, it promotes angiogenesis, and when VEGF decreases below this level, TSP1 induces endothelial cell apoptosis. Therefore, an important role of TSP1 during vascular development is in vascular pruning and remodeling to adjust the oxygen need of the tissue by eliminating excess blood vessels. If oxygen level is high, because of excess perfusion, VEGF levels will decrease, and those vessels will be eliminated by the action of TSP1 on endothelial cells.

14.2.3 ATTENUATION OF RETINAL VASCULAR DEVELOPMENT AND NEOVASCULARIZATION IN MICE OVEREXPRESSING TSP1 IN THE LENS

To further demonstrate the importance of TSP1 in retinal vascular homeostasis and neovascularization, we hypothesized that aberrant expression of TSP1 would severely compromise these processes. To test this hypothesis, we generated transgenic mice that expressed TSP1 in their lens using αA-crystallin promoter (27). The expression of αA-crystallin normally begins around embryonic day 12, prior to initiation of postnatal retinal vascularization. Therefore, expression of TSP1 prior to postnatal vascularization of the retina could severely compromise this process. We showed that mice overexpressing TSP1 in their lens are defective in the postnatal development of retinal vasculature and neovascularization during OIR (27). We observed a significant delay in the development and maturation of retinal vasculature, which was associated with a significant increase in apoptosis of retinal endothelial cells and decrease in retinal vascular density. Transgenic mice expressing increased levels of TSP1 in their lens were also protected from ischemia-mediated neovascularization during OIR. Together, these studies further confirmed the important role of TSP1 during retinal vascular development and neovascularization. Therefore, pathological conditions that affect the threshold levels for VEGF and/or TSP1 will severely impact retinal vascular homeostasis.

14.3 ESTABLISHMENT OF MOUSE RETINAL ENDOTHELIAL CELLS

Our in vivo studies demonstrated that TSP1 is an important modulator of retinal vascular homeostasis. To gain further insight into the molecular and cellular mechanisms that mediate TSP1 function, we established retinal endothelial cells from wild-type and TSP1$^{-/-}$ mice (28). Our initial characterization of these cells indicated that the TSP1$^{-/-}$ retinal endothelial cells are more migratory and express increased amounts of fibronectin compared to wild-type cells, characteristics of proangiogenic endothelial cells. We also showed increased fibronectin expression in retinal vasculature of TSP1$^{-/-}$ mice (29). We later showed that lack of TSP1 in retinal endothelial cells affects expression of specific cell cycle regulators, including cyclin A and cyclin D1, and cdk2 (29). The cyclin D1 has been demonstrated to modulate cell migration through inhibition of TSP1 expression and Rho-activated kinase II (30). This is consistent with our observation that retinal endothelial cells lacking TSP1 express increased levels of cyclin D1. Therefore, a reciprocal relationship may exist between cyclin D1 and TSP1 that impacts cell migration.

PROTOCOL 2: ISOLATION OF MOUSE RETINAL ENDOTHELIAL CELLS

Isolation of retinal endothelial cells from mouse has been very difficult, and to my knowledge has not been previously reported. We took advantage of the Immortomouse, which expresses a temperature-sensitive large T antigen. We have crossed our wild-type and transgenic mice to Immortomice and have successfully cultured retinal endothelial cells and astrocytes (28,31).

1. Immortomice (express a temperature-sensitive SV40 large T antigen) are obtained from Charles River Laboratories (Wilmington, Massachusetts). TSP1$^{-/-}$ mice in the C57BL/6J background were generated as previously described (32). TSP1$^{-/-}$ mice were crossed with Immortomouse, and the immorto/TSP1$^{-/-}$ mice were identified by PCR analysis of DNA isolated from tail biopsies. The PCR primer sequences were as follows: immorto-forward: 5'- CCT CTG AGC TAT TCC AGA AGT AGT G -3', immorto-reverse: 5'-TTA GAG CTT TAA ATC TCT GTA GGT AG 3'; Neo-forward: 5'-TGC TGT CCA TCT GCA CGA GAC TAG -3', Neo-reverse: 5'-GAG TTT GCT TGT GGT GAA CGC TCA G-3'; TSP1-forward: 5'AGG GCT ATG TGG AAT TAA TAT CGG 3', and TSP1-reverse: 5'-GAG TTT GCT TGT GGT GAA CGC TCA G- 3'.

2. Antibody-coated magnetic beads for endothelial cell isolation are prepared as follows. Sheep antirat Dynabeads (Dynal Biotech, Lake Success, New York) are washed thrice with serum-free DMEM (Dulbecco's Modified Eagle's Medium; Invitrogen, Carlsbad, California) and then incubated with rat antimouse PECAM-1 monoclonal antibody MEC13.3 (BD Pharmingen, San Diego, California) overnight at 4°C (10 µL/50 µL beads in DMEM). Following incubation, beads are washed thrice with DMEM containing 10% fetal bovine serum (FBS) and resuspended in the same medium.

3. Eyes from one litter (6 to 7 pups) of 4-week-old wild-type or TSP1$^{-/-}$ immortomice are enucleated and hemisected. The retinas are dissected out aseptically under dissecting microscope and kept in serum-free DMEM containing penicillin/streptomycin (Sigma, St. Louis, Missouri).

4. Twelve to fourteen retinas (from one litter) are pooled together, rinsed with serum-free DMEM, minced into small pieces in a 60 mm tissue culture dish using sterilized razor blades, and digested in 5 mL of collagenase type I (1 mg/mL in serum-free DMEM, Worthington, Lakewood, New Jersey) for 30–45 min at 37°C.

5. Following digestion, DMEM with 10% FBS is added, and cells are pelleted. The cellular digests then are filtered through a double layer of sterile 40 µm nylon mesh (Sefar America Inc., Fisher Scientific, Hanover Park, Illinois), centrifuged at 400 × g for 10 min to pellet cells, and cells are washed twice with DMEM containing 10% FBS.

6. The cells are resuspended in 1.5 mL medium (DMEM with 10% FBS), and incubated with sheep antirat magnetic beads precoated with anti-PECAM-1 as described earlier for 3 h at 4°C.

7. After affinity binding, magnetic beads are washed six times with DMEM with 10% FBS, and bound cells in endothelial cell growth medium are plated into a single well of a 24-well plate precoated with 2 µg/mL of human fibronectin (BD Biosciences, Bedford, Massachusetts). Endothelial cells were grown in DMEM containing 20% FBS, 2 mM L-glutamine, 2 mM sodium pyrovate, 20 mM HEPES, 1% nonessential amino acids, 100 µg/mL streptomycin,

100 U/mL penicillin, freshly added heparin at 55 U/mL (Sigma, St. Louis, Missouri), endothelial growth supplement 100 µg/mL (Sigma, St. Louis, Missouri), and murine recombinant interferon-γ (R&D, Minneapolis, Minnesota) at 44 units/mL.

8. Cells are maintained at 33°C with 5% CO_2. Cells will be progressively passed to larger plates, maintained, and propagated in 1% gelatin-coated 60 mm dishes.

14.3.1 ENHANCED PROANGIOGENIC SIGNALING IN TSP1⁻/⁻ RETINAL ENDOTHELIAL CELLS

Increased fibronectin expression in endothelial cells is generally associated with a proangiogenic phenotype (33). The increased fibronectin expression in TSP1⁻/⁻ retinal endothelial cells is consistent with their enhanced proliferative and migratory phenotype (28,34). Fibronectin interactions with its cellular receptor (integrins) trigger intracellular signaling pathways, including Src/PI3-kinase, which impact endothelial cell proliferation, migration, and survival (35). We observed TSP1⁻/⁻ retinal endothelial cells have increased levels of active Src kinase compared to wild-type cells (29). However, this was not associated with increased levels of MAPK/ERKs.

Activation of the Src signaling pathway may result in Src-dependent activation of the PI3-kinase pathway, and its downstream effectors including MAPK/P38, Rac/Cdc42, and Akt/PKB. TSP1⁻/⁻ retinal endothelial cells expressed increased levels of active MAPK/P38 compared to wild-type cells without a significant change in the total levels of MAPK/P38. The levels of Rac1-GTP and Cdc42-GTP were also higher in TSP1⁻/⁻ retinal endothelial cells. We observed a significant increase in the levels of active Akt/PKB in TSP1⁻/⁻ retinal endothelial cells compared to wild-type cells, as well as in retinal vasculature in vivo (29). Thus, TSP1 expression impacts endothelial cell phenotype through the modulation of proangiogenic signaling pathways. Inhibition of these signaling pathways using specific pharmacological inhibitors demonstrated that the enhanced migratory phenotype of TSP1⁻/⁻ retinal endothelial cells was dependent on these proangiogenic signaling pathways.

14.3.2 DECREASED PROAPOPTOTIC SIGNALING IN TSP1⁻/⁻ RETINAL ENDOTHELIAL CELLS

The biological consequences of TSP1 interaction with its angioinhibitory receptor CD36 is the recruitment of Fyn (a member of the Src family of kinases) and activation of downstream events that lead to activation of MAPK/JNK and caspase 3, promoting apoptosis in microvascular endothelial cells (11,12). We observed decreased activation of Fyn in TSP1⁻/⁻ retinal endothelial cells compared to wild-type cells. This was also associated with decreased active JNK2 in TSP1⁻/⁻ retinal endothelial cells. However, the lack of TSP1 did not affect the levels of pro- and active caspase 3 in retinal endothelial cells (29). Thus, TSP1 expression impacts retinal endothelial cells apoptotic pathways through activation of the Fyn-JNK MAPK pathway. Therefore, lack of TSP1 enhances cell survival and proliferation signaling, and it eliminates cell apoptotic signaling, promoting angiogenesis as observed in the TSP1$^{-/-}$ mouse retina and retinal endothelial cells.

14.3.3 Downregulation of TSP1 Expression in Wild-Type Retinal Endothelial Cells

The majority of studies of TSP1 function in endothelial cells have focused on its role as an inducer of apoptosis. Our studies demonstrated that TSP1 is also a major suppressor of endothelial cell proliferation and migration through modulation of fibronectin expression and limiting Src/PI3-kinase signaling pathways. Although the role of CD36 in inhibition of angiogenesis has been demonstrated both in vivo and in vitro, our in vivo evaluation of retinal vascular development in TSP1$^{-/-}$ and CD47$^{-/-}$ suggests an important role for CD47 in angioinhibitory activity of TSP1. How signaling from CD36 and CD47 may be integrated to modulate different aspects of TSP1 angioinhibitory function remains elusive. However, our observations reemphasize the role of TSP1 as an endogenous inhibitor of angiogenesis and a tumor suppressor, whose downregulation in a variety of tumors promotes the proangiogenic phenotype and tumor progression.

To demonstrate that lack of TSP1 is solely responsible for the phenotype of TSP1$^{-/-}$ retinal endothelial cells, we knocked down TSP1 expression in wild-type retinal endothelial cells. Mouse TSP1-specific siRNAs were synthesized and cloned into a retroviral vector as recommended by the supplier (Invitrogen) and used to infect wild-type retinal endothelial cells. Stable clones of retinal endothelial cells expressing a specific TSP1 siRNA were evaluated for their TSP1 expression and their migratory properties. We observed that downregulation of TSP1 expression in wild-type cells resulted in their enhanced migration, as we observed in TSP1$^{-/-}$ retinal endothelial cells (figure 14.3). Therefore, lack of TSP1 is sufficient to promote a proangiogenic phenotype in retinal endothelial cells.

Protocol 3: Construction of Mouse TSP1-Specific siRNAs

TSP1-specific siRNA were utilized to knock down TSP1 in wild-type retinal endothelial cells and evaluate its effects on endothelial cell migration, adhesion, proliferation, apoptosis, and capillary morphogenesis.

1. Scan the mouse cDNA sequence for TSP1 for putative target sequences for siRNA using the siDESIGN center (www.dharmacon.com). The identified sequences are blasted against the mouse genome database to exclude sequences that match nonspecific mRNA.
2. An oligonucleotide containing the 19-base specific sequence (table 14.1), its reverse complementary sequence, a 9-base spacer sequence, and terminal sequences for ligation to the BglII and HindIII restriction sites of pSUPER-retro vector is synthesized (Invitrogen, Carlsbad, California). This oligonucleotide, and another with a complementary sequence, are annealed by heating to 90°C for 4 min, then to 70°C, and slowly cooling to 10°C in 100 mM NaCl and 50 mM HEPES, pH 7.4.

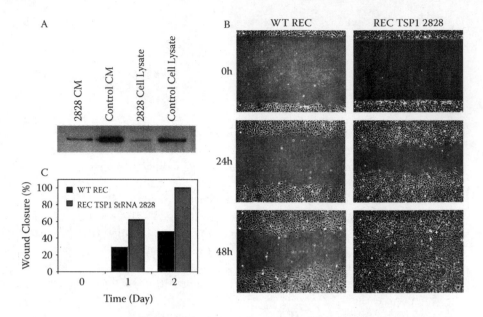

FIGURE 14.3 Enhanced migration of wild-type retinal endothelial cells expressing a TSP1-specific siRNA. (A) Western blot analysis of conditioned medium (CM) or lysates from wild-type retinal endothelial cells expressing a specific TSP1 siRNA (2828) or control virus. (B) Wound migration of wild-type retinal endothelial cells infected with a control or TSP1-specific siRNA (2828) virus. Wound closure was monitored by photography in digital format after 24 or 48 h. (C) The quantitative assessment of migration assay as percentage distance migrated.

TABLE 14.1
Murine TSP1-Specific siRNAs

siRNAs	DNA Sequence
TSP1ORF2828	5′-CCAGGCCGACCATGATAAA-3′
TSP1ORF3547	5′-TCAGAGTGGTGATGTATGA-3′
TSP1ORF3138	5′-GAACTTGTCCAGACTGTAA-3′
TSP1ORF2831	5′-GGCCGACCATGATAAAGAT-3′
Control	5′-TTCTCCGAACGTGTCACGT-3′

3. The duplexes are then cloned into the pSUPER-retro vector. Recombinant pSUPER-retro vector is then transformed into *E. coli* DH5α (Invitrogen), and clones containing vector with insert are selected on 100 µg/mL ampicillin plates, screened by digestion with BglII, and isolated using the plasmid midi kit (QIAGEN, Valencia, California). The successful ligation of BglII-HindIII insert with the vector digested with BamHI and HindIII will result in loss of BamHI and BglII recognition sites. BglII site is compatible with BamHI, but their ligation results in loss of both recognition sites. Therefore, the plasmid with insert will not be digested with BglII, indicating successful ligation.

4. Generally, four such oligonucleotides are synthesized for siRNA-mediated knockdown of gene expression by 70–90%. An oligonucleotide that lacks homology with mouse sequences is used as a control.

5. The retroviral packaging cell line Phi-NX (Nolan Lab, Stanford, California) will be grown in Dulbecco's modified Eagle's medium (DMEM) with 10% FBS supplemented with 1% penicillin-streptomycin and 1% glutamine. Cells (4×10^6) will be seeded in 100 mm dishes.

6. The next day, cells are transfected with 5 µg pSUPER-retro control siRNA vector or vector containing TSP1-specific siRNA insert using Lipofectin (Invitrogen). The cell medium, which contains retroviral particles, will be harvested and filtered through a 0.45 µm filter 48 h post transfection.

7. Filtered virus-containing medium (1 mL) is added to retinal endothelial cells (1×10^6 cells per 60 mm dish) in the presence of polybrene (4 µg/mL) to improve viral infection for 6 h. The virus-containing medium is then replaced with fresh medium, and 24 h later puromycin (2 µg/mL) is added (the pSUPER-retro vector contains a puromycin resistance gene to allow for selection). The puromycin-resistant cells are expanded and used for analyses. Generally, the efficiency of infection is around 80–90% in our laboratory.

14.4 TSP1 AND MODULATION OF NITRIC OXIDE SYNTHASE ACTIVITY

The predominant nitric oxide synthase (NOS) expressed in endothelial cells is eNOS. It catalyzes the formation of NO from L-arginine, which results in the activation of soluble guanylyl cyclase and initiation of various signaling cascades (36). NO produced in the endothelium is involved in signaling for vasorelaxation, platelet aggregation, proliferation of vascular smooth muscle cells, and other mechanisms of cardiovascular homeostasis. The eNOS is constitutively expressed in endothelial cells, as well as in some neurons and cardiac myocytes. In resting cells, myristoylation and palmitoylation cause eNOS to localize primarily to the plasma membrane, but upon cellular activation it undergoes a unique and dynamic cycle of translocation. Within the plasma membrane, eNOS is targeted to caveolae, where its interactions with the scaffolding protein caveolin inhibit its activity. Increased levels of intracellular Ca^{2+} cause calmodulin (CaM)-mediated disruption of the eNOS–caveolin complex and activation of membrane-associated eNOS. Depalmitoylation of eNOS results in its translocation to the cytoplasm, where phosphorylation may attenuate its activity. The cycle is completed with the eventual return of eNOS to the cell membrane by mechanisms not yet fully understood.

This cycle of redistribution is thought to be a method of regulating eNOS activity. Other possible regulatory mechanisms include interaction with B2 bradykinin receptor, which appears to inhibit eNOS, and HSP90, which may enhance eNOS activation. The eNOS is recently shown to associate with platelet-endothelial cell adhesion molecule-1 (PECAM-1/CD31) and localize to cell–cell junctions. This association results in inhibition of eNOS activity (37,38). However, phosphorylation of PECAM-1 upon shear stress leads to dissociation of eNOS from PECAM-1, thus activating eNOS (39). The molecular and signaling mechanism responsible for regulation of eNOS activity by PECAM-1 requires investigation, and PECAM-1 may be involved in relocalization of inactivated eNOS to the plasma membrane, completing its redistribution cycle to the plasma membrane.

The angiogenic effect of VEGF is largely mediated via eNOS, and NO itself is an upstream promoter of VEGF expression (40). VEGF activates eNOS through activation of the PI3K-Akt/PKB pathway in a calcium-independent manner (41,42). Akt1 activates eNOS through phosphorylation of eNOS on S1176 in mice (equivalent S1179 and S1177 for bovine and human, respectively), augmenting NO release. NO can then modulate blood flow and many aspects of angiogenesis, including cell migration, cell proliferation, and capillary morphogenesis.

Sustained activation or absence of Akt1 has been associated with pathological angiogenesis, and leaky and immature blood vessels (43,44). The changes in vascular permeability and angiogenesis in Akt1$^{-/-}$ mice were linked to reduced expression of TSP1 and TSP2 (44). Therefore, regulation of TSP expression by Akt1 may impact endothelial cell function through modulation of extracellular matrix composition (44). However, the lack of pericytes recruitment and vascular maturation observed in newly formed blood vessels of Akt1$^{-/-}$ mice might result from a lack of eNOS activity in Akt1$^{-/-}$ endothelial cells (44–47).

We observed increased levels of active Akt1 in retinal endothelial cells prepared from TSP1$^{-/-}$ mice consistent with their promigratory and proliferative phenotype (29). Figure 14.4 shows that retinal vasculature of TSP1$^{-/-}$ mice have increased amounts of eNOS compared to wild-type mice. Furthermore, we observed increased amounts of active eNOS (S1176) in TSP1$^{-/-}$ retinal EC compared to wild-type cells. TSP1 was recently shown to antagonize proangiogenic signaling mediated by NO in endothelial cells in a cGMP-dependent manner (48). Our studies suggest that TSP1 blocks NO-mediated

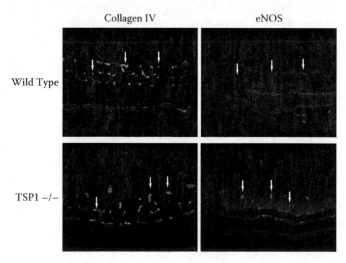

FIGURE 14.4 (See accompanying color CD.) Frozen eye sections prepared from 6-week-old wild-type or TSP1$^{-/-}$ mice and stained with an antibody to collagen IV (Chemicon) or eNOS (Santa Cruz). Collagen IV stains the retinal blood vessels (arrows). Note persistent staining of retinal blood vessels with eNOS antibody from TSP1$^{-/-}$ mice.

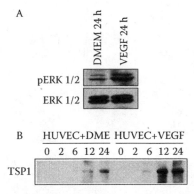

FIGURE 14.5 Increased expression of TSP1 in endothelial cells incubated with VEGF. (A) Sustained activation of MAPK/ERKs in human umbilical vein endothelial cells (HUVECs) incubated with VEGF (20 ng/mL) for 24 h or serum-free medium (DMEM, control). (B) Increased secretion of TSP1 in serum-free medium conditioned by HUVEC incubated with VEGF or serum-free medium (DMEM, control). Note sustained activation of MAPK/ERKs in HUVEC incubated with VEGF for up to 24 h and increased production of TSP1.

activities through the action of TSP1 on Akt1, thus preventing eNOS activation. How TSP1 effects on Akt1 activation are mediated requires further investigation.

A reciprocal relationship between VEGF and TSP1 has been observed in many studies (49–53). However, it is not known whether VEGF or TSP1 can directly modulate each other's expression. A direct inhibitory effect of TSP1 on VEGF in the ovary has been recently demonstrated, whereby TSP1 binds and promotes internalization of VEGF through the low-density lipoprotein receptor-related protein-1 (54). Our studies indicate that VEGF can also directly upregulate TSP1 expression, perhaps through sustained activation of MAPK/ERKs (figure 14.5). We have also previously shown that sustained activation of MAPK/ERKs in proangiogenic, transformed brain endothelial cells induces TSP1 expression in a MAPK/ERKs-dependent manner (55). Thus, TSP1 may act as a feedback regulator of proangiogenic signaling and directly impact Akt1 activation through modulation of VEGF expression, a known activator of eNOS, in endothelial cells.

14.5 SUMMARY

TSP1 plays an important role in retinal vascular development. In its absence, the developing retinal vasculature fails to undergo appropriate pruning and remodeling, resulting in increased vascular density. This was mainly attributed to increased number of endothelial cells that are protected from apoptosis in the absence of TSP1. These activities of TSP1 are mediated through multiple intracellular signaling pathways that are integrated with those activated by VEGF to maintain the differentiated, quiescent state of the endothelium (figure 14.5). Interaction of TSP1 with CD36 is believed to recruit the Src kinase family member Fyn and activate the JNK/MAPK pathway, resulting in endothelial cell apoptosis. However, this may depend on specific signaling through CD47 that dampens the proliferative and migratory activity of endothelial cells. These activities of TSP1 are contradicted through activation

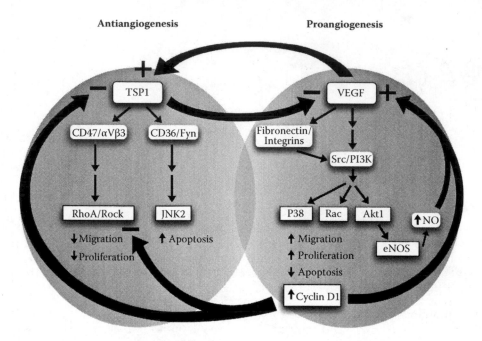

FIGURE 14.6 (See accompanying color CD.) Pro- and antiangiogenesis signaling pathways initiated by VEGF and TSP1 in endothelial cells, and potential feedback mechanisms, which act to maintain the quiescent, differentiated state of endothelium. Alterations in any of the components of these intracellular signaling pathways will disturb the angiogenic balance, resulting in regression or growth of new blood vessels.

of promigratory and prosurvival signaling pathways, such as the P38/MAPK, Rac, and Akt1. Activation of Akt1 can directly activate eNOS, producing the proangiogenic factor NO. These signaling events are kept in check by a number of feedback mechanisms, as outlined in figure 14.6. Alterations in any of the components of these pathways will have severe consequences on retinal vascular homeostasis.

REFERENCES

1. Adams, J.C. and Lawler, J.: The thrombospondins, *Int J Biochem Cell Biol* 2004, 36: 961–968.
2. Bornstein, P.: Thrombospondins as matricellular modulators of cell function, *J Clin Invest* 2001, 107: 929–934.
3. Iruela-Arispe, M.L., Liska, D.J., Sage, E.H., and Bornstein, P.: Differential expression of thrombospondin 1, 2, and 3 during murine development, *Dev Dyn* 1993, 197: 40–56.
4. Baenziger, N.L., Brodie, G.N., and Majerus, P.W.: A thrombin-sensitive protein of human platelet membranes, *Proc Natl Acad Sci U.S.A.* 1971, 68: 240–243.
5. Christopherson, K.S., Ullian, E.M., Stokes, C.C., Mullowney, C.E., Hell, J.W., Agah, A., Lawler, J., Mosher, D.F., Bornstein, P., and Barres, B.A.: Thrombospondins are astrocyte-secreted proteins that promote CNS synaptogenesis, *Cell* 2005, 120: 421–433.
6. Sheibani, N. and Frazier, W.A.: Thrombospondin-1, PECAM-1, and regulation of angiogenesis, *Histol Histopathol* 1999, 14: 285–294.

7. Good, D.J., Polverini, P.J., Rastinejad, F., Le Beau, M.M., Lemons, R.S., Frazier, W.A., and Bouck, N.P.: A tumor suppressor-dependent inhibitor of angiogenesis is immunologically and functionally indistinguishable from a fragment of thrombospondin, *Proc Natl Acad Sci U.S.A.* 1990, 87: 6624–6628.

8. Rastinejad, F., Polverini, P.J., and Bouck, N.P.: Regulation of the activity of a new inhibitor of angiogenesis by a cancer suppressor gene, *Cell* 1989, 56: 345–355.

9. Lawler, J.: The functions of thrombospondin-1 and-2, *Curr Opin Cell Biol* 2000, 12: 634–640.

10. Jimenez, B. and Volpert, O.V.: Mechanistic insights on the inhibition of tumor angiogenesis, *J Mol Med* 2001, 78: 663–672.

11. Jimenez, B., Volpert, O.V., Crawford, S.E., Febbraio, M., Silverstein, R.L., and Bouck, N.: Signals leading to apoptosis-dependent inhibition of neovascularization by thrombospondin-1, *Nat Med* 2000, 6: 41–48.

12. Jimenez, B., Volpert, O.V., Reiher, F., Chang, L., Munoz, A., Karin, M., and Bouck, N.: c-Jun N-terminal kinase activation is required for the inhibition of neovascularization by thrombospondin-1, *Oncogene* 2001, 20: 3443–3448.

13. Tolsma, S.S., Volpert, O.V., Good, D.J., Frazier, W.A., Polverini, P.J., and Bouck, N.: Peptides derived from two separate domains of the matrix protein thrombospondin-1 have anti-angiogenic activity, *J Cell Biol* 1993, 122: 497–511.

14. Dawson, D.W., Pearce, S.F., Zhong, R., Silverstein, R.L., Frazier, W.A., and Bouck, N.P.: CD36 mediates the in vitro inhibitory effects of thrombospondin-1 on endothelial cells, *J Cell Biol* 1997, 138: 707–717.

15. Simantov, R. and Silverstein, R.L.: CD36: a critical anti-angiogenic receptor, *Front Biosci* 2003, 8: s874–882.

16. Febbraio, M., Hajjar, D.P., and Silverstein, R.L.: CD36: a class B scavenger receptor involved in angiogenesis, atherosclerosis, inflammation, and lipid metabolism, *J Clin Invest* 2001, 108: 785–791.

17. Fruttiger, M.: Development of the mouse retinal vasculature: angiogenesis versus vasculogenesis, *Invest Ophthalmol Vis Sci* 2002, 43: 522–527.

18. Wang, S., Wu, Z., Sorenson, C.M., Lawler, J., and Sheibani, N.: Thrombospondin-1-deficient mice exhibit increased vascular density during retinal vascular development and are less sensitive to hyperoxia-mediated vessel obliteration, *Dev Dyn* 2003, 228: 630–642.

19. Ito, M. and Yoshioka, M.: Regression of the hyaloid vessels and pupillary membrane of the mouse, *Anat Embryol* 1999, 200: 403–411.

20. Short, S.M., Derrien, A., Narsimhan, R.P., Lawler, J., Ingber, D.E., and Zetter, B.R.: Inhibition of endothelial cell migration by thrombospondin-1 type-1 repeats is mediated by {beta}1 integrins, *J Cell Biol* 2005, 168: 643–653.

21. Iruela-Arispe, M.L., Lombardo, M., Krutzsch, H.C., Lawler, J., and Roberts, D.D.: Inhibition of angiogenesis by thrombospondin-1 is mediated by 2 independent regions within the type 1 repeats, *Circulation* 1999, 100: 1423–1431.

22. Kanda, S., Shono, T., Tomasini-Johansson, B., Klint, P., and Saito, Y.: Role of thrombospondin-1-derived peptide, 4N1K, in FGF-2-induced angiogenesis, *Exp Cell Res* 1999, 252: 262–272.

23. Freyberg, M.A., Kaiser, D., Graf, R., Vischer, P., and Friedl, P.: Integrin-associated protein and thrombospondin-1 as endothelial mechanosensitive death mediators, *Biochem Biophys Res Commun* 2000, 271: 584–588.

24. Freyberg, M.A., Kaiser, D., Graf, R., Buttenbender, J., and Friedl, P.: Proatherogenic flow conditions initiate endothelial apoptosis via thrombospondin-1 and the integrin-associated protein, *Biochem Biophys Res Commun* 2001, 286: 141–149.

25. Isenberg, J.S., Ridnour, L.A., Dimitry, J., Frazier, W.A., Wink, D.A., and Roberts, D.D.: CD47 is necessary for inhibition of nitric oxide-stimulated vascular cell responses by thrombospondin-1, *J Biol Chem* 2006, 281: 26069–26080.

26. Smith, L.E., Wesolowski, E., McLellan, A., Kostyk, S.K., D'Amato, R., Sullivan, R., and D'Amore, P.A.: Oxygen-induced retinopathy in the mouse, *Invest Ophthalmol Vis Sci* 1994, 35: 101–111.

27. Wu, Z., Wang, S., Sorenson, C.M., and Sheibani, N.: Attenuation of retinal vascular development and neovascularization in transgenic mice over-expressing thrombospondin-1 in the lens, *Dev Dyn* 2006, 235: 1908–1920.

28. Su, X., Sorenson, C.M., and Sheibani, N.: Isolation and characterization of murine retinal endothelial cells, *Mol Vis* 2003, 9: 171–178.

29. Wang, Y., Wang, S., and Sheibani, N.: Enhanced proangiogenic signaling in thrombospondin-1-deficient retinal endothelial cells, *Microvascular Res* 2006, 71: 143–151.

30. Li, Z., Wang, C., Jiao, X., Lu, Y., Fu, M., Quong, A.A., Dye, C., Yang, J., Dai, M., Ju, X., Zhang, X., Li, A., Burbelo, P., Stanley, E.R., and Pestell, R.G.: Cyclin D1 regulates cellular migration through the inhibition of thrombospondin 1 and ROCK signaling, *Mol Cell Biol* 2006, 26: 4240–4256.

31. Scheef, E., Wang, S., Sorenson, C.M., and Sheibani, N.: Isolation and characterization of murine retinal astrocytes, *Mol Vis* 2005, 11: 613–624.

32. Lawler, J., Sunday, M., Thibert, V., Duquette, M., George, E.L., Rayburn, H., and Hynes, R.O.: Thrombospondin-1 is required for normal murine pulmonary homeostasis and its absence causes pneumonia, *J Clin Invest* 1998, 101: 982–992.

33. Kim, S., Bell, K., Mousa, S.A., and Varner, J.A.: Regulation of angiogenesis in vivo by ligation of integrin alpha5beta1 with the central cell-binding domain of fibronectin, *Am J Pathol* 2000, 156: 1345–1362.

34. Wang, Y., Wei, X., Xiao, X., Hui, R., Card, J.W., Carey, M.A., Wang, D.W., and Zeldin, D.C.: Arachidonic acid epoxygenase metabolites stimulate endothelial cell growth and angiogenesis via mitogen-activated protein kinase and phosphatidylinositol 3-kinase/akt signaling pathways, *J Pharmacol Exp Ther* 2005, 314: 522–532.

35. Wilson, S.H., Ljubimov, A.V., Morla, A.O., Caballero, S., Shaw, L.C., Spoerri, P.E., Tarnuzzer, R.W., and Grant, M.B.: Fibronectin fragments promote human retinal endothelial cell adhesion and proliferation and ERK activation through alpha5beta1 integrin and PI 3-kinase, *Invest Ophthalmol Vis Sci* 2003, 44: 1704–1715.

36. Forstermann, U. and Munzel, T.: Endothelial nitric oxide synthase in vascular disease: from marvel to menace, *Circulation* 2006, 113: 1708–1714.

37. Govers, R., Bevers, L., de Bree, P., and Rabelink, T.J.: Endothelial nitric oxide synthase activity is linked to its presence at cell-cell contacts, *Biochem J* 2002, 361: 193–201.

38. Dusserre, N., L'Heureux, N., Bell, K.S., Stevens, H.Y., Yeh, J., Otte, L.A., Loufrani, L., and Frangos, J.A.: PECAM-1 interacts with nitric oxide synthase in human endothelial cells: implication for flow-induced nitric oxide synthase activation, *Arterioscler Thromb Vasc Biol* 2004, 24: 1796–1802.

39. Bagi, Z., Frangos, J.A., Yeh, J.C., White, C.R., Kaley, G., and Koller, A.: PECAM-1 mediates NO-dependent dilation of arterioles to high temporal gradients of shear stress, *Arterioscler Thromb Vasc Biol* 2005, 25: 1590–1595.

40. Papapetropoulos, A., Garcia-Cardena, G., Madri, J.A., and Sessa, W.C.: Nitric oxide production contributes to the angiogenic properties of vascular endothelial growth factor in human endothelial cells, *J Clin Invest* 1997, 100: 3131–3139.

41. Fulton, D., Gratton, J.P., McCabe, T.J., Fontana, J., Fujio, Y., Walsh, K., Franke, T.F., Papapetropoulos, A., and Sessa, W.C.: Regulation of endothelium-derived nitric oxide production by the protein kinase Akt, *Nature* 1999, 399: 597–601.

42. Dimmeler, S., Fleming, I., Fisslthaler, B., Hermann, C., Busse, R., and Zeiher, A.M.: Activation of nitric oxide synthase in endothelial cells by Akt-dependent phosphorylation, *Nature* 1999, 399: 601–605.
43. Phung, T.L., Ziv, K., Dabydeen, D., Eyiah-Mensah, G., Riveros, M., Perruzzi, C., Sun, J., Monahan-Earley, R.A., Shiojima, I., Nagy, J.A., Lin, M.I., Walsh, K., Dvorak, A.M., Briscoe, D.M., Neeman, M., Sessa, W.C., Dvorak, H.F., and Benjamin, L.E.: Pathological angiogenesis is induced by sustained Akt signaling and inhibited by rapamycin, *Cancer Cell* 2006, 10: 159–170.
44. Chen, J., Somanath, P.R., Razorenova, O., Chen, W.S., Hay, N., Bornstein, P., and Byzova, T.V.: Akt1 regulates pathological angiogenesis, vascular maturation and permeability in vivo, *Nat Med* 2005, 11: 1188–1196.
45. Ackah, E., Yu, J., Zoellner, S., Iwakiri, Y., Skurk, C., Shibata, R., Ouchi, N., Easton, R.M., Galasso, G., Birnbaum, M.J., Walsh, K., and Sessa, W.C.: Akt1/protein kinase Balpha is critical for ischemic and VEGF-mediated angiogenesis, *J Clin Invest* 2005, 115: 2119–2127.
46. Yu, J., deMuinck, E.D., Zhuang, Z., Drinane, M., Kauser, K., Rubanyi, G.M., Qian, H.S., Murata, T., Escalante, B., and Sessa, W.C.: Endothelial nitric oxide synthase is critical for ischemic remodeling, mural cell recruitment, and blood flow reserve, *Proc Natl Acad Sci U.S.A.* 2005, 102: 10999–11004.
47. Hatakeyama, T., Pappas, P.J., Hobson, Ii. R.W., Boric, M.P., Sessa, W.C., and Duran, W.N.: Endothelial nitric oxide synthase regulates microvascular hyperpermeability in vivo, *J Physiol* 2006, 574: 274–281.
48. Isenberg, J.S., Ridnour, L.A., Perruccio, E.M., Espey, M.G., Wink, D.A., and Roberts, D.D.: Thrombospondin-1 inhibits endothelial cell responses to nitric oxide in a cGMP-dependent manner, *Proc Natl Acad Sci U.S.A.* 2005, 102: 13141–13146.
49. Sheibani, N. and Frazier, W.A.: Thrombospondin 1 expression in transformed endothelial cells restores a normal phenotype and suppresses their tumorigenesis, *Proc Natl Acad Sci U.S.A.* 1995, 92: 6788–6792.
50. Greenaway, J., Gentry, P.A., Feige, J.J., LaMarre, J., and Petrik, J.J.: Thrombospondin and vascular endothelial growth factor are cyclically expressed in an inverse pattern during bovine ovarian follicle development, *Biol Reprod* 2005, 72: 1071–1078.
51. Zhang, Y.W., Su, Y., Volpert, O.V., Vande and Woude, G.F.: Hepatocyte growth factor/scatter factor mediates angiogenesis through positive VEGF and negative thrombospondin 1 regulation, *Proc Natl Acad Sci U.S.A.* 2003, 100: 12718–12723.
52. Vikhanskaya, F., Bani, M.R., Borsotti, P., Ghilardi, C., Ceruti, R., Ghisleni, G., Marabese, M., Giavazzi, R., Broggini, M., and Taraboletti, G.: p73 Overexpression increases VEGF and reduces thrombospondin-1 production: implications for tumor angiogenesis, *Oncogene* 2001, 20: 7293–7300.
53. Filleur, S., Courtin, A., Ait-Si-Ali, S., Guglielmi, J., Merle, C., Harel-Bellan, A., Clezardin, P., and Cabon, F.: SiRNA-mediated inhibition of vascular endothelial growth factor severely limits tumor resistance to antiangiogenic thrombospondin-1 and slows tumor vascularization and growth, *Cancer Res* 2003, 63: 3919–3922.
54. Greenaway, J., Lawler, J., Moorehead, R., Bornstein, P., Lamarre, J., and Petrik, J.: Thrombospondin-1 inhibits VEGF levels in the ovary directly by binding and internalization via the low density lipoprotein receptor-related protein-1 (LRP-1), *J Cell Physiol* 2007, 210: 807–818.
55. Wu, J. and Sheibani, N.: Modulation of VE-cadherin and PECAM-1 mediated cell-cell adhesions by mitogen-activated protein kinases, *J Cell Biochem* 2003, 90: 121–137.

6

*Lipid Mediators and
Signaling in the RPE*

15 Lipidomic Approaches to Neuroprotection Signaling in the Retinal Pigment Epithelium

Nicolas G. Bazan, Victor L. Marcheselli,
Yan Lu, Song Hong, and Fannie Jackson

CONTENTS

15.1 OVERVIEW

Lipidomics is the identification and characterization of multiple complex lipids from a cell, tissue, or organ. Also, lipidomics can be applied to a part of a cell (e.g., photoreceptor outer segment). Lipidomics includes interactions between lipids and interactions of lipids with other molecules (e.g., lipid–protein interactions). Lipid–protein interactions are particularly important from a functional point of view when the protein in question is, for example, an ion channel, receptor, enzyme, or transporter. Lipidomics-based analysis of specific stereochemical lipid structures has become important for the elucidation and bioactivity of signaling lipids. Lipidomics approaches are analogous to the definition of gene expression profiles—genomics—and of the identification and characterization of proteins—proteomics. Thus, the genome, the proteome, and the lipidome are distinct areas of knowledge. Moreover, lipidomics belongs to the larger domain of metabolomics, which aims to attain a complete definition and understanding of the metabolome. Several recent reviews deal with various aspects of lipidomics (1–5).

A major leap in the initiation of lipidomics was the development and application of electrospray ionization (ESI/MS). This advancement, with other tandem mass spectrometry (MS) software improvements, gave rise to new technologies. With these tools, the exhaustive definition and quantification of heterogeneous lipid classes, molecular species, and small transiently appearing lipids in a cell or part of a cell can be pursued. Several years ago, massive efforts using, for example, thin-layer chromatography (TLC) (including silver ion, to separate molecular species of lipid classes according to the degree of unsaturation), gas–liquid chromatography, and high-pressure liquid chromatography (HPLC) in combination with other chemical procedures, were necessary. These techniques were laborious, lengthy, with several shortcomings (e.g., loss of labile lipids), and also required relatively large samples to be able to detect the smaller lipidic pools.

In this chapter, we describe methodological approaches to study omega-3 and omega-6 essential fatty acids as well as some of their derivatives. Several of these derivatives are biologically active lipids—that is, mediator/messengers. Arachidonic acid, the major omega-6 fatty acid of the body, is ubiquitously distributed throughout tissues. Docosahexaenoic acid (DHA), the major omega-3 fatty acid, on the other hand, attains its highest concentration in the central nervous system, mainly in photoreceptor outer segments (6). DHA is esterified to phospholipids, and it is recycled continuously during the daily renewal of the outer segments. This process is circadian in mammalians and involves the shedding of the tip of photoreceptors by the retinal pigment epithelial cells (RPEs) in the morning. Then the tips of photoreceptors are phagocytized and degraded by the RPE cells. At the same time, outer segment membrane biogenesis is activated in the inner segments of the photoreceptors. The outcome of these daily events is an outer segment of exactly the same length as the day before. Membrane biogenesis engages active synthesis of phospholipids containing DHA (7). This renewal involves an efficient retrieval of DHA from the RPE cell to the base of the photoreceptor through a "short loop" involving the interphotoreceptor matrix (figure 15.1). Likewise, vitamin A is retrieved. The dietary supply of omega-3 fatty acids converges on the liver, where elongation and desaturation of 18:3 omega-3 (linolenic acid) takes place, prior to its secretion into the bloodstream for its delivery through lipoproteins to the RPE (8,9). Thus, the RPE sits at the center stage of DHA arrival to the retina from the liver and on its processing of outer segment phospholipids and retrieval to the photoreceptor cells. The molecular nature of these processes is not completely understood.

The RPE cells are critical for the survival of photoreceptors. It has been demonstrated that in several retinal degenerative diseases photoreceptors die because of a failure in RPE cell survival. Therefore, a quest has been undertaken in many laboratories to identify neuroprotective signaling in the RPE cell. This chapter describes lipidomic approaches used to identify and study neuroprotectin D1 and related lipids that have been shown to exert potent and specific neuroprotection in the RPE cells (10–13) as well as in the brain (14–16). We also describe experimental approaches for the study of other lipid mediators with pro- and anti-inflammatory bioactivity. Inflammatory lipid signaling has become a major focus in neuroprotection mechanisms (7,14,17). The methods described here highlight the power of lipidomics in increasing our understanding of the bioactivity of DHA oxygenation derivatives. Although it has been extensively reported that DHA participates in photoreceptor and brain damage, our laboratory, in contrast,

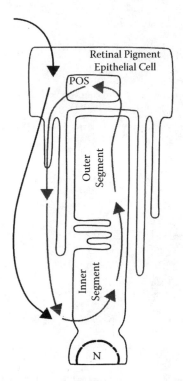

FIGURE 15.1 (See accompanying color CD.) Docosahexaenoic acid trafficking in RPE and photoreceptors. Arrow indicates DHA trafficking. At the top is depicted the arrival of DHA from the choriocapillarias to the RPE cell and then its movement through the interphotoreceptor matrix to the inner segment of the photoreceptor. Red arrows represent the DHA "short loop" that accompanies the photoreceptor's outer segment renewal. POS, photoreceptor outer segment after phagocytosis.

suggested that the retina forms mono-, di-, and tri-hydroxyl derivatives of DHA that may be neuroprotective (18). Because lipoxygenase inhibitors were found to block this synthesis, an enzymatic process catalyzed by a lipoxygenase was suggested (18), and the name *docosanoids* was proposed because they are derived from the 22C DHA. Eicosanoids are derived from the 20C arachidonic acid. Because lipidomics technology was not available at that time, the mass spectrometry methods available did not allow us to define the detail structure and stereochemistry of these DHA-oxygenated derivatives. Upon the advent of lipidomic analysis based on LC-PDA-ESI-MS-MS, the identification and detailed characterization of NPD1 was performed.

15.2 PROTOCOLS

PROTOCOL 1: LIPID EXTRACTION

Lipid extraction from tissue samples

Lipids from retina, RPE, and from subcellular fractions, in general, are extracted following modifications of the procedures described by Folch et

TABLE 15.1

Parent and Product Ions of Various Lipids for Single Reaction Monitoring (SRM) MS/MS

Analyte	Parent Ion	Product Ion	Collision Energy	Ionization
NPD1	359.2	153.1	22	Negative
RvD1	375.1	141.1	20	Negative
RvE1	349.1	195.1	22	Negative
15S-HETE	319.1	219.1	18	Negative
12S-HETE	319.1	179.1	20	Negative
PGE2	351.1	189.1	26	Negative
PGD2	351.1	233.2	16	Negative
PGD2-d4	355.1	275.3	22	Negative
Lipoxin-A4	351.1	217.2	26	Negative
DHA	327.1	283.2	16	Negative
Arachidonic acid	303.1	259.3	18	Negative
C16-PAF	526.4	184.2	38	Positive
C18-PAF	557.6	184.1	42	Positive
C16-Lyso-PAF	482.5	184.2	36	Positive
C18-Lyso-PAF	510.7	184.2	44	Positive

al. (19), and then purified by solid-phase extraction (SPE). Different extraction techniques have been compared, using radioactive or deuterated lipids, and the extraction technique described here resulted in the highest yields (table 15.1).

1. Homogenize retinal tissue samples (glass/glass homogenizer #21, Kontes Glass Co., Vineland, New Jersey) in 1 mL of Chloroform:Methanol (2:1, v/v).
2. Wash the homogenizer twice with 1 mL C:M 2:1.
3. Collect both C:M extracts and add 5 µL of internal standards (Mixture of PGD_2-d4 and PAF-C16-d4, 0.01 µg/µL) and vortex thoroughly.
4. Cap samples under nitrogen and sonicate for 5 min in a water bath (Branson 3510).
5. Centrifuge the samples at 1000 × g at 4°C for 10 min or more, if needed.
6. Gently aspirate the lipid extract from the pellet and transfer to a 16 × 100 mm glass test tube with Teflon lined screw caps.
7. Wash the pellet with 1 mL of C:M 2:1, vortex, centrifuge again, and pool extracts together.
8. Add 0.2 volumes of 0.05% $CaCl_2$ to C:M extracts. Vortex extensively until a milky emulsion is formed (30 s). Then centrifuge the extracts for 15 min for phase separation.
9. Save the lower lipidic layer, and discard the upper aqueous layer (carefully avoid disturbing the interphase).

10. Evaporate the organic solvent using a nitrogen evaporator, and resuspend the lipids in desired volume of methanol (usually 1 mL) for further purification by SPE.

11. Save the pellet for protein analysis to normalize lipidic data. Allow the proteins to dry overnight. Then add 1 N NaOH to hydrolyze the proteins. Next assay proteins using the Lowry Proteins Assay.

Lipid extraction from cell culture plates

Procedures for lipid extraction in cell cultures were also followed according to the Folch technique, except for skipping the SPE purification; lipid extracts were directly prepared for mass spectrometric analysis.

When cell cultures are used (usually 6-well plates), media (3 mL) is removed and saved for analysis, after which wells are washed with 2 mL ice-cold 1 × PBS pH 7.4. Then 1 mL methanol is added, and cells carefully scraped off with 0.6 cm disposable cell scrapers and transferred to screw cap glass culture tubes. Plates are washed once with 1 mL methanol, and methanolic extracts pooled. Four mL of chloroform is added to all extracts, sonicated shortly in a bath sonicator, and then flushed with N_2 and stored at $-80°C$.

Lipid extraction on tissues or cultured cells is performed as described by Folch et al. (19). In short, C:M (2:1) extracts are centrifuged at 1000 × g at 4°C to separate insoluble precipitates, which are separated and saved for protein content analysis (usually used to normalize data). Pellets are washed once with 1 mL C:M (2:1) and extracts pooled. Volumes of organic extracts are normalized with more C:M (2:1), then 0.2 volumes of aqueous 0.05% $CaCl_2$ are added and vortexed extensively until a milky emulsion is formed (30 s). The use of $CaCl_2$ in this step is important to unify and generate only Ca salts, thereby improving chromatographic profiles and also improving the yield of highly polar lipids (amphiphilic) into the organic phase. Aqueous and organic phase separation is obtained by centrifugation at 1000 × g at 4°C. Aqueous (upper) and chloroform (lower) layers are carefully separated. Avoid disturbing the interphase, which stays with the organic phase. Lipids are then collected by drying the organic phase in a N_2 stream evaporator, then resuspended in methanol for further purification or analysis. The resuspension volume depends on application. Tissue culture cell extracts are resuspended in 100 μL methanol for direct mass spectrometric analysis.

Lipid extraction from cell culture media

Culture cell media collected as described in the previous paragraph are centrifuged at 1000 × g to pellet cellular debris, and 1 mL media is removed and mixed with 9 mL of C:M (1:1). Then the media is flushed with N_2 and stored at $-80°C$ until the lipid extraction procedure is performed.

To each of the samples, 5 μL of deuterated internal standard mix solution (PGD_2-d_4/PAF-C16-d_4 (0.01 μg/μL of each standard, (Cayman Chemical Company, Ann Arbor, Michigan)) is added before starting the lipid extraction.

Lipid extraction on cell media is performed on C:M (1:1) extracts (10 mL in total) by addition of 3.5 mL of 0.05% $CaCl_2$ (ph 3.5 with acetic acid) then vortexed extensively, centrifuged, and phase-separated as described in the previous paragraph. Lipid extracts are also dried in a N_2 stream evaporator, and then typically resuspended in 100 μL of methanol for mass spectrometric analysis.

PROTOCOL 2: LIPID EXTRACT PURIFICATION

Purification is performed only on tissue extracts by SPE, a modified version of a technique previously described by Gronert et al. (20). This procedure is performed on SPE columns (Bond Elut LRC, C_{18} INT 500 mg, Varian, Inc., Palo Alto, California) with the help of an SPE vacuum manifold (Visiprep DL™, Supelco, Bellafonte, Pennsylvania).

1. Samples in 1 mL methanol are pre-equilibrated at pH 3.5 with 9 mL of water.
2. Activate C_{18} columns with 10 mL of methanol followed by 10 mL water, pH 3.5 (do not allow the columns to dry), and load the sample to the column.
3. Wash with 6 mL of 15% methanol in water pH 3.5, followed by 6 mL of water pH 3.5 and allow the column to dry for 5 min.
4. Wash with 6 mL hexane to remove neutral lipids and allow the column to dry for 2 min.
5. Add the 16 × 100 mm glass test tubes into the chamber to collect the lipids.
6. Elute polar lipids from SPE columns with 10 mL of chloroform:methanol (C:M 2:1).
7. Dry down the elutant under nitrogen.
8. Resuspend samples in methanol, usually 200 μL for each sample.

PROTOCOL 3: LC-PDA-ESI-MS-MS-BASED LIPIDOMIC ANALYSIS

Equipment. The methodological steps described in this chapter are based on the use of the following equipment:

1. For quantitative analysis, a LC-TSQ (Thermo Fischer Scientific Co., Waltham, Massachusetts), composed by a LC Surveyor System (Thermo Fischer Scientific Co., Waltham, Massachusetts) containing an autosampler, LC pump, and PDA detector, a 20 μL injection loop, and a C_{18} Pursuit column with 100 mm length × 2.1 mm OD, loaded with 5 μm stationary phase (Varian, Inc., Palo Alto, California).
2. For qualitative analysis, or shotgun lipidomics, by intrasource separation and multidimensional mass spectrometry to elucidate the structural composition of lipid mediators as well as their metabolic intermediates, we used a

LC-UV-ESI-Linear Ion Trap LTQ (Thermo Fischer Scientific Co., Waltham, Massachusetts). The system utilized a C_{18} reverse-phase discovery column (Supelco, Bellafonte, Pennsylvania), 100 mm length × 2.5 mm OD, and 5 μm stationary phase.

Quantitative analysis of specific analytes

Quantitative Analysis. For quantitative analysis, usually 20 μL of purified lipid extracts are injected into the LC-PDA Surveyor System. Elution on a linear gradient [100% solution A (40:60:0.01 methanol/water/acetic acid) to 100% solution B (99.99:0.01 methanol/acetic acid)], running at a flow rate of 300 μL/min for 45 min, is performed. LC effluents are monitored in a PDA detector and then diverted to the mass spectrometer for quantitative analysis.

Lipids are subjected to ionization in an electro-spray-ionization (ESI) probe mounted on a TSQ Quantum (Thermo Fischer Scientific Co., Waltham, Massachusetts), a triple quadrupole mass spectrometer capable of performing on positive or negative ion detection mode MS-MS detection.

This type of instrument is highly recommended for quantitative analysis. It can be configured for different scan modes, but for quantitative purposes we suggest using selected reaction monitoring (SRM), which is explained in figure 15.2, for detection of NPD1 from RPE tissular lipid extract. In short, panel A shows a full scan (range m/z 110 to 1000) of negative ions obtained by electrospray ionization of a lipidic effluent monitored at a retention time of 19.30 min. Between all the peaks, we can detect (highlighted) a minor ion at m/z 359. This scan was obtained on the Q_1 quadrupole of the TSQ Quantum. Panel B shows a full scan of all the product ions (m/z 100 to 359) on the Q_3 quadrupole after the parent ion, selected in Q_1 was subjected to collision-induced fragmentation on the Q_2 quadrupole at 1.5 mTorr of argon gas and 22 eV collision energy. Highlighted product ion m/z 153 was our selected ion for SRM detection. It is important to point out that the majority of unlabeled ions shown here are background peaks that will offset quantitative data analysis if single ion monitoring is applied. Panel C shows a typical SRM detection for NPD1, where only product ion m/z 153 was monitored on Q_3 quadrupole after ion m/z 359 was selected and fragmented, as shown in panels A and B.

Table 15.1 shows the pairs of parent/product ions and collision energies for selected lipids obtained in negative or positive ionization modes.

It is important also to point out that in a single run, up to eight analytes can be measured simultaneously on this instrument; see figure 15.3 for an example of detecting seven analytes, including the internal standard PGD_2-d_4. It should be taken into consideration that the amount of time the instrument spends on each analyte is reduced proportionally as we increase the number of analytes, because the instrument scans all the selected analytes before starting a new scan; therefore, sensitivity is decreased. To understand this better, consider how long it takes an analyte peak to elute (in seconds), and factor this value by the scan time (in seconds, usually 0.2 s) times by the number of analytes. This is the number of readings we will detect for that peak; therefore, there is a better chance of profiling the peak when a higher number of readings is obtained.

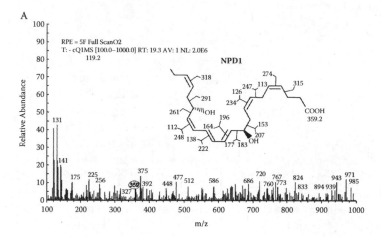

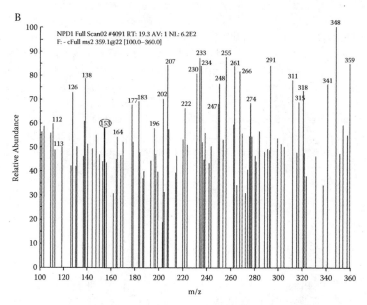

FIGURE 15.2 (See accompanying color CD.) Analytic approach to detect a minor lipid derivative in a complex matrix such as retinal tissular extract. Panel A: Full scan of ions detected at RT: 19.3 min on Q1 quadrupole in the m/z 100–1000 range. NPD1, our analyte of interest at m/z 359, is shown highlighted. Insert shows the molecular structure of NPD1 and its pattern of fragmentation. Panel B: Fragmentation pattern of ion m/z 359 at 22 eV collision energy in presence of 1.5 mTorr of Argon gas, as detected in Q3 quadrupole in the m/z 100–360 range. Ions shown labeled are specific product fragments from parent ion m/z 359, which are shown on Panel A indicated on the molecular structure. Most unlabeled ion peaks are background. Product ion m/z 153 was highlighted because it was selected for SRM monitoring.

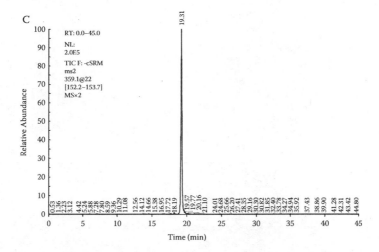

FIGURE 15.2 (continued) (See accompanying color CD.) Analytic approach to detect a minor lipid derivative in a complex matrix such as retinal tissular extract. Panel C: TIC elusion profile by SRM analysis for m/z 153 product ion, after fragmentation of m/z 359 parent ion under same conditions shown in Panel B. The chromatogram display is very clean, and its specificity has increased enormously; this is the best approach for quantitative analysis in complexed matrix.

Quantitative analysis. The data analysis involves the integration of the peak area. This value needs to be processed on a calibration curve obtained from the analyte (figure 15.4). Therefore, it is critical to use high-quality calibration standards and to run the calibration curves under the same conditions as unknown samples. Also, it is highly recommended that a deuterated internal standard be added so it can be resolved from the endogenous compounds, and to permit correction for multiple errors that could have been carried over from the sample lipid extraction, injection, matrix and coelutant interference, and system and source instabilities.

For calculation purposes, the first step is to correct the measured area counts for the analyte using a correction factor obtained by factoring the total area counts measured for the internal standard added to the sample, and the area counts measured for the internal standard during sample analysis. Then the corrected value should be interpolated to the calibration curve for that particular compound. An example of a Microsoft Excel worksheet and the respective standardized formulas used for sample calculations is shown in figure 15.5. This worksheet can be expanded to incorporate all the analytes measured in a single run for each sample. For example, in figure 15.3 up to seven analytes were quantified simultaneously.

Shotgun lipidomics or structural identification of novel lipid mediators

Here we provide a protocol using liquid-chromatography-ultraviolet-tandem mass spectrometry (LC-UV-MS/MS) with atmospheric pressure ionization (1,21,22). This is a sensitive and specific means for chemical identification in

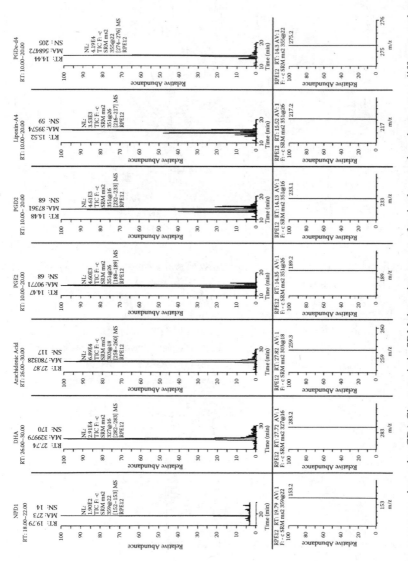

FIGURE 15.3 (See accompanying color CD.) Characteristic SRM detection report of a single sample, where seven different analytes were simultaneously scanned. For each analyte, the upper cell indicates a total ion current (TIC) chromatogram, showing the retention time (RT) for the analyte, mass area (MA) of the integrated peak, and signal-to-noise (SN) reports. The lower panel shows the spectral analysis for the specific product ion quantified. All analytes are endogenous products, with the exception of PGD_2-d4, which was the added internal standard.

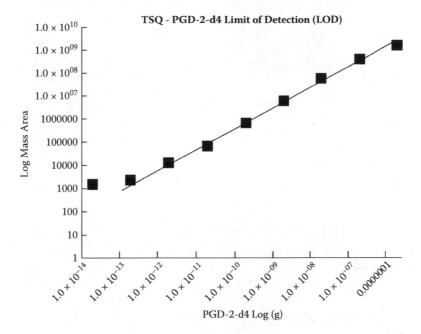

FIGURE 15.4 Example of a calibration standard curve obtained for the PGD2-d4 that was used also as internal standard. This curve is linear in the range of 1 μg to 0.1 pg. Calculations for unknown data could be obtained by simple interpolation within the linear range of the curve.

lipidomic analysis. Low-energy ionization or electrospray ionization (ESI) of lipids generates primarily molecular (or pseudomolecular) ions for low- or intermediate-energy collision-induced dissociation tandem MS/MS analysis and can avoid unwanted degradation products (1).

Sample preparation is performed as in protocol 1, followed by SPE purification protocol as described in protocol 2. Chromatograms and spectra of samples and standards were acquired in negative-ion mode using a Finnigan LC-UV-ESI-linear ion trap LTQ and triple quadruple TSQ Quantum tandem mass spectrometer (Thermo Fischer Scientific Co., Waltham, Massachusetts). The system utilized a reverse-phase C18 Discovery column (Supelco, Bellafonte, Pennsylvania), 100 or 150 mm long × 2 mm of OD. The column was kept in a column heater at 35°C. The LC system used a Surveyor MS quaternary LC pump (Thermo Fischer Scientific Co., Waltham, Massachusetts). When a 100-mm-long column was used, the mobile phase A (methanol:water:acetic acid 65:35:0.01) flowed isocratic for 8 min at 0.2 mL/min, then ramped to 100% mobile phase B (methanol) from 8.01 to 30 min at the same flow rate, eluted isocratically for 5 min in mobile phase B, and finally returned to 100% mobile phase A using a 10 min gradient. The photodiode-array UV detector was set to scan from 200 to 400 nm. Conditions for MS-MS were: electric spray voltage at 4 kV; heating capillary at 250°C and −45 V; tube lens offset at 55 V; sheath N_2 gas at 40 units (near 0.6 L/min); and auxiliary N_2 gas at 3 units (0.045 L/min). Quantification was based on the peak areas from selective ion monitoring (SIM) chromatograms. Calibration

	A	B	C	D	E	F	G
1							
2			Docosanoids Calculations				
3			Integrated Areas From Mass Spect.				
4			µg/injected to instrument	NPD-1	PGD2-d4		
5				0.001 µg/µl	0.01 µg/µl		
6	NPD-1		0.02	2948686			
7							
8	PGD2-d4		0.2		56559473		
9							
10							Calculated Protein Content
11	Sample #	Treatment		NPD-1	Ist:PGD2-d4	Samples	mg Proteins/ Total Sample
12							
13	C - 55	Control		370	3758199	C - 55	0.1969
14							
15				NPD-1			
16	Sample #	Treatment	µg/total sample				
17							
18	C - 55	Control	=(D13*E8*C$6*5)/(E13*$D$6*20)	9.441E-06			
19							
20	Sample #	Treatment		NPD-1			
21			pmole/mg Protein				
22							
23	C - 55	Control	=(D18*1000000)/G13*360	0.133			

FIGURE 15.5 Example of Excel calculation worksheet. Data input shown in column D, row 7, and column E, row 9 are integrated areas for external standards injected to the mass spectrometer. Data input on columns C and D, row 14, are the analytes' integrated areas for the sample. The formula in column C, row 19, automatically converts analytes and internal standard readings from the mass spectrometer into corrected µg analyte in the whole sample. The factor 5 shown is the volume (µL) of internal standard (PGD2-d4) added to the whole sample. Also there is 20 which is the volume (µL) of PGD2-d4 injected to the mass spectrometer to obtain the reading shown on column E, row 8, which is used in the formula to correct to the whole sample. The formula shown in column C, rows 24, is to calculate the analyte in pmole/mg of protein. In column G, row14, was entered the amount of proteins measured for the whole sample.

curves for each corresponding standard were performed. A typical example for PGD_2 calibration curve is shown in figure 15.6.

Analysis of lipid mediators via LC-MS/MS based on spectra and fragmentation mechanisms are described for LXA_4 in figure 15.7. The specific molecular ions generated by electrospray of LC effluent are selected and fragmented in the ion trap via collision-induced dissociation (CID), this generates the MS/MS product ions spectra or MS^2. If we select some of the specific MS/MS ions and fragment them again in the ion trap, we obtain the MS^3; following this approach, we can obtain the MS^4, up to MS^n.

Functional groups and their specific position in the lipid molecular structure are the fingerprint information for structure elucidation and identification. The multiple stage tandem mass spectrometry provides a step-by-step fragmentation tool for revealing the fingerprint information of the molecular ions (23). This approach is demonstrated by LC-UV-MS^n analysis of lipoxin A_4 (figure 15.7). Lipoxin A_4 was eluted out of the C18 column and injected into the ESI-MS/MS^n after scanning by

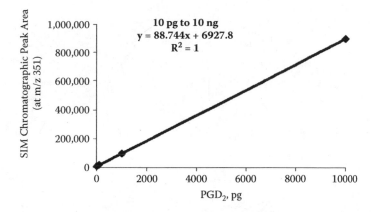

FIGURE 15.6 Standard calibration curve. SIM chromatographic areas (at m/z 351) is linearly responsive to the amount of PGD2 injected to LC-MS/MS.

the UV detector. ESI deprotonation formed the molecular ions m/z 351, which was selected and fragmented in the linear ion trap. The fragmented ions pattern was detected, generating the MS/MS or MS^2 spectrum shown for molecular ion m/z 351, in Panel A, figure 15.7. Afterward, by selecting MS/MS product ions m/z 333 and 251 in the ion trap, and then fragmented again in separate phases, the respective MS^3 product ion spectra were obtained and shown on Plates B and C of figure 15.7.

The MS/MS ion m/z 251 in figure 15.7 is proposed to form via the cleavage of bond at C14-C15 along the pathways from (a) to (c) shown in Panel E with the structure as the insert in Panel C of m/z 251, where ions m/z 114, 134, 189 ($251-H_2O-CO_2$), 207 ($251-CO_2$), 233 ($251-H_2O$) lead to the structure assignment of ion m/z 251. The MS/MS ion m/z 333 is expected to be generated via pathways from (d) to (f) of Panel E of figure 15.7 forming the structure shown in Panel E (f). This is consistent with MS^3 spectrum of product ion m/z 333, shown in Panel B of figure 15.7. The ions m/z 113, 189 ($233-CO_2$), 219, 233, 261, 289 ($333-CO_2$), 315 ($333-CO_2$) lead to the structure assignment of ion m/z 333. As well as the ions m/z 163, 189, and 248 in MS^4 of m/z 289 in Panel D of figure 15.7 lead to the structure assignment of ion 289. When combined, these ions confirmed the structure of ion m/z 333 and, subsequently, the structure of LXA_4 anion m/z 351.

We usually run up to n = 3 for MS^n because when n level increases, the signal intensities of the ions decrease rapidly. The approach with n higher than 4 does not seem to be practical for regular applications unless the duty cycle for gas-phase ion manipulation by the mass spectrometer is further improved. However, higher n should provide more definitive "fingerprints" for the structure when there are many functional groups in the lipids that need to be elucidated.

Examples of LC-UV-MS-MS-based lipidomic analysis for docosanoids such as RvD1, NPD1, and their precursor DHA are shown in figure 15.8. Lipids were separated chromatographically and their respective chromatographic retention times (CRT) (Panel A), and MS/MS spectra and breakdown ion profiles are shown in panels B and C. The eicosapenaenoic acid (EPA) and derived lipid mediators Resolvin

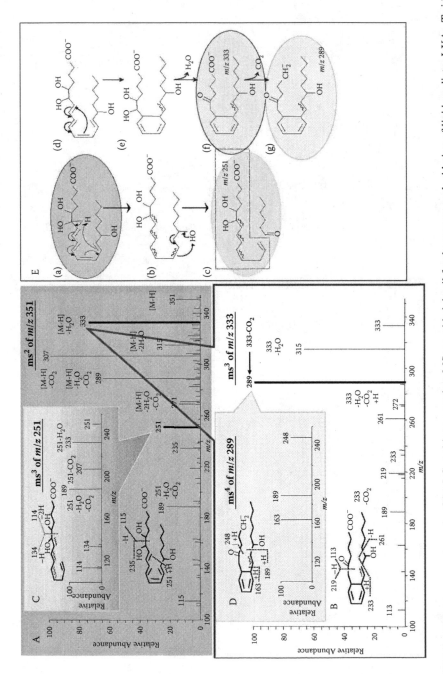

FIGURE 15.7 (See accompanying color CD.) LC-multistage tandem MS^n provided detail for the structure elucidation of lipid mediator LXA$_4$. To identify the structure of lipids, we need to elucidate functional groups and the specific position in the molecular structure throughout of MS fragmentation patterns.

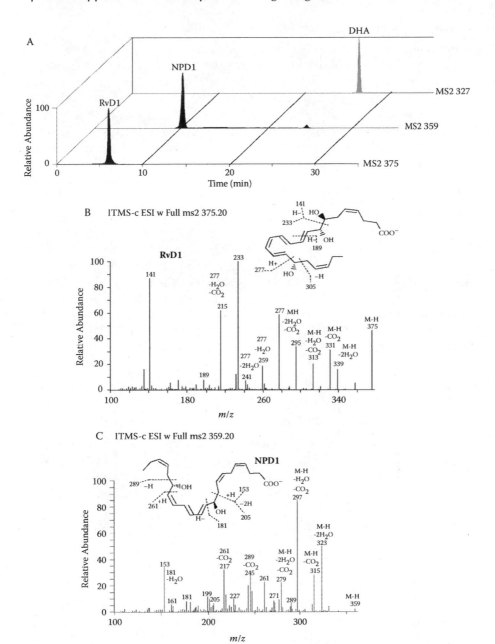

FIGURE 15.8 (See accompanying color CD.) LC-UV-MS/MS-based lipidomics for doco-sanoids derived from docosahexaenoic acid. Analysis was conducted using a LC-UV-linear ion trap MS/MS (LTQ, Thermo Fischer Scientific Co., Waltham, Massachusetts). (A) SIM chromatograms showing triHDHA (m/z 375), diHDHA (m/z 359), and DHA (m/z 327). (B) MS/MS wideband spectrum (m/z 375) for RvD1 with structure inset where the fragment ions are interpreted. (C) MS/MS wideband spectrum (m/z 359) for NPD1.

E1, PGE$_3$, and LTB$_5$, as well as lipidomic profiles of their isomers can be acquired at once from one single sample injection as shown in figure 15.9. Profile of respective CRTs are shown on Panel A, and the MS-MS spectra obtained with this method are shown on panels B and C. Following this approach we can also provide the lipidomic profile identification for eicosanoids, such as 20-LTB$_4$, PGE$_2$, LXA$_4$, 5,15-dihydroxy eicosatetraenoic acid, and their precursor arachidonic acid, shown in figure 15.10 (Panels A, B, C, D, and E). The quantification can also be conducted using the chromatographic peak areas in SIM, MS/MS, of the selected ions from MS/MS. The linear response of the LC-UV- MS/MS for the PGD2 standard is shown in figure 15.6, where SIM chromatographic peak areas were found linear for 10 pg to 10 ng injected amounts.

Lipidomic databases were constructed with MassFrontier™ software (developed by Mistrik, R., version 4.0, HighChem, Ltd., San Jose, California). The characteristic UV spectra of standard lipids were written into the subdatabase names, as well as their chromatographic retention times, or CRTs and product ion mass spectra. Therefore, MassFrontier™ could handle the acquired UV spectral results and CRTs for the identification of the unknown lipid mediators' mass spectra. The efficacy of this database was assessed with the identification of neuroprotectin D1 biosynthesized by retinal pigment epithelial cells using LC-UV-MS/MS (shown in figure 15.11). A subdatabase "mTOz 359 UV270nm" was selected with the molecular ion of interest and matched UV λ_{max} 270 nm.

PROTOCOL 4: OTHER PROCEDURES

Alkaline hydrolysis of esterified fatty acids

To analyze the fatty acid composition of complex lipids (e.g., phospholipids), hydrolysis of the esterified fatty acids is conducted. This quantitative hydrolysis should produce an end product in the form of a fatty acid or a salt that will be suitable for mass spectrometric analysis (readily ionizable). For this purpose we proceed as follows:

1. Suspend lipid extract in 50 µL of methanol or ethanol in a 13 × 100 mm screw top test tube.
2. Add 8 µL of 1 N NaOH.
3. Add 42 µL H$_2$O.
4. Vortex for about 20 s.
5. Flush N$_2$ on the test tube, cap, and wrap in aluminum foil to protect from light.
6. Incubate at 42°C for 3 h.
7. Add 100 µL H$_2$O, then adjust pH to 3.5 with 0.05 N HCl. Try not to run below pH 3.0 range.
8. Extract aqueous phase with hexane:isopropanol (3:2 V/V) 2 mL, vortex 20 s, then centrifuge at 3000 rpm for 5 min.
9. Remove organic phase from the top of the aqueous phase, and then repeat procedure to wash with 1 mL more of solvent mixture.
10. Then dry organic extracts and resuspend in methanol for analysis.

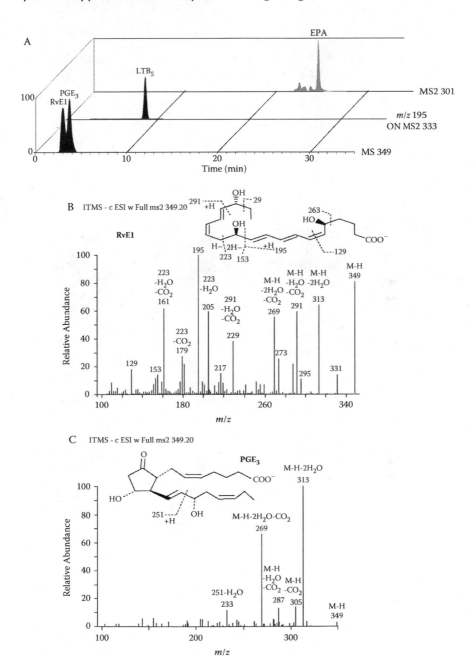

FIGURE 15.9 (See accompanying color CD.) LC-UV-linear ion trap MS/MS-based lipidomics for eicosanoids derived from eicosapentaenoic acid. Analysis was conducted using a LC-UV-linear ion trap MS/MS. (A) SIM chromatograms showing triHEPEs (m/z 349), diHEPE (m/z 195 on MS^2 333 for LTB$_5$), and eicosapentaenoic acid (m/z 301). (B) MS/MS wideband spectrum (m/z 349) for RvE1 with structure inset where the fragment ions are interpreted. (C) MS/MS wideband spectrum (m/z 349) for PGE$_3$.

Analysis of phospholipids and fatty acids

To analyze by gas liquid chromatography (GLC) the content and composition of the fatty acids free or esterified to glycerol in neutral lipids or phospholipids, we convert them into methyl esters, to decrease the volatility and increase stability at the temperature range used during GLC run (70 to 230°C) and make them suitable for interaction with the stationary phase. The longer the carbon chain, the less volatile is the methyl ester; thus, there is a stronger interaction with the stationary phase. This indicates you need

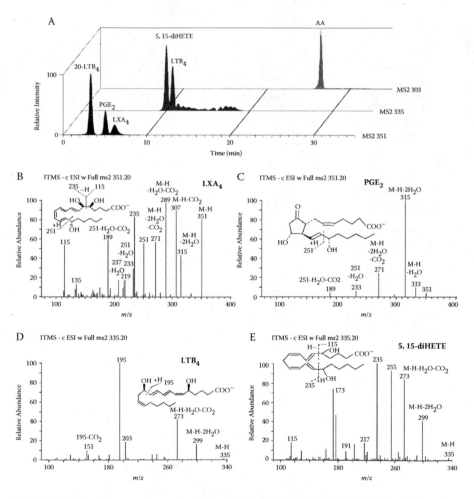

FIGURE 15.10   (See accompanying color CD.) LC-UV-MS/MS-based lipidomics for eicosanoids derived from arachidonic acid. Analysis was conducted using a LC-UV-linear ion trap MS/MS. (A) SIM chromatograms showing triHETEs (m/z 351), diHETEs (m/z 335), and arachidonic acid (m/z 303). (B) MS/MS wideband spectrum (m/z 351) for LXA$_4$ with structure inset where the fragment ions are interpreted. (C) MS/MS wideband spectrum (m/z 351) for PGE$_2$. (D) MS/MS wideband spectra m/z 335 for LTB$_4$. (E) MS/MS wideband spectra at m/z 335 for 5, 15-diHETE.

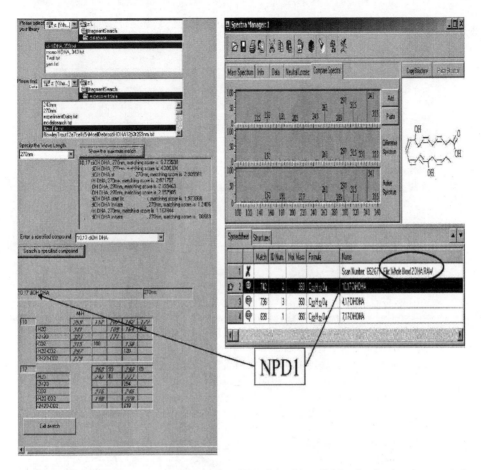

FIGURE 15.11 (See accompanying color CD.) Searching lipidomic databases revealed NPD1 in retinal pigment epithelial cells; the left is our theoretical database, and the right is the database built with data from standards using Mass Frontier Software.

to increase the temperature to minimize the retention times. This technique permits methyl ester separation of fatty acid in liquid phase (SP-2330) on the base of the carbon chain length and the number of double bonds.

Procedure

1. The sample (aliquot of C:M extract or silicagel spot from TLC plate) is resuspended in 1.96 mL of toluene:methanol (1:1) (v:v). Store samples under N_2 at −40°C until use.
2. Add 40 μL concentrate sulfuric acid (final concentration 0.38 M), flush with N_2 and vortex well.
3. A modification you would consider to speed up the assay that produces the same results is to add at once toluene:methanol:sulfuric Acid (100:100:4) 2 mL per reaction tube, flush with N_2 and cap tight. Here, procedure can be

      stopped or continued with incubation. If stopped, samples have to be kept at
      −20 to −40°C until use.
4. Let samples reach room temperature; vortex each sample.
5. Incubate in a shaking water bath for 4 h at 65°C.
6. Cool to room temperature, add 1 mL of water, vortex the mixture, and add
    3 mL of hexane (or petroleum ether), vortex for an extended period (30 s).
7. Centrifuge to separate the aqueous layer from the hexane layer, remove the
    hexane layer, add 3 mL more of hexane to the aqueous layer, and repeat the
    procedure. Collect together both hexane extracts.
8. Evaporate the hexanes under N_2 current.
9. Add adequate amount of internal standard mixture containing fatty acid
    methyl esters 13:0, 15:0, 17:0, 19:0, and 21:0 (Standard 14A, Nu-Check Prep
    Inc., Elysian, Minnesota).
10. Resuspend in adequate volume of hexane and inject into GLC.

GLC analysis was performed on a CP-3800 gas chromatograph (Varian, Inc.,
Palo Alto, California) mounted with an autosampler and FID detector; it used
helium as carrier gas, and hydrogen and compressed air for the flame ionization detector. The instrument was loaded with a SP™ 2330 fused silica capillary column (Supelco, Bellafonte, Pennsylvania) 30 m × 0.25 mm × 0.2 μm
film thickness. The instrument was run using a column temperature program
at 70°C, then ramped to 150°C at 20°C/min, held for 8 min, then ramped again
to 210°C at 5°C/min, held for 6 min, then ramped again to 230°C at 10°C/min,
held for 5 min, then returned to 70°C at 20°C/min. The injector temperature
was 220°C, and used on split/less–split (50%) mode. FID detector was set at
250°C. The helium carrier gas was flowing at 1.5 mL/min.
    The Star Chromatography Workstation (Varian, Inc., Palo Alto, Califor-
nia) was capable of controlling the GLC run program, as well as the autosampler injection program, and data collection and peak area analysis. Samples
after methanolysis procedures were loaded with an internal standard mixture
containing fatty acid methyl esters 13:0, 15:0, 17:0, 19:0, and 21:0 (Standard
14A, Nu-Check Prep Inc., Elysian, Minnesota). Peak elusion profile identification was performed on a mixture of saturated and polyunsaturated fatty acid
methyl esters (Standard 489, Nu-Check Prep Inc., Elysian, Minnesota).

Assessment of neurotrophin bioactivity on lipid mediators. Neurotrophins
modulate photoreceptor survival (24–28). Therefore, it is of interest to define
neurotrophin-mediated signaling. Here, we provide an experimental design for
the study of neurotrophin signaling using a lipidomic approach. For this purpose, human RPE cells grown to confluence and a high degree of differentiation displaying apical-basolateral polarization are used (28). These RPE cells
have prominent apical microvilli, zonae occludens, positive immunoreactivity,
and transepithelial resistance of at least 400 Ω × cm^2. The use of human RPE
cells in barrier-forming monolayers allows addressing of the issue of "sid-
edness" in lipid mediator release. We have recently shown that several neurotrophins with bioactivities that promote neuronal and/or photoreceptor cell

survival are agonists of NPD1 synthesis (15). Neurotrophins are added to the upper incubation chamber (apical cellular surface), and then the upper chamber media and the lower chamber media (basal cellular surface) are collected separately, and subjected to lipidomic analysis. Among several neurotrophins recently studied, PEDF (pigment epithelium-derived factor) was by far the most potent stimulator of NPD1 synthesis (15). PEDF, a member of the serine protease inhibitor (serpin) family, was initially identified in the same human retinal pigment epithelial cells preparation described here (29).

PROTOCOL 5: FORMATION OF CONFLUENT MONOLAYERS OF HUMAN RPE CELLS

RPE cells are seeded onto Millicell-HA culture plate inserts (Millipore, Bradford, Massachusetts), placed in 24-well plates, and allowed to reach confluence. Cultures are maintained in Chee's essential replacement medium (28) until the experiments were performed. The medium includes Eagle's minimal essential medium with calcium (MEM, Irvine Scientific, Irvine, California), 1% heat-inactivated calf serum (JRH Bioscience, Lennexo, Kansas), amino acid supplements, and 1% bovine retinal extract. The Millicell-HA filter inserts allow separate manipulation of the culture media bathing the apical or basal surfaces of the RPE monolayer and measurement of the transepithelial resistance (TER), which provides a measure of cell differentiation and confluence. Cultures are used for experiments once they developed a TER of at least 400 Ω $\times$ cm^2, as measured by an Epithelial Volt-ohmmeter (EVOM, World Precision Instruments, New Haven, Connecticut).

Characterization of this preparation in addition is made by EM to show well-differentiated human RPE with prominent apical microvilli, and zona occludens-1 (ZO-1) antibody immunoreactivity to illustrate the polyhedral shape of the cells. One experimental design is to add growth factors (20 ng/ mL or varying concentrations) to the apical medium and 72 h later. Apical and basal media are collected separately and subjected to lipidomic analysis.

Procedure: Exposure of ARPE-19 cells to growth factors

Cells at 75–80% confluence (72 h growth in DMEM/F12 + 10% FBS) are serum-starved for 2 h before exposure to growth factors. The serum-starved cells are treated with TNF-α (10 ng/mL) and H$_2$O$_2$ (400 to 800 μM) to induce oxidative stress and challenged with increasing concentrations of DHA (10, 30, 50, 100, and 200 nM) and PEDF (50 ng/mL) simultaneously with oxidative stress for 4 h before harvesting. Cell extracts are made and protein concentrations adjusted by Bio-Rad protein reagent and used for Western blot analysis. To study neuroprotection by DHA and PEDF in the oxidative stress-induced ARPE-19 cells, 72 h cells were serum-starved for 8 h before the introduction of oxidative stress and challenged with DHA and PEDF for 15 h. Cells were analyzed to detect Hoechst-positive apoptotic cells.

The upper chamber compartment is filled with 500 μL medium (bathing the apical cell monolayer surface) containing 0.1% HAS (human serum albumin), and 50 nM DHA, or 50 nM DHA plus added neurotrophins (10 to 200 ng PEDF or CNTF, or 20 ng of BDNF, Cardiotrophin, CNTF, FGF, GDNF, LIF, NT3, or Persephin). The lower chamber is filled with 500 µL media (bathing the basal cell monolayer surface) containing 0.1% HAS. Cells were incubated for 72 h, then apical and basal media are removed and collected for analysis. After allowing the cells to rest for at least 72 h on fresh media, the experiments can be repeated subsequently as many times as necessary. In our laboratory, we have run 28 sets of experiments during 9 months using the same cells.

PROTOCOL 6: ASSESSMENT OF NPD1-MEDIATED CYTOPROTECTION AGAINST A2E

To explore protection signaling, here we provide as an example of experimental design for the study of A2E actions on RPE cells in culture. A2E, a lipofusin component, accumulates in the RPE during aging (30–33) in AMD, in Stargardt macular dystrophy (an early-onset form of AMD), and in animal models of this disease (31–33). As a consequence, RPE apoptosis takes place, which precedes photoreceptor cell degeneration and death (33).

Our experimental design involved the treatment of RPE cells in culture with A2E, which resulted in rapid apoptotic cell death. However, NPD1 [a recently discovered neuroprotective mediator in RPE cells (12) and in brain and retina tissue (13)] was found to be cytoprotective against A2E-induced apoptosis.

PROTOCOL 7: HOECHST STAINING

NPD1 inhibits A2E-induced RPE cell damage in ARPE-19 cultures. Phase contrast and the calcein AM (green) and ethidium homodimer (red) cell viability assay can be used (12), as well as fluorescence microscopy by Hoechst staining (10), which is described as follows.

ARPE-19 cells are washed once with 1 mL PBS, after which they are fixed in 1 mL methanol for 15 min at room temperature. Cells are washed again with 1 mL PBS. Then perform the staining procedure that follows. ARPE-19 cells are incubated with 2 µM Hoechst 33258 reagent dissolved in Lock's solution (Promega Corporation, Madison, Wisconsin) at room temperature for 15 min. Cells are washed once with PBS and photographed using a Nikon DIAPHOT 200 microscope and UV fluorescence. Images are recorded by a Hamamatsu Color Chilled 3CCD camera and Photoshop 5.0 software (Adobe Systems, Mountain View, California). Hoechst-positive nuclei are identified by their strong bluish fluorescence and condensed or granular characteristics. This differs from normal nuclei, which are large, pale, and relaxed. This assay has also been used to assess outer segment phagocytosis in RPE cells under oxidative stress conditions (34).

ACKNOWLEDGMENTS

This research was supported by the National Institutes of Health (NIH), National Eye Institute (NEI) grant R01 EY005121; NIH, National Center for Research Resources (NCRR) grant P20 RR016816; and a grant from the American Health Assistance Foundation.

GLOSSARY OF TERMS

PDA: Photodiode array detector, capable of detecting a wide range of wavelengths. The characteristics of a good detector performing on LC-MS-MS are: wide UV; visible range, usually 200–800 nm; sensitivity, which is the minimum chemical load to be detected, and the absorbance range; a very small flow cell, to minimize peak distortion, with the increase of use of micro or nano-LC, the use of reduced LC flow rates, and column sizes, call for carefully planning the size of a flow cell to prevent excessive dead volume, without resting sensitivity.

ESI: Electro spray ionization. This is a technique that transfers ions in liquid phase to the gas phase so that they can be conveyed into the ion source of the mass spectrometer. This is obtained through a device installed in the head of the mass spectrometer, generating a fine mist from LC effluents, within a strong electrical field; therefore, the charged ions migrate into the instrument, leaving behind solvents and other ions of opposing polarity. The operator has the option to select ions with specific charges. This is a mild ionization, so sensitive molecules can be ionized without disturbing structures. This approach has many advantages also over other ionization techniques such as derivatization (to facilitate ion formation) or Atmospheric Pressure Chemical Ionization (APCI).

SRM: Selected reaction monitoring. This is an analytical tool where quantitative analysis is performed on a selected parent/product ion monitoring under specific fragmentation conditions. The fact that two ions (parent/product) under specific conditions are selected means that the specificity and sensitivity are enormously increased.

SIM: Single ion monitoring. This technique is based on monitoring of a single ion, increasing sensitivity but reducing specificity. For example, in table 15.1 ion 351.1 m/z is common for PGE_2, PGD_2, and Lipoxin A_4; therefore, analysis is performed on SIM mode. Those ions will be detected and quantified together because they are isomers or structural analogs and unless special LC columns are used, they will be eluted at the same retention time.

Full Scan: This detection modality on a triple quadrupole mass spectrometer will provide the full profile of product ions in the sample when a parent ion is selected on the Q1 quadrupole. Then the full scan is detected on the Q3 quadrupole after fragmentation on the collision cell (Q2).

Amphifilic: Refers to the chemical behavior of a molecule in a milieu. Such a molecule could behave as an anion or cation depending on the pH, temperature, organic composition, or aqueous phases. These behaviors will affect chemical solubility and will facilitate the transition from one to the other phase quantitatively.

REFERENCES

1. Murphy, R.C., Fiedler, J., and Hevko, J., Analysis of nonvolatile lipids by mass spectrometry, *Chem. Rev.* (Washington, D.C.). 101, 479, 2001.
2. Wenk, M.R., The emerging field of lipidomics, *Nat. Rev. Drug. Discov.*, 4, 594, 2005.
3. Han, X. and Gross, R.W., Global analyses of cellular lipidomes directly from crude extracts of biological samples by ESI mass spectrometry: a bridge to lipidomics, *J. Lipid. Res.*, 44, 1071, 2003.
4. Rapaka, R.S. et al., Targeted lipidomics: signaling lipids and drugs of abuse, *Prostaglandins Other Lipid Mediat.*, 77, 223, 2005.
5. Serhan, C.N. et al., Resolution of inflammation: state of the art, definitions and terms, *FASEB J.*, 21, 325, 2007.
6. Bazan, N.G., Synaptic lipid signaling: significance of polyunsaturated fatty acids and platelet-activating factor, *J. Lipid Res.*, 44, 2221, 2003.
7. Rodriquez de Turco, E.B. et al., Post-Golgi vesicles cotransport docosahexaenoylphospholipids and rhodopsin during frog photoreceptor membrane biogenesis, *J. Biol. Chem.*, 272, 10491, 1997.
8. Scott, B.L. and Bazan N.G., Membrane docosahexaenoate is supplied to the developing brain and retina by the liver, *Proc. Natl. Acad. Sci. U.S.A.*, 86, 2903, 1989.
9. Bazan, N.G. and Rodriguez de Turco, E.B., Alterations in plasma lipoproteins and DHA transport in progressive rod-cone degenerations (PRCD), in *Retinal degeneration and regeneration*, Kato, S. and Osborne, N.N., eds., Kugler Publications, Amsterdam/New York, 1996, 89.
10. Mukherjee, P.K. et al., Neuroprotectin D1: a docosahexaenoic acid-derived docosatriene protects human retinal pigment epithelial cells from oxidative stress, *Proc. Natl. Acad. Sci. U.S.A.*, 101, 8491, 2004.
11. Bazan, N.G., Cell survival matters: docosahexaenoic acid signaling, neuroprotection and photoreceptors, *Trends. Neurosci.*, 29, 263, 2006.
12. Mukherjee, P.K. et al., Neurotrophins enhance retinal pigment epithelial cell survival through neuroprotectin D1 signaling, *Proc. Natl. Acad. Sci. U.S.A.*, 104, 13152, 2007.
13. Bazan, N.G., Neuroprotectin D1 (NPD1): a DHA-derived mediator that protects brain and retina against cell injury-induced oxidative stress, *Brain Pathol.*, 15, 159, 2005.
14. Lukiw, W.J. et al., A role for docosahexaenoic acid-derived neuroprotectin D1 in neural cell survival and Alzheimer disease, *J. Clin. Invest.*, 115, 2774, 2005.
15. Belayev, L. et al., Docosahexaenoic acid complexed to albumin elicits high-grade ischemic neuroprotection, *Stroke*, 36, 118, 2005.
16. Marcheselli, V.L. et al., Novel docosanoids inhibit brain ischemia-reperfusion-mediated leukocyte infiltration and pro-inflammatory gene expression, *J. Biol. Chem.*, 278, 43807, 2003. Erratum in: *J. Biol. Chem.*, 278, 51974, 2003.
17. Bazan, N.G., Omega-3 fatty acids, pro-inflammatory signaling and neuroprotection, *Curr. Opin. Clin. Nutr. Metab. Care*, 10, 136, 2007.
18. Bazan, N.G., Birkle, D.L., and Reddy T.S., Docosahexaenoic acid (22:6, n-3) is metabolized to lipoxygenase reaction products in the retina, *Biochem. Biophys. Res. Commun.*, 125, 741, 1984.

19. Folch, J. et al., A simple method for the isolation and purification of total lipids from animal tissues, *J. Biol. Chem.*, 226, 497, 1957.
20. Gronert, K. et al., Transcellular regulation of eicosanoid biosynthesis, *Methods Mol. Biol.*, 120, 119, 1999.
21. Griffiths, W.J. et al., Electrospray-collision-induced dissociation mass spectrometry of mono-, di- and tri-hydroxylated lipoxygenase products, including leukotrienes of the B-series and lipoxins, *Rapid Commun Mass Spectrom.*, 10, 183, 1996.
22. Balazy, M., Eicosanomics: targeted lipidomics of eicosanoids in biological systems, *Prostaglandins Other Lipid Mediat.*, 73, 173, 2004.
23. Hong, S. et al., Resolvin D1, Protectin D1, and related docosahexaenoic acid-derived products: Structure analysis via electrospray/low energy tandem mass spectrometry based on spectra and fragmentation mechanisms. *J. Am. Soc. Mass Spectrom.*, 18, 128, 2007.
24. LaVail, M.M. et al., Multiple growth factors, cytokines, and neurotrophins rescue photoreceptors from the damaging effects of constant light, *Proc. Natl. Acad. Sci. U.S.A.*, 89, 11249, 1992.
25. Faktorovich, E.G. et al., Photoreceptor degeneration in inherited retinal dystrophy delayed by basic fibroblast growth factor, *Nature*, 347, 83, 1990.
26. Valter, K. et al., Time course of neurotrophic factor upregulation and retinal protection against light-induced damage after optic nerve section, *Invest. Ophthalmol. Vis. Sci.*, 46, 1748, 2005.
27. Politi, L.E., Rotstein, N.P., and Carri, N.G., Effect of GDNF on neuroblast proliferation and photoreceptor survival: additive protection with docosahexaenoic acid, *Invest. Ophthalmol. Vis. Sci.*, 42, 3008, 2001.
28. Hu, J. and Bok, D., A cell culture medium that supports the differentiation of human retinal pigment epithelium into functionally polarized monolayers, *Mol. Vis.*, 7, 14, 2001.
29. Tombran-Tink, J. et al., PEDF and the serpins: phylogeny, sequence conservation, and functional domains, *J. Struct. Biol.*, 151, 130, 2005.
30. Sparrow, J.R. et al., A2E-epoxides damage DNA in retinal pigment epithelial cells. Vitamin E and other antioxidants inhibit A2E-epoxide formation, *J. Biol. Chem.*, 278, 18207, 2003.
31. Radu, R.A. et al., Light exposure stimulates formation of A2E oxiranes in a mouse model of Stargardt's macular degeneration, *Proc. Natl. Acad. Sci. U.S.A.*, 101, 5928, 2004.
32. Bui, T.V. et al., Characterization of native retinal fluorophores involved in biosynthesis of A2E and lipofuscin-associated retinopathies, *J. Biol. Chem.*, 281, 18112, 2006.
33. Cideciyan, A.V. et al., Mutations in ABCA4 result in accumulation of lipofuscin before slowing of the retinoid cycle: a reappraisal of the human disease sequence, *Hum. Mol. Genet.*, 13, 525, 2004.
34. Mukherjee, P.K. et al., Photoreceptor outer segment phagocytosis attenuates oxidative stress-induced apoptosis with concomitant neuroprotectin D1 synthesis, *Proc. Natl. Acad. Sci. U.S.A.*, 104, 13158, 2007.

Index

A

B

C